Food Geographies

Social, Political, and Ecological Connections

Pascale Joassart-Marcelli

San Diego State University

ROWMAN & LITTLEFIELD

Lanham • Boulder • New York • London

Executive Editor: Michael Kerns
Assistant Acquisitions Editor: Haley White
Sales and Marketing Inquiries: textbooks@rowman.com

Credits and acknowledgments for material borrowed from other sources, and reproduced with permission, appear on the appropriate pages within the text.

Published by Rowman & Littlefield
An imprint of The Rowman & Littlefield Publishing Group, Inc.
4501 Forbes Boulevard, Suite 200, Lanham, Maryland 20706
www.rowman.com

86-90 Paul Street, London EC2A 4NE

British Library Cataloguing in Publication Information Available

Library of Congress Cataloging-in-Publication Data
Names: Joassart, Pascale, author.
Title: Food geographies : social, political, and ecological connections / Pascale Joassart-Marcelli.
Description: Lanham, Maryland : Rowman & Littlefield, 2022. | Series: Exploring geography | Includes bibliographical references and index.
Identifiers: LCCN 2021050223 (print) | LCCN 2021050224 (ebook) | ISBN 9781538126646 (cloth) | ISBN 9781538126653 (paperback) | ISBN 9781538126660 (epub)
Subjects: LCSH: Food habits—Social aspects. | Food habits—Political aspects. | Food habits—Environmental aspects. | Cultural geography.
Classification: LCC GT2850 .J63 2022 (print) | LCC GT2850 (ebook) | DDC 394.1/2—dc23/eng/20211130
LC record available at https://lccn.loc.gov/2021050223
LC ebook record available at https://lccn.loc.gov/2021050224

Contents

Textboxes, Figures, and Tables

Textboxes

Figures

Tables

Preface

Over the past two decades, food has become a serious topic, taking up more space in academic scholarship, popular media, and everyday conversation. Driven by social anxieties caused by a seemingly endless list of problems linked to food, including food insecurity, obesity, cultural loss, animal abuse, labor exploitation, environmental degradation, and climate change, research on food has exploded. Publications, conferences, and centers dedicated to food multiplied quickly. Students' growing interest in these issues prompted the creation of courses and curricula examining food from new angles, emphasizing its social, cultural, political, economic, and environmental aspects. Amid the growth of interdisciplinary work on food, geography emerged as a unifying discipline, well positioned to connect environment and society and to offer a systemic and integrative approach to the study of food that builds bridges with anthropology, sociology, history, political science, environmental studies, economics, and other disciplines. Indeed, as a discipline, geography sits comfortably at the crossroads of the humanities and social and environmental sciences. Moreover, food is intensely geographical, it connects us to the earth and to each other though webs of relationships that are political, economic, and social and unfold at many scales, ranging from the body to the globe.

In 2008, as a new professor of geography at San Diego State University, I got caught in this whirlwind of excitement about food. Having had lifelong interests in cooking, gardening, and eating, I realized that food did not have to be a hobby, separate from my professional life and academic interests. Indeed, food was revealing itself as a fascinating topic of research, giving me a unique window to address questions that have always animated my work about urban inequality, poverty, environmental justice, grassroots activism, and their relationships to immigration, race, and gender.

Food also turned out to be an ideal topic to introduce students to the breadth and power of geography. After taking a few years to develop the course and get it approved by my university, I had the pleasure to teach *Geography of Food* for the first time in 2011, attracting just a dozen adventurous upper-division students. Since then, the course has become a permanent feature of our departmental schedule, enrolling much larger groups of students from geography and many other disciplines and reflecting the tremendous interest among students to learn more about where our food comes from, how it is produced, what its impacts on people and the environment are, and what it says about the diversity of human life. My experience teaching the geography of food for a decade has prompted me to write this textbook in the hope of filling a gap and offering a unique and inclusive perspective inspired by geography.

Over the past decade, more teaching materials have become available, including a number of excellent books such as *Taking Food Public* (Williams-Forson and Counihan 2012), *Food* (Clapp 2016), *Food and Society* (Guptill, Copelton, and Lucal 2013), *Environment and Food* (Sage 2011), and multiple new editions of *Food and Culture* (Counihan and Van Esterik 2012), adding to classics such as *Consuming Geographies* (Bell and Valentine 1997), *Food in Society* (Atkins and Bowler (2001), *Geographies of Agriculture* (Robinson 2014), *Food Politics* (Nestle 2013), and *Food Nations* (Belasco and Scranton (2002). Many of these, including *Food and Place*, which I coedited with my San Diego State University colleague Fernando Bosco (Joassart-Marcelli and Bosco 2018), are edited volumes that gather chapters written by multiple authors working from a variety of perspectives, underscoring the interdisciplinary nature of the field but often lacking a unifying voice. In addition, most emphasize one aspect of the complex relationships between food, people, and the earth, whether it is culture, agriculture, consumption, or the environment. These left me wanting for a textbook that would bring these different aspects together and draw connections between them in a way that geography is known for.

At the same time, several new books were published taking a critical stance on the food system and suggesting ways to make it more just and sustainable, underscoring the deep structural inequalities that shape how/where food is produced and what people eat. Such works include *Cultivating Food Justice* (Alkon and Agyeman 2011), *Food Justice* (Gottlieb and Joshi 2013), and *The New Food Activism* (Alkon and Guthman 2017), alongside a host of more narrowly focused monographs such as the recent *Black Food Geographies* (Reese 2019). These books make invaluable contributions to the field of food studies and geographies of food by offering a critical perspective that is sometimes lacking in textbooks.

Food Geographies builds on this large body of work and offers a unique perspective to study food that is grounded in geography, integrative, and decidedly critical. The book stands out for at least four reasons. First, it intentionally takes a systemic, integrative, and multi-scalar approach in which food is always intertwined in the environmental, social, political, and personal. You will learn about transnational corporations as well as the gut biome and discover surprising connections between the two. Second, its empirical focus spans the globe. Although many examples come from the United States, the book also includes numerous case studies from other countries, highlighting similarities, differences, and connections between the Global North and the Global South and emphasizing spatial connections through food. Through many textboxes and illustrations, including photos, graphs, and maps, you will travel around the world—from Peru's Ica Valley to Indonesian ancient forests, Tanzanian collective farms, Filipino aquaculture pounds, and French food halls. Third, the book offers a critical perspective influenced by political ecology, feminism, Marxism, and anti-racist scholarship, shedding light on food-related social and environmental injustices. You will question the current system and examine ways to enact meaningful change to build a better future by producing, consuming, and engaging with food differently. Finally, while paying attention to urgent and complex problems and asking challenging questions, you will never lose track of the pleasure and joy that food may offer as a conduit to self-expression, comfort, well-being, livelihood, and social connections.

Food Geographies

Learning Objectives

- Recognize the multiple meanings of food and its complex geographies.
- Appreciate food's significance to social and environmental health and the urgency of better understanding the relationships between food, environment, and society.
- Conceptualize food as a geographic fact that connects people and places.
- Become familiar with the interdisciplinary breadth of the growing field of food studies.
- Begin to think about food in terms of interdependent social, economic, political, biological, and health systems.
- Get introduced to the key theoretical concepts, tools, and methods used by geographers to make sense of how people produce and consume food.
- Understand the structure of the book, its key themes, and primary objectives.

We all need to eat. Around the world, food provision activities consume much of our time, whether it is growing food, cooking it, selling it, eating it, or just thinking about it. If you are reading this book, you are probably among the growing number of people who spend time thinking about food. You may even be among those who call themselves "foodies"; you enjoy reading about food, trying new recipes, sharing meals with others, and discovering authentic restaurants, and you likely invest time and resources in these activities. For you, food may be more than sustenance. Beyond eating, you may engage with food as a cultural phenomenon, by posting photos on social media, reading food memoirs and lifestyle magazines, collecting cookbooks, visiting specialty markets, or watching food films and television shows. In many ways, food tells the story of who we are and, if you are like me, you probably get great pleasure from food. Indeed, my interest in food as a scholarly subject of inquiry stems first and foremost from my nearly insatiable appetite for "good" food.

Yet, food is not all pleasure. Historically, it has been a major source of struggle; entire populations have disappeared because of insufficient food, and many wars have been waged over agricultural resources such as land and water. Today, over nine hundred million people are undernourished and do not have enough food to lead a healthy active lifestyle. While most hungry people live in the Global South, approximately fifty million people in the United States suffer from food insecurity. Thus, for many, food equates to hard work and agonizing stress.

In the past decade, our food has come under increased scrutiny as we began realizing the social and environmental costs generated by the modern food system.

An explosion of popular books and documentaries have described problems with the ways we produce and consume food and proposed solutions. These works reflect and fuel social anxieties regarding the safety, health, and environmental sustainability of food. You too may be concerned about your food choices—or lack thereof—and wonder about their impacts on bodies, local communities, distant places, and the planet. Perhaps you worry about the cost of food, your body weight and image, food allergies and illnesses, animal abuse and labor exploitation, genetic engineering, climate change, chemicals used in food production, proper food etiquette, authenticity, or culinary technique.

This book provides concepts and tools to help us understand the significance of food in our everyday life and its relationship to the well-being of consumers, workers, animals, communities, regions, and the planet as a whole. It does so by focusing on *food geographies*, emphasizing connections, similarities, and differences in the meaning of food across places. As it shall become clear within the next few pages, the geography of food cannot easily be summed up by a theoretical approach, substantive area, or methodology. Hence, I have adopted the plural form of the term—food geographies—to emphasize the pluralism that characterizes this growing subdiscipline of geography.

1.1. Food as a Geographic Fact

Food must be produced and consumed *somewhere* and, as geographers believe, that somewhere matters. The physical geography of place—its climate, natural resources such as soil and water, and accessibility—shapes food production, availability, taste, and diets. Similarly, the human geography of place—its economic, political, social, and cultural characteristics—influences what is produced, how, and for whose consumption. At the same time, food-related activities constantly reshape places, putting stress on environmental resources, influencing economic growth, creating new food cultures, and prompting state intervention. The relationship between food, people, and environments is dynamic and mutually constitutive. In a global context where population continues to grow, economies are increasingly connected, and environmental pressures are mounting, these spatial relationships keep changing and adapting, with some people and regions benefiting, while others are hurting. A poignant illustration of the dynamic bond between food, people and environments is provided by Native people's relationship to salmon in the Klamath river basin as described in textbox 1.1.

As this example illustrates, food is intensely geographic. It is deeply entangled in the social, cultural, and environmental geographies that define places. First, food is a central element of how societies, in particular places and times, have organized themselves *economically* and *politically* to provide for the most basic of human needs and support populations at various scales, from rural communities to the global economy. Tribes and nations would not be able to survive and cities would not have been created without the existence of a social structure organizing the utilization and distribution of scarce resources to feed their population. This may involve locally adapted methods of irrigating, cultivating, and preserving food, as well as reliance on trade and food imports. As such, food is important in shaping political economic networks connecting places, including relationships between regions that produce raw food (e.g., rural areas and the Global South) and those that import and process these foods (e.g., urban areas and the Global North). By definition, these relations are imbued with power and often rife with conflict. In the case of the

Box 1.1   Indigenous Fishing

The Karuk, Hupa, and Yurok people have lived in the Klamath basin since time immemorial, long before the arrival of European Americans in this mountainous region of northwestern California. They have relied on the river for salmon, steelhead trout, and other local species of fish (see figure 1.1) and gathered acorns, berries, and medicinal plants in the surrounding forest. Their diet was diverse yet limited by what nature provided, with salmon and acorns being the most significant sources of protein and energy. To care for these valuable resources, Native people engaged in forest and fisheries management practices that were fused with rituals, including ceremonies and festivals. For them, a well-managed forest understory is "a supermarket, where Karuk people can gather berries and acorns; a pharmacy, where they can find herbs to treat coughs and inflammation; and a hardware store, with hazel and bear grass for making baskets" (Braxton Little 2018).

(continued)

FIGURE 1.1 A Hupa fisherman on the Klamath River, circa 1923. Native people often used spears to fish. Today, in an era of commercial fishing, these practices have practically disappeared and are reserved for ceremonies or recreation.

Source: Edward Curtis, "A Smoky Day at the Sugar Bowl," The North American Indian. (Library of Congress): https://picryl.com/media /a-smoky-day-at-the-sugar-bowl-hupa.

Box 1.1   *Continued*

However, Native people's access to these natural resources and their ability to manage them according to traditional knowledge has been under constant threat for the past several centuries. The Gold Rush migration, which increased competition for resources, led to the extermination and displacement of Native people. More recently, in the 1960s, the federal and state governments authorized utility companies to build dams on the Klamath river, causing Native people to experience new and severe threats to their ancestral foodways and social fabric. By blocking the upstream run of salmon, the dams prevented spawning and led to a drastic decline in fish population, especially Chinook salmon (Norgaard et al. 2011). Water pollution also increased and negatively affected fish populations. At the same time, commercial fishing expanded in the region, adding further pressure on salmon fisheries. In addition, the privatization of land ownership and subsequent changes in forest management led to the devaluation of tribal knowledge and the abandonment of ancestral practices of low-intensity burning that for centuries helped maintain oak stands, understory fungi, and associated wildlife by removing competing conifer trees and fuel for larger and much more destructive fires (Long et al. 2017). Fishing and hunting regulations by the California Department of Fish and Game is further limiting Native people's rights to access resources upon which they have historically depended.

Together, these recent changes have denied the Karuk, Hupa, and Yurok people access to their traditional food, adding hunger to a long history

of structural violence against Native people that amounts to genocide (Norgaard et al, 2011). With the disappearance of subsistence food provisioning mechanisms, people have become increasingly dependent on commercial food mass-produced elsewhere. Consequently, diets have changed dramatically in the past thirty years; highly processed food has replaced smoked salmon and acorn mush. This has led to a rise of food-related health problems such as diabetes, heart disease, and other chronic illnesses. Today, the official poverty rate among the Karuk, Hupa, and Yurok people living on the reservation ranges from 33 to 52 percent, well above the California average of 14 percent but below tribal estimates. Less than 10 percent of the population works in fishing, forestry, or agriculture and many households depend on food assistance and commodity food from the US Department of Agriculture. These consequences have been documented in a series of recent technical reports by the US Department of the Interior (2012) to help decide whether the four utility dams built on the Klamath river should be removed. Tribes themselves have been actively involved in protesting the dams and drawing attention to the significance of the river and forest in supporting their foodways, health, and culture. Today, several projects including the Western Klamath Restoration Partnership (2018) seek to restore agroforestry and traditional fire management practices through collaboration between Native people, the scientific community, and nonprofit and government agencies, such as the US Forest Service.

Klamath people, it is their interaction with European American settlers and eventually corporate utility companies that dismantled their ancestral food system, by robbing them of access to land and water and increasing their dependence on processed food. Their food economies in the past two centuries have been shaped by their evolving relationship to the outside world.

Second, food is one of our most primal connections to the earth and its *biophysical environment*. While food production relies on natural resources like soil and water, it also leaves a drastic mark on the environment and severely alters landscapes, causing ecological and climatic changes which are experienced and addressed differently across places. For example, Native people of northwestern California have a unique relationship to the Klamath basin both as a source of food and a culturally significant and sacred place. They consider themselves care-takers of the river and the forest, not owners. As such, they have been invested in sustaining the environment by only taking what they need for subsistence and preserving resources for future generations. The building of the dam signaled a new era in which tribal knowledge and sustainable food provisioning practices were devalued and supplanted by modern industrial values of productivity, efficiency, and profit. These changes have had significant impacts not just on the foodways, livelihoods, and well-being of local tribes, but also on the collective natural environment by increasing the risk of fire, decimating oak forests, polluting the water, and contributing to climate change, with potential consequences well beyond the reservations.

Finally, food is a powerful way for people to define themselves *culturally* and relate to each other *socially*. As a cultural symbol and practice, food is a powerful agent in place-making. Food has the capacity to distinguish, exclude, and bring people together. In other words, much can be revealed about culture through the lens of food. For example, salmon has been an integral part of Native people's culture in the Pacific Northwest; it has informed spiritual ceremonies, community events, culinary traditions, and everyday life for centuries. Thus, the changing food practices of Native people along the Klamath river symbolize cultural and spiritual losses. They also reflect racist cultural hierarchies in which white practices are privileged at the expense of traditional knowledge. For instance, contemporary narratives of health on the reservations of northwestern California tend to portray Native people as uneducated and therefore likely to engage in unhealthy behavior such as smoking tobacco, drinking alcohol, and eating high-calorie and low-nutrient food. This perspective suggests a need for education and public health programs to help Native people make better food choices, instead of acknowledging that these behaviors are the result of government sponsored disruptions of their subsistence economies. By holding on to traditional food practices and organizing for the removal of dams to "bring the salmon home" (Hormel and Noorgard 2009), Native people are actively revitalizing cultural ecologies and resisting oppressive economic, political, and social forces threatening them.

Whether we live in large cities, suburbs, or small rural towns, in the Global North or in the Global South, food ties us in myriad ways to the political, economic, social, cultural, and environmental relations that unfold at multiple scales, ranging from the micro-local to the global. In short, it connects us spatially to the rest of the world. As we discuss in the next section, studying food from a geographic perspective puts these spatial connections at the center of our analysis.

Food Studies: An Interdisciplinary Field 1.2.

Scholars from a variety of disciplines have contributed to the relatively new field of food studies, asking a wide range of questions and developing theoretical concepts to think about food in systemic, analytically rigorous, and critical ways. What is common among these various disciplines is the belief that food is more than a biological matter. Although long regarded as irrelevant, feminine, and mundane, the study of food is now firmly established as an important, rigorous, and interdisciplinary area of scholarly work characterized by a desire to approach food as a "social fact." This means that we understand food as shaped by our social environment, including economic circumstances, moral values, social norms, politics, and fashion that vary across space and time. Similarly, we believe that food plays a central role in transforming social environments. These recognitions have led many scholars to approach their study of food from a systemic perspective that pays attention to the connections between the production, distribution, marketing, regulation, and consumption of food

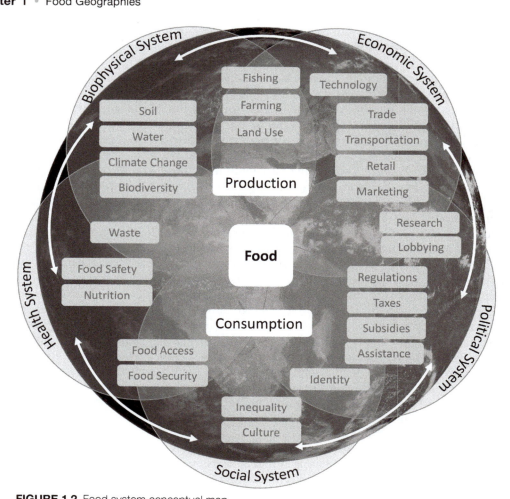

FIGURE 1.2 Food system conceptual map.

Source: Author, inspired by Nourishlife (2020) with background globe from https://pixy.org/372102/.

and how the economic, political, environmental, and social comes together through food (see figure 1.2). This systemic and interdisciplinary approach distinguishes food studies from more technical disciplines such as health, nutrition, agronomy, and food science that tend to focus on one specific aspect of food.

Still, despite its interdisciplinary nature, the study of food remains influenced by rigid disciplinary boundaries. For instance, the humanities and social and environmental sciences only occasionally converse with each other and most research and teaching are done in disciplinary silos. Those within the *humanities*, including history, philosophy, and literary, cultural, and media studies, primarily study the significance of food in shaping history and symbolizing human experiences. For historians, the domestication of plants and animals is particularly important because it marks the beginning of what we call civilization. Ironically, it also coincides with the spread of wars and the rise of imperialism that, even in our current era, represents violent attempts to control land and populations for economic gains, including a reliable food supply. Many historians, such as Mintz (1986), Laudan (2013) and Kipple (2007), have documented the profound impacts of colonialism on the global production, exchange, and consumption of food commodities like sugar, tea, and spices, highlighting the economic advantages it gave colonial powers as well as the destruction and exploitation it caused in colonized areas.

Some philosophers study food as a powerful window to explore what it means to be a person and what our obligations to others are (see Curtin and Heldke 1992, Telfer 2012, and Kaplan 2012). Conceptualizing humans as "eaters"—not just "thinkers"—opens new ways to think about human agency and responsibility and raises important ethical questions about eating animals, the role of biotechnology, body politics, and global hunger. Philosophy has also informed work on food justice, including discussions about good food and the right to food, and the aesthetics of food as art and taste.

Scholars working with literary texts and other media often approach descriptions of food and food practices as symbolic of culture. For them, films, classic literature, cookbooks, magazines, and social media reveal the meaning of food to particular social groups in certain times and places and highlight social and cultural dynamics and structures (see Appadurai 1988, Pazo 2016, Rousseau 2012). The exponential growth of food media in the past two decades has prompted a growing number of scholars to apply their analytical skills to "reading between the lines" and identifying underlying ideologies and social biases in this important realm of popular culture.

Food has also animated much research in the *social sciences*, where economists, political scientists, anthropologists, sociologists, and others have sought to answer pressing questions regarding hunger, poverty, health disparities, and environmental stress associated with contemporary food production and consumption. Development economists have long been concerned with agrarian questions, including our ability to feed growing populations, the trade-offs between agriculture and industry, and the effects of trade and foreign investment on subsistence agriculture (see Schultz 1964, Boserup 2017). They have also contributed to research on hunger and famines shedding light on the role of markets and other social institutions in explaining uneven access to food (Sen 1982). More recently, environmental economists have begun studying the hidden costs of food production on nature and comparing the effectiveness of various policies aiming at reducing these costs.

Political scientists have focused their attention on food politics—the ways that food production, distribution, and consumption are negotiated and regulated by political actors at various scales, including non-governmental organizations, local governments, nation states and international bodies (see Paarlberg 2013). Many emphasize the role of subsidies and trade agreements in supporting the creation of a global corporate food regime that favors the Global North. Others study how public policy may help create a healthier, more equitable, and more sustainable food system.

There is a long tradition of studying food within anthropology—a discipline concerned with the origins of humans and how they have evolved over time culturally, socially, and physically. Food, of course, is essential to the survival of humans and, according to anthropologists, reveals much about their cultures and shared values (see Lévi-Strauss 1964 and Douglas 1984). Similarly, sociologists consider how food creates and reflects social structures, including differences based on class, race, and gender (DeVault 1994). For example, sociologists have been influential in thinking about taste as socially constructed (Bourdieu 1984). This means that the food we like reflects our socioeconomic status and class position, more than our actual taste buds. Psychologists have also examined how food preferences, desires, and disgusts are influenced by mental and emotional factors, which may be unconscious (Connor and Armitage 2002, Lyman 2012).

Recent years have witnessed an explosion of public health research on food motivated in great part by fears of an obesity epidemic. Most scholars within this field focus on explaining health-related food behaviors or dietary choices as a function of individual, social, and environmental factors. For example, much has been written about the negative impacts of living in a neighborhood where the supply of affordable healthy food is limited while fast food is abundant (Walker et al. 2010).

As it should become clear by now, there is significant overlap across these disciplines, with many scholars approaching similar problems from different perspectives. Geographers too have their own approach to studying food. Yet, the integrative nature of geography—a discipline that sits comfortably at the crossroads of the humanities and social and environmental sciences—makes it particularly well suited for promoting systemic approaches and transdisciplinary connections.

1.3. Geographies of Food

Geographers have made important contributions to food studies, adding to the work of sociologists, anthropologists, and other scholars in the humanities, social sciences, and environmental sciences. They have explored a wide range of topics related to the production, transformation, distribution, representation, communication, consumption, disposal, and recycling of food. Yet, as Mandelblatt (2012, p. 154) argues, "separating geographers from the rest of scholars who work in the field of food studies is somewhat awkward and indeed artificial." Rather than their unique substantive contributions, the most important influence and distinguishing characteristics of geography within food studies have been its capacity to bring diverse disciplines into conversations by using space, place, and scale as convening and unifying concepts.

1.3.1. The Geographic Perspective on Food

Geography is typically defined as the study of the interactions between people and their environments, including the physical features of the earth's surface (*physical geography*) as well as the cultural and social arrangements of human settlements (*human geography*). While most geographers specialize in either human or physical geography, they firmly believe in the interconnections between both and uniquely focus much of their efforts in understanding *nature-society interactions*. Instead of being defined by a substantive topic of inquiry, such as the economy, social structures, culture, power relations, plants, water, or climate, geography is more accurately defined by its approach, which consists of investigating how things relate to each other spatially. In short, geography is less about *what* we study than it is about *how* we study things.

We geographers definitely do not have a monopoly on the study of food, but we offer a unique perspective focused on understanding where things are, why particular spatial patterns exist, how the social and environmental are related, how places change over time, and what this means for people inhabiting them. These spatial questions are at the core of geographic inquiry and can be applied to many different topics, including food. In fact, geographers believe that everything—people, natural objects, human-made objects, and social structures—exists somewhere in space. Therefore, every academic subject can be approached geographically by considering it in terms of location, shape, scale, network, and distance. This means that geographers are often involved in interdisciplinary collaboration with researchers from other fields. For instance, economic geographers may work with economists to understand where food commodities are produced and how they are traded globally, political geographers may engage with political scientists to explore the changing role of the state in regulating food, cultural geographers may collaborate with anthropologists to investigate how particular places shape food cultures, and environmental geographers may interact with ecologists to study how agricultural and fishing practices impact ecosystems.

The integrative and interdisciplinary nature of geography is particularly useful in the study of food—a complex and multifaceted object that cannot be summed up to one function or relegated to one discipline alone. For example, studying a single fruit as common as a tomato from a geographic perspective raises and sheds light on a number of interesting and related questions that span multiple topics and disciplines (see textbox 1.2).

The usefulness of the geographic approach is attested by the emergence, in the past two decades, of geography of food as a new subfield of the discipline. In the United States and in Europe, specialty groups have been formed in the last ten years by the main professional associations to promote collaboration among academics and increase visibility. The number of published articles and conference presentations focused on food geographies have exploded in the past few years.

To think spatially, geographers rely on a few important and unique *theoretical concepts* and *analytical tools*. This may be a good point for us to become acquainted with some of these ideas to which we will return throughout the book.

1.3.2. Key Theoretical Concepts

Theoretically, geographers tend to frame their study of food with one or more key concepts, including space, place, network, scale, and nature-society interaction. Those who emphasize *space* typically focus on questions such as where things are and how they relate to each other. There is an extensive theoretical literature in the discipline—and a healthy level of debate—regarding the nature of space. Generally, space is understood as the multidimensional area in which particular phenomena of interest occur. This could be a physical surface defined by its attributes and boundaries or a discursive/symbolic space shaped by the way people describe and imagine the area where activity takes place. In an era of increased mobility, linked to advances in transportation and communication technologies, spatial relations have been dramatically altered, spanning much of the globe and connecting people virtually through newly configured virtual spaces.

In the tomato example provided in textbox 1.2, researchers who focus on the "tomato trail" privilege space as an important component of understanding the food system. Examining where tomatoes are grown, packaged, processed, and consumed and how these locations relate to each other frames a broader analysis of the political, economic, and environmental features of the modern food system. This approach is similar to what Ian Cook (2006) describes as "following" in the first of a series of influential articles seeking to define the field of geography of food.

The spatial approach of "following" foods also relates to the concepts of scale and network, which have been equally important in geography. *Scale* describes the size of a feature, phenomenon, or the level of analysis. When considering food spaces, scholars have focused their attention on a variety of scales from the molecular to the global. Although the contemporary food system is typically viewed as operating at the global scale, in recent years, scholars and activists have become especially interested in understanding and promoting local food systems as an alternative to the global structure and its associated ills. This has led to exciting work on the relationships between the global and the local and raised interesting questions regarding the meaning of "local," including how it is defined and produced. Indeed, recent work in geography shows that scale, like space, is social and political.

Box 1.2

Studying the Tomato from a Geographic Perspective

As a geographer or spatial thinker, you may look at a tomato in your grocery store and start wondering about several issues: Where did the tomato originate? Where was this tomato grown? Was it on a small farm or in a large industrial greenhouse? Whose hands planted, cultivated, and picked it? How much did the farm workers earn? Was it enough to support their families and communities? Am I endorsing labor exploitation and environmental destruction on the other side of the planet? Was this tomato grown sustainably? What does sustainability mean anyway? How much water and chemical inputs did it require? How far did the tomato travel? What will I prepare with it? Could I make a "traditional" marinara sauce? Will my friends or family like this? Wouldn't it be more fun to go out to eat at the new café that opened in the neighborhood? Why are tomatoes so predominant in Mediterranean cuisine? Will this help me stay fit and healthy? Should I care? I could go on and on.

You may be surprised to learn that scholars have actually tried to answer most of these questions. Although not all formally trained as geographers, these researchers illustrate the power and breadth of geographic thinking in approaching food. For example, in her book *Tangled Routes: Women, Work, and Globalization on the Tomato Trail* (2007), Deborah Barndt investigates the journey of the corporate tomato in North America. Focusing on women workers in various places along the tomato supply chain, including fields, factories, roads, warehouses, borders, and fast-food restaurants in Mexico, Canada, and the United States, she shows how globalization works on the ground and what it means for the everyday lives of people across the north-south divide. She articulates how the Indigenous *tomatl*, originally grown and consumed by Maya and Aztec peoples, has been replaced by the genetically engineered modern tomato through a series of historical events and political economic decisions. Beginning with European colonization, she details specific moments and pinpoints key actors leading to the demise of subsistence agriculture and the rise of the global, industrial, corporate food system dominated by low-wage labor. Barndt also highlights ways that women along the tomato trail resist, organize and advocate simultaneously for biodiversity, cultural diversity, healthier food, and better livelihoods. Her rich multi-locale ethnographic study is framed and held together by the spatial relationships that connect women along the tomato route and exemplifies the organizing and integrating capacity of the geographic perspective.

Similarly, in *Tomatoland*, Barry Estabrook (2012) explores the true cost of the modern tomato in the United States, including rampant labor exploitation, environmental degradation, and loss of nutrients and flavor. He traces the tomato back to the town of Immokalee, Florida, where Latino immigrant tomato pickers live in dilapidated and overcrowded encampments, earn less than $10,000 a year, and are exposed to numerous toxic chemical fertilizers and pesticides. Along the way, he takes us through detours via the Atacama desert in southern Peru (where the ancestors of the tomato originated before moving to central America and eventually Europe and the rest of the world through the Colombian exchange); labs on the campus of the University of California, Davis (where researchers are trying to develop new seeds that withstand harsh environmental conditions, resist pests and disease, and produce more nutritious fruits); the produce section of a supermarket (where tasty and affordable tomatoes are almost impossible to find); a small Pennsylvania farm (where a passionate farmer grows tomato from heirloom seeds); and a New York City farmers market (where those tomatoes are sold to affluent consumers). His emphasis on the power of those who control the knowledge and technology associated with farming the "best" tomato echoes some of the criticism that Jim Hightower directed to the university agricultural research complex in his notorious report *Hard Tomatoes, Hard Times* (1972). Like Brandt, Estabrook shows us how workers, growers, and consumers in seemingly disconnected places are parts of a larger food system and highlights how they are organizing and working together to resist labor exploitation, toxic chemical use, and biodiversity loss.

Others have addressed the cultural significance of the tomato as an indicator of place, a symbol of national cuisine, and a type of healthy food. It occupies a privileged place in Italian culture and food identity, even though it was not cultivated in Italy before the sixteenth century, when it was introduced by the Spanish who brought it from America (see figure 1.3). The Marzano tomato, which is often considered the ideal fruit to be canned for authentic Neapolitan pizza and traditional sauces, was recently attributed a specific DOP (protected designation of origin), which officially codifies its

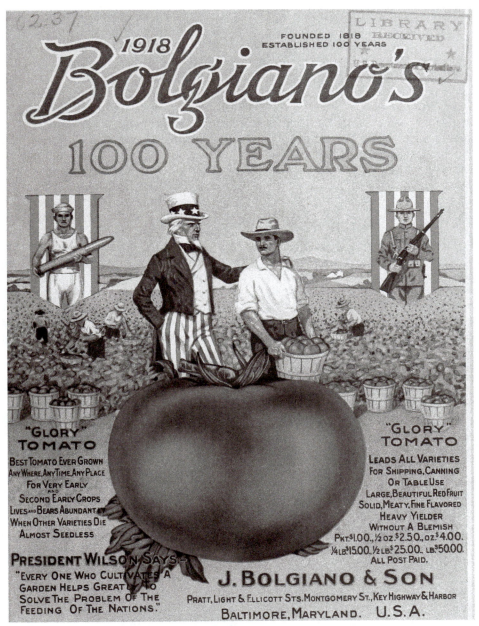

FIGURE 1.3 Tomato seed packet from Bolgiano and Son, 1918. The geographic history of seeds, including their travels around the world, first from America to Europe via the Columbian Exchange and then from Italy back to the United States, illustrates the global political economic nature of food production.

Source: Bolgiano's 100 Years "Glory" Tomato, J. Bolgiano & Son, Baltimore, MD. Special Collections, USDA National Agricultural Library. Accessed October 12, 2018 at https://www.nal.usda.gov/exhibits/speccoll/items/show/8745.

(continued)

Box 1.2 *Continued*

territorial identity as belonging to the Campania region between Naples and Salerno. According to several observers, this label has led to a price premium and massive fraud in the global trade of the prized tomato, which ultimately hurts the small farmers in Campania and benefits those able to market "fake Marzanos" that are grown and processed elsewhere. In a context where people are trying to reconnect with local food producers and seeking seasonal and locally grown produce, geographic origin has become an important political, economic, and cultural issue.

In California, the same tomato tells a different story. There, as Ken Albala (2010) recounts in his article "The Tomato Queen of the San Joaquin," it helped Tillie Lewis build a massive canning business and profoundly transform food in America. Through a number of controversial business deals, Lewis launched the American canned "pomodoro" industry, effectively replacing most imports from Italy. Her product line also included canned juice, spaghetti sauce, and pasta—a precursor to the infamous Chef Boyardee—produced in a large industrial cannery, where most of the labor was provided by women (figure 1.4). Eventually, Lewis became a major producer of C rations for American soldiers in the Korean War, which spearheaded her transition into the first mainstream commercial diet food industry with the Tasti-Diet. Similar to C rations,

the diet was based on packaged food that was "sweet to the taste, but kind to the waist," including canned fruits, vegetables, and meat, as well as artificially sweetened snacks adding up to about 1,200 calories per day. The enterprise received the endorsement of the American Medical Association and was incredibly successful, dramatically influencing the way Americans eat by equating processed food with health and convenience.

Other interesting topics that may be investigated through the tomato are "food miles" and "carbon footprints." Traditionally, tomatoes have been grown in open fields in regions with a favorable Mediterranean climate. More recently, however, they have been produced in high-tech conditioned greenhouses in cooler climates, such as in Canada and the Netherlands. Questions have risen regarding the environmental impact of this new type of production, particularly the contribution to greenhouse gases through additional energy requirements and the trade-offs with transportation-related emissions. In a farm-to-shop study of tomatoes in Australia, Roggeveen (2014) shows that emissions related to transportation and storage are lower than those associated with production, giving support to traditional farming in appropriate climate zones. This of course has significant economic implications for those interested in competing on the global tomato market with new technologies.

FIGURE 1.4 Woman working at an assembly line in a tomato canning plant, circa 1940.

Source: https://commons.wikimedia.org/wiki/File:Working_at_the_cannery_(3126975216).jpg.

The idea of *network* has been helpful in thinking about these spatial relationships, with many scholars having adopted the notion of space as "relational"—meaning that space is not a fixed entity, an abstract notion, an empty container, or a mere surface, but instead is made and shaped relationally through the *movements* of information, goods, and people. Relational space is fluid and created by layers of interrelations—a complex web of networks that come together "under the surface" and may not be immediately visible. Massey (2005) used the phrase "meeting place" to describe how space emerges through the temporary consolidation of competing relations. For example, urban agriculture spaces in the United States and Western Europe can be understood as emerging out of political, economic, social, and cultural networks that shape flows of capital, people, ideas, and practices. Although these spaces tend to be seen as micro-local, it is useful to recognize how they are shaped by external relationships, including urban disinvestment and corporate relocation strategies, increased migration from the Global South, consumer practices, global activist networks, national social policies on food security, and so forth. At different times and under different circumstances, such spaces would not exist.

Place is another extremely important and foundational concept in geographic inquiry. It typically refers to space that is made meaningful by human activity. It could be a nation or a city, but it could also be a garden, a kitchen, a street, or any significant lived space. It includes physical and human characteristics. The resurgence of place as a key concept is tied to a shift of interest among researchers in the social sciences and humanities toward the "everyday" and the diversity of human experiences. This is often referred to as the "cultural turn"—a moment in the 1980s when social scientists, particularly geographers, turned their attention to cultural processes as central to understanding and shaping the human experience. This ushered a shift in both substantive interests and methods that brought credibility to previously ignored topics and approaches. Food is one such topic, along with other everyday practices linked to identity and difference. Within this context, place is seen as a constitutive element of cultural processes to the extent that it shapes our everyday activities and is simultaneously transformed by them. Even in a highly globalized era, place remains significant in influencing how the global movements of people, capital, and ideas unfold differently across space. Equally important is the realization that places are experienced differently by different people. Place, in that sense, is more than its natural and human-made physical attributes, it also encompasses the social and cultural processes that affect belonging and inclusion as well as the human-environment interactions that are central to the geographic approach.

Human-environment or *nature-society interactions* are viewed by most geographers as the "synthetic-core of geography" (Peet 1998), which, as defined above, is about the relationships between people and their environments. On the one hand, society shapes and alters the natural environment by creating landscapes transformed by human settlements. On the other hand, nature provides society with the resources and materials needed to establish livelihoods and cultures. These multilateral relationships are rife with tensions and contestations to the extent that natural resources are limited and social power is uneven. Thus, in studying nature-society interactions, human geographers tend to emphasize the social, political, economic, and cultural forces that lead to environmental change at multiple scales.

Together, these concepts inform the geographic perspective, including the kind of questions we ask as described in Table 1.1.

| | | TABLE 1.1 | Key Concepts and Sample Questions in Geography of Food |

Concept	Definition	Sample Questions	Ch.
Space	The multi-dimensional area in which particular phenomena of interest occur	• How did colonialism influence global food trade?	3
		• Where is food insecurity most prevalent?	6
		• How does climate change impact food yields around the world?	8
		• What is the environmental footprint of food?	8
		• What factors influence food access within cities?	11
Place	Space that is made meaningful by human activity	• What is the impact of biotechnology on Indigenous communities?	2
		• Why are farm workers invisible in rural places?	4
		• How are food practices influenced by local culture?	10
		• How do food festivals transform cities?	11
		• What is the role of food in creating home, especially among immigrants?	12
		• How do ethnic restaurants support immigrant communities?	11
		• Why have food trucks become so popular in cities around the world?	11
		• Does where you live make you fat?	13
Network/ Movement	Interconnections that facilitate the movement of people, things, ideas, and capital between places	• What type of foods are imported from where?	3
		• How does migration support the global food system?	4
		• How does the flow of agrifood technology reflect power?	3
		• Does fair trade help farming communities in the Global South?	5
		• Does globalization threaten regional food cultures?	10
		• How do social movements gain strength from networks?	14
Scale	Size of a feature, phenomenon, or level of analysis	• Why has average farm size increased?	2
		• Should fisheries be managed locally, nationally, or internationally?	9
		• Are local food systems more sustainable?	7
		• How does the body relate to larger scales?	13
Nature-Society Interaction	Relationships between human activity and nature, including the ways humans use and modify nature	• How have agricultural landscapes changed over time?	2
		• Does urban agriculture improve sustainability?	8,11
		• How are land and water resources allocated to different social groups?	3,7
		• How does industrial farming impact the natural environment and climate?	7,8
		• How are Indigenous coastal communities affected by climate change?	9

1.3.3. Geographic Methods

Geographers have a large toolbox at their disposal to collect and analyze data to answer these "where" and other related questions. Unlike other disciplines that rely primarily on secondary sources such as texts or large quantitative data sets, geography places much importance on *fieldwork*—the process of collecting data about people, cultures, and the natural environment. Instead of working in laboratories or controlled environments, geographers tend to immerse themselves in particular places in order to experience and observe what they study firsthand. In fact, the origins of geography are associated with the work of explorers who discovered distant places and gathered information to describe them, often in service of imperial powers. Although the discipline has distanced itself from this troubling past, many geographers continue to see their work as a form of spatial exploration. Fieldwork does not necessary require travel to faraway places; it can take place on your campus or in your neighborhoods. For example, geographers interested in urban agriculture may travel to Cuba to learn about the very extensive networks of gardens used by Havana residents as their primary source of food. Alternatively, they may work in their own communities, including college campuses, to observe how people come together to grow food. Regardless of where it takes place, fieldwork necessitates a commitment to spending time in the very places we seek to understand. Without this spatial immersion, we may miss very important aspects of the relationships between people and their environment, which are only revealed through carefully planned observation as well as casual interactions with people and nature that take time to develop. In fact, we learn *in* and *from* the environment in fieldwork. Although this can be difficult, uncomfortable, and challenging, it is always interesting and typically very rewarding.

The example provided in textbox 1.1 illustrates the difference between secondary and fieldwork data. Publicly available data, such as the US Census, may provide evidence of an alarming level of food insecurity among Native people in Northern California and its relationship to poverty. Yet, it is only by spending time in these places, learning about daily food practices, traditional methods of agroforestry, local plant and animal ecology, and sense of place, that one can begin to make connections between historical dispossession, contemporary energy policy, devaluation of traditional forest management approaches, and food insecurity.

Various types of data can be gathered through fieldwork, including quantitative and qualitative data. Things that can be counted or enumerated, such as population size, income, number of plants, gallons of water used, and crop yields, constitute *quantitative* data. These are often gathered through surveys, audits, and other forms of direct observation where the researcher records numerical values. In contrast, *qualitative* data consist of information that describe or characterize the "qualities" of people, things, practices, and places in ways that cannot be easily quantified. Such data are typically collected through interviews, focus groups, participant observation, or other forms of interaction with the subjects of research. For example, researchers may be interested in perceptions about ethnic food in an increasingly diverse city or the everyday experiences associated with cooking at home. Interviewing ethnic entrepreneurs, conducting a group activity at an immigrant center, going shopping with fathers, or sharing meals with families have helped researchers gain new knowledge on the topics mentioned above.

Geographers interested in culture often engage in qualitative research known as *ethnography*—an approach traditionally based in anthropology that emphasizes the study of people's culture and worldviews from the perspective of an insider, requiring extensive immersion and detailed descriptive work. Today, ethnographic work has become more reflexive based on the recognition that researchers come to the field with personal histories and biases making it difficult to represent an "insider" perspective. Instead, ethnographic work is understood as a negotiated relationship between insiders and outsiders that yields biased yet valuable information. Although many researchers engage in ethnographic work, geographers tend to pay more attention to context than those in other disciplines.

Participatory methods, in which the researcher engages directly with the research subjects through a variety of activities, have become common in the discipline. Such approaches aim to give research participants a more active role in the ways they are being represented. These methods also allow researchers to observe directly how people interact with their environment. *Visual methods*, in which data are gathered from photographs, films, and drawings, have also grown in popularity. Advances in *mobile technologies*, including smart phones, tablets, and other GPS-enabled devices, have also created new opportunities for qualitative and quantitative data collection, including citizen engagement with science through social media or shared applications. It is important to note, however, that not all geographers engage in fieldwork; many rely on data collected by others and, even when they collect their own data, they often supplement and validate them with secondary sources of information.

There are important debates within geography and in the social sciences regarding the validity of qualitative and quantitative approaches. Some argue that qualitative data are approximations that cannot be generalized. In contrast, others claim that quantitative data are reductive in the sense that they simplify the human experience by ignoring factors that cannot be measured. Yet, both can be valuable in their own ways and, in recent years, many scholars have engaged in *mixed methods* that combine both types of data. Ultimately, the value of particular data is linked to how rigorously they were collected and what kind of research question is being asked. Those interested in spatial patterns are more likely to privilege quantitative data, while those interested in places tend to favor qualitative data. Still, this distinction is blurred and has become increasingly so with the expansion of new and mixed methods.

Data collection is only the first part of research, the next step consists of making sense of the data: synthesizing, analyzing, and interpreting. Here, too, geographers have unique tools at hand. Perhaps, the most well-known of such tools is *mapping*. A map is a two-dimensional representation of the earth-surface on which selected information is displayed. Because cartographers make many choices regarding what variables to include, colors, labels, projections, and so on, maps are not neutral representations of facts. Indeed, like other abstract representations, maps are simplifications of reality, which emphasize the spatial features of one or more variables of interest. Although they can be powerful depictions of various social and environmental phenomena, maps need to be interpreted carefully (see textbox 1.3). Acknowledging the limitations of mapping and the power relations embodied in maps is an important aspect of *critical geography*—a branch of the discipline concerned with issues of power and inequality in both real-world spatial arrangements and their representations.

The potential of maps to illuminate spatial patterns and the advance of computer technology have led to the development and growing use of *Geographic Information Systems* or GIS—computer systems used for capturing, storing, displaying, and analyzing data based on their geographic location on the earth's surface. GIS allow researchers to store and manage large datasets about all kinds of phenomena, including natural features such as water availability, soil characteristics, and deforestation; political territories and jurisdictions; population characteristics such as race, income, education, and employment; the location of farms, factories, markets, transportation routes, and so on. Once organized, these data can then be mapped together by combining "layers" to identify relationships between places and/or between variables within a particular place. For example, maps have been used to show the spatial relationship between supermarkets and race, revealing greater access to supermarkets in white neighborhoods.

| Box 1.3 | **The Power and Danger of Maps** |

Maps are very effective tools to display information succinctly and visualize important relationships. Yet, if designed or used poorly, they may be misleading and have severe unintended (and in some cases intended) consequences. For instance, geographers have contributed to research on community food insecurity by mapping access to food, including the location of supermarkets within cities (see chapter 11). Such maps are used to define "food deserts" as those areas with no supermarkets nearby. Further, when juxtaposed with data on race and income, they illustrate the spatial relationship between the location of food deserts and socioeconomic factors. These are then used as a potential explanation for health disparities by race and income. Although this type of map reveals an association between food access and race or income, it does not provide an explanation of how these variables may have contributed to the emergence of food deserts. More analysis is needed to understand the complex and dynamic processes in which retailers have historically abandoned low-income communities of color while new investments in food retail make former food deserts unaffordable to older residents and causes displacement of low-income people and racial minorities. In addition, the maps do not provide any information about the everyday experience of residents and their health conditions. We may question whether the absence of supermarkets necessarily leads to food insecurity; not all residents of food deserts experience the same level of food insecurity, and other important factors might come into play in shaping people's ability to obtain nutritious food. Those factors, however, cannot be shown on maps.

Despite these limitations, food desert maps have informed policy interventions to attract new markets in areas lacking them. For example, unique tax incentives have been given to investors willing to build supermarkets in zones delineated on food desert maps. However, there is now growing recognition that these maps may be misleading to the extent that they do not show all available sources of food such as community garden or small ethnic markets, which may reflect their lack of value in the eyes of investors and policy makers. As a result, policy interventions might contribute to the decline of existing food businesses that are often better positioned to serve the local community. Building on existing community assets may be a more efficient and equitable way to improve community food security. Yet, to the extent that these are "off the map" and therefore invisible, they are less likely to receive policy attention.

Spatial analysis is a branch of statistics focused on finding meaning in spatial data. The goal of spatial analysis is to provide statistical evidence of spatial associations and patterns, including clustering, dispersion, correlation, and so on. For example, environmental geographers may study crop resilience to climate change by analyzing the spatial relationships between the occurrence of droughts, changing temperatures, and crop yields and computing statistical indictors of the strength of these relationships. In the study of food deserts, they could statistically test the relationship between various measures of access to supermarkets and the average income of residents across neighborhoods within a metropolitan area. Spatial analysis has been greatly enhanced by GIS and new geo-computational technologies that allow researchers to work with ever larger datasets, extending over longer time periods.

Beyond traditional cartography, geographers value *visualization* to display, read, and synthesize complex data. They are constantly developing new ways to "map" information, including word clouds, multidimensional mapping, and other conceptual graphic representations.

Geographers working with qualitative data also depend on rigorous methods to analyze and interpret their data. A common approach consists of "coding" textual data, such as field notes, transcribed interviews, and documents, for relevant themes— an approach known as *content analysis*. These themes can be defined a priori based on knowledge of the literature. Preferably, however, they would emerge organically as recurring topics or phrases raise interesting and relevant issues. They would be refined over time as existing data are revisited or new data are collected in response to early observations. Ultimately, patterns, associations, and explanations would surface, substantiated by quotes or other evidence. Because qualitative research is more concerned with generating hypotheses than it is with testing pre-established hypotheses, qualitative analysis is often defined as *inductive*. Instead of starting with a theory and looking for specific data to support or reject it, it begins with a question and leads to multiple stages of data collection and analysis through which a hypothesis emerges. This process is known as *grounded theory* because the theory is "grounded" in field observation and analysis. Such an approach requires a constant conversation between theory, data, and interpretations, which involves a great deal of writing. Indeed, many experts consider *writing* to be an essential component of qualitative method. By taking notes, first in the field and later in research memos, and eventually preparing manuscripts for publications, researchers work hands-on with data and develop what Rapley (2011, 274) calls "a qualitative analytic attitude."

As noted above, critical geographers have often condemned the work of cartographers and statisticians for failing to "go under the surface." For this reason, they tend to favor qualitative approaches that do not focus on specific variables, but instead consider the totality of evidence, including contextual and presumably tangential information that may turn out to be much more important than anticipated by theory. This is not to say that qualitative research is atheoretical. In fact, both qualitative and quantitative approaches are shaped by theoretical foundations that frame the types of questions being asked and the methods being used.

1.3.4. Foundational Approaches

Geography, like most disciplines, is characterized by healthy debates about the best way to approach the relationship between people and their environment. There are foundational differences between how researchers conceptualize the spatiality of human experience and frame their research. These meta-theoretical notions usually rest on broad assumptions about the significance and meaning of space, place, scale, networks, and other key geographic concepts introduced above. Over time, new approaches have emerged, displacing older ones without replacing them entirely.

Because geography of food is a relatively new field, it is primarily influenced by recent perspectives in human and environmental geography, including political economy, political ecology, cultural geography, and feminist geography, which tend to be critical of older perspectives. Yet, the first studies of food are found in regional geography—the most traditional and long-lasting approach in the discipline.

Regional geography, as its name indicates, studies the world's regions. It tends to be very descriptive since the goal is to distinguish regions by identifying their unique characteristics, including social, economic, political, cultural, and environmental. Because food is such an important aspect of economic activities, social relations, cultural identities, and interactions with nature, regional geographers have produced detailed descriptions of the unique food and agricultural practices of various regions. These approaches often emphasize the role of heritage and terroir. For example, Barbara and James Shortbridge's work on American foodways (1998) identifies significant regional differences in how food is grown and prepared across the United States, with agriculture, group identity, and place attachment playing an important role in shaping regional cuisines such as the loco moco in Hawaii, lobster rolls in Maine, chili in Cincinnati, and crawfish in Louisiana. These types of distinctions have led to somewhat humorous—albeit reductive and formulaic—maps depicting American food traditions by state (see figure 1.5).

FIGURE 1.5 Map of iconic foods of the United States.

Source: Author.

The regional approach draws heavily on the concept of place and emphasizes its impact on food culture. However, it has been criticized for a variety of reasons in the past several decades. A common criticism was leveled by quantitative geographers, especially those interested in spatial analysis, who viewed this approach as too descriptive and therefore not useful in producing generalizable predictions. Beginning in the mid-1950s, they sought to make geography more "scientific" by promoting the use of statistics to test hypotheses and advance theory. Geographers who are more open to place-based descriptive approaches have also criticized the regional approach for relying on rigid notions of region or nation and failing to acknowledge the importance of factors diminishing the significance of fixed geographic boundaries, including globalization, mobility, changing cultural economies and identity politics, and the relational qualities of place. These themes are central to geographic approaches that have grown in popularity in recent decades and have had the most influence on the study of food.

Political economy is often credited for putting geography of food on the map (Winter 2003). As Young (2012, 2) argues "the food system has become the most important globally embedded network of production and consumption." The idea of *commodity chains* has been particularly influential in showing these transnational connections between places and actors as agricultural products are transformed into food commodities for consumption (Le Heron et al. 2001, Fine 1994, Goodman and Watts 1997). Early political economic perspectives emphasized the spatiality of the commodity chain and the uneven power relations between producers and consumers in different places. Inspired by Marxism, they drew attention to the exploitation of labor and nature for the benefit of transnational corporations. They also showed how neoliberal policy reforms gave disproportionate power to seed and chemical companies, food corporations, and large supermarket chains. In its early stage, this approach focused primarily on the political and economic forces shaping agriculture and trade. Significant work has also been dedicated to understanding hunger and the uneven distribution of food (Watts 1983, Young 1997).

In contrast, *cultural geographers* study food consumption as the primary realm where culture, identity, and difference are constructed and negotiated. This emphasis, inspired by the "cultural turn" that swept geography and other social sciences in the 1980s, was partly a reaction to the perceived rigidity and determinism of the political economic approach that viewed global capitalism and neoliberalism as the primary forces structuring the food system. Instead, cultural geographers focus on how food acquires meaning through everyday consumption practices that may challenge, destabilize, or reproduce social structures (Jackson 2002, Cook and Crang 1996). Many have been influenced by post-structuralism—a perspective that rejects the essentialism of categories like class, race, and gender and instead views them as fluid, interconnected, and always in the making. They acknowledge the agency of consumers and other actors in food networks and recognize the diversity and place-contingency of human experiences with food. David Bell and Gill Valentine's 1997 book *Consuming Geographies* was extremely influential in shaping this approach by drawing attention to the various interconnected scales at which food and eating acquires cultural meaning from the body to the globe.

Recent work in political economy has reconnected production with consumption, emphasizing the circulation of money, food, and knowledge at various scales. *Circuits of knowledge* have become an important topic as researchers focus on understanding how meaning and value are created and altered at various stages and places along the food chain. For example, Ian Cook (2004) "followed" a papaya through a multi-local ethnography to understand its production as a commodity, not only through the labor of farm workers and packers, but also in the marketing strategies of supermarkets and consumption narratives of elite consumers. Susanne Friedberg (2004) studied how consumer tastes in Paris influence the livelihoods of French bean farmers in Burkina Faso and Zambia.

While acknowledging the global nature of our contemporary food system, many cultural geographers work at more proximate scales including the rural, the urban, the domestic, and the body. Those interested in the rural tend to focus on livelihoods and landscapes, including research on *geographic imaginaries* of rurality that erase economic hardship and environmental destruction. *Political ecologists*, who study the links between people and nature through the power dynamics involved in controlling, managing, and knowing the environment, have played a key role in this area of geography of food. Their work, which intersects with political economy, shows how capitalism and the modern agrifood system dominate and abuse nature, including through deforestation and land-grabs (Jarosz 2009). In addition, they have examined how nature and rural livelihoods are represented in food narratives underlying ethical consumerism (Bryant and Goodman 2004). One of the strengths of political ecology is its ability to attend to the material and symbolic qualities of nature—a hybridity that also describes food.

Urban geographers have produced numerous studies of access to food within cities. In recent years, they have progressively turned their attention to how food is being used to rebrand, label, and otherwise transform urban places. For example, they have studied the role of food in urban contexts such as new development, commercial districts, ethnic neighborhoods, restaurants, and so-called alternative spaces such as farmers markets and community gardens (Joassart-Marcelli and Bosco 2018).

Feminist geographers have had a profound influence by challenging the ways we think about and approach food. While the improvement of women's lives may motivate their work, the contributions of feminist geographers extend beyond gender issues by encouraging us to rethink difference more broadly as something that is experienced, lived, and constructed spatially across various scales and domains. Doing so requires that we consider previously ignored topics and scales of analysis, such as the domestic and the everyday, and engage with new research methods, including participatory and performative methods, to unpack dominant narratives. For example, feminist geographers have helped shift our attention to the gendered nature of feeding and eating (Valentine 1999). They have been influential in rejecting the conceptual boundaries that separate food from the eating body and challenging us to think about eating as an embodied process in which we materially engage with food and connect viscerally and emotionally to the human and more-than-human (Hayes-Conroy and Hayes-Conroy 2008, Longhurst and Johnston 2012). Focusing on the *body* draws attention to how difference is both materialized (e.g., when a lack of healthy food leads to malnutrition and unhealthy bodies) and socially constructed (e.g., when specific bodies or bodily practices are labeled as unhealthy).

These contemporary foundational approaches share a common desire to understand power relations and the role of food in reproducing difference, whether it is about economic inequality and labor exploitation, vulnerabilities and capabilities related to nature and rural livelihoods, disparities in access to food, or stigma linking food to race, ethnicity, class, gender, or other labels. This concern is manifest in the prevalence of *food justice* and food activism as a central theme within geographies of food, including related work on the *right to food* and *food sovereignty*. In addition to identifying spatial patterns associated with the "causes, symptoms, processes, and outcomes of food inequality" (Agyeman and McEntree 2014), many geographers study and participate in various forms of resistance to oppression associated with food. By seeking to bring about positive change in the food system and engaging actively with marginalized communities, they support meaningful collaborative efforts to address injustices in the food system, but also inform how we do food research and what questions we ask (Alkon and Guthman 2017).

This volume addresses multiple geographies of food, including the diversity of substantive issues, methodological approaches, theoretical foundations, and alternative visions that characterize our subfield.

1.4. Goals and Organization of the Book

The primary goal of *Food Geographies* is to equip students with geographic concepts (e.g., space, scale, place, network), analytical tools (e.g., spatial analysis, mapping, ethnographic research, representational analysis), and theoretical perspectives (e.g., political economy, post-structuralism, feminism, political ecology) to examine key food issues. Specific goals include:

1. conceptualize food from a systemic perspective that connects the *global* to the *local*, emphasizing how global food production and consumption networks are *spatially* organized and shape local livelihoods and experiences in uneven ways;
2. study the relationship between food and the natural environment, drawing attention to *environment-society interactions,* including the environmental effects of food production and consumption and their spatial patterns;
3. explore the cultural significance of food and the relationships between food, identity, and *place*, highlighting how race, ethnicity, nationalism, gender, and class interact with place-based food practices, experiences, and representations.
4. think critically about food; envision and evaluate solutions to foster food security, justice, sovereignty, and sustainability; understand the role of food in resisting oppression; develop a compassionate moral vision that includes an appreciation of geographic difference and the humanity of those struggling with hunger, poor health, poverty, and social stigma.

To achieve these objectives, the book is organized in three parts related to the three geographic themes outlined above: political economy, nature-society interactions, and food cultures. In the first part, we focus on the political economic geographies of food and investigate the spatial organization of the food system, emphasizing its global, industrial, corporate, capitalist, and neoliberal nature and examining its alternatives. The central questions of this section are where food is produced, why it is produced there, and what the social and economic consequences are for people and regions involved in this process.

Part II centers on the environmental geographies of food: the nature-society relationships and political ecologies in which food is imbricated. We investigate the environmental pressures that the contemporary food system generates, including stress on natural resources, ecosystems, and climate. We also draw attention to the role of policy and governance in shaping the use of nature and discuss opportunities and challenges to feeding the world sustainably. This section also addresses how the environmental burdens of food are unevenly distributed and affect livelihoods differently. We ask where environmental burdens are concentrated, why, and to what effects.

The third part of the book is dedicated to food and culture. It focuses on the spaces and scales where food becomes meaningful and meshes with cultural identity and difference, including race, ethnicity, and gender. Our approach emphasizes how place and scale shape the way food is experienced and affect health and well-being. Our questions are related to where food is consumed, prepared, enjoyed, imagined, and ingested, with particular attention to the significance of national, urban, domestic, and body scales.

POLITICAL AND ECONOMIC GEOGRAPHIES OF FOOD

Agriculture and Farming 2

Learning Objectives

- Appreciate the significance of agriculture in human history and the evolving relationship between people and the earth.
- Understand agrarian landscapes and agroecosystems as shaped by political and economic forces that influence how people control and use land, water, labor, capital, and technology to produce food.
- Examine the various ways in which people around the world organize themselves socially and politically to produce food.
- Draw attention to the importance of physical, social, cultural, political, and economic geographies in shaping food production.

Most historians agree that agriculture—often viewed as a "prerequisite to civilization"—is one of the most defining contribution to human history. Known as the Neolithic revolution, the domestication of plants and animals that began around 10,000 BCE catalyzed tremendous societal change, including religion, government, economy, arts, and culture. Because it allowed for a more consistent supply of food than was afforded by hunting and gathering, agriculture freed up labor for other purposes, triggering the development of tools, weapons, and writing among other innovations. With those came urbanization and trade, but also conflict and war.

From a geographic perspective, the impact of agriculture on the environment and the spatial organization of society cannot be understated. The beginning of agriculture signals a new relationship between people and nature—a relationship that differs across time and space and continues to evolve. In this chapter, we review the history of farming and examine the various ways in which people around the world organize themselves socially and politically to harness their environment to obtain food. Conceptualizing the farm as a social, political, economic, and cultural place, we trace the history of food provisioning to today's industrial, large-scale, intensive farming operations. In other words, we pay attention to the place-based social structures that underlie farming, including the institutional arrangements involving the ownership and use of land, labor, and capital. We show how different organizations of food provisioning both influence and reflect social and environmental relations that are visible and readable in agrarian landscapes.

We begin this chapter with the foraging model of food provisioning and subsistence, which predates agriculture but remains an important source of food today in several places. We then turn to various forms of agriculture around the world and compare their key characteristics, paying particular attention to the nature-society interaction on the role of physical, economic, political, and social characteristics in shaping agroecosystems.

2.1. The Birth of Agriculture

2.1.1. Food Provisioning before Agriculture: Hunting and Gathering

For most of human history, food has been obtained through foraging, which consists of hunting or fishing wild animals and gathering wild fruits, nuts, and other edible plants. Today, most hunter-gatherers have been displaced by the spread of agriculture. Yet there remain a few places where foraging is still an important mode of subsistence, mostly in areas unsuited for agriculture like deserts, arctic tundra, and tropical rain forests. Examples include the San (or Bushmen) of Botswana and Namibia (see textbox 2.1), the Batak of Northern Palawan in the Western Philippines, the Sentinelese of the Andaman Islands in the Bay of Bengal, the Spinifex (or Pila Nguru) of the Great Victorian Desert of Australia, the Hadza of East Africa, and the Pirahã of the Maici River in the Amazon.

Many myths and controversies surround foraging societies, resulting primarily from the Eurocentric perspective that has attempted to explain history through Western values. For instance, Victorian scholars who first observed and labeled "hunter-gatherers" in nineteenth-century British colonies, described them as "primitive and lazy savages" and assumed that their living conditions could only be improved by the adoption of Western techniques of agriculture. Such perception provided a justification for colonialism, which led to the demise of many hunter-gatherer societies around the world. A major source of misunderstanding regarding foraging societies is also due to the lack of acknowledgment of their diversity and a desire among scholars to identify universal features of hunter-gatherer societies. In reality, past and present foragers vary widely in their environments and social structures and, in many cases, foraging was combined with some animal and plant domestication. A final source of confusion regarding foraging is its romanticization and recent popularization through trends such as the so-called "paleo" diet and the use of foraged food by celebrity chefs.

First among foraging's misconceptions is the idea that the hunter-gatherer life was harsh, short, and miserable. Several anthropological studies have challenged this notion, suggesting that hunter-gatherers lived long, healthy, and relatively non-stressful lives. Because population densities were low and natural resources plentiful, most groups were able to find food easily under normal circumstances. According to some studies, people only needed a few hours of work each day to meet their basic needs and therefore enjoyed more leisure time than peasants did in agrarian societies (Lanchester 2017). This hotly contested idea is summed up in anthropologist Marshall Sahlins's (1968) description of hunter-gatherers as the "original affluent society." Although many scholars reject this view (Kaplan 2000), ongoing debates indicate that there are different ways to assess quality of life. In these assessments, Western standards put a disproportionate emphasis on individual material possession for future consumption, while foragers seem to value the collective satisfaction of basic needs in the present, what Suzman (2017) calls "affluence without abundance."

Another misconception is the assumption that hunter-gatherers were passively dependent on their environment and did not attempt to control or manage nature.

Box 2.1

The San Hunter Gatherers of Southern Africa

The San, who were named "Bushmen"—meaning bandits—by Dutch colonizers, have been hunting and gathering food in southern Africa for millennia. Persecution by Europeans who settled in the area starting in the mid-1600s led to their near extinction. Today there remain about one hundred thousand San who live primarily in the Kalahari Desert of Namibia and Botswana. Their nomadic lifestyle, however, has been threatened by vagrancy laws, hunting restrictions, modern development, and even land conservation projects that make it illegal for them to hunt and forage on their ancestral land. As a result, the majority have taken jobs on commercial farms and ranches, where labor exploitation is widespread.

Building on efforts at the United Nations to promote the rights of Indigenous people, the San have become more politically active since the 1980s and have created organizations to represent themselves in political settings and manage the natural resources on which their livelihoods depend (Hitchcock 2020). Many have adopted the name of First People to emphasize that they were the first inhabitants of southern Africa. Nevertheless, they continue to face discrimination and multiple challenges. In 1997, more than one thousand San were removed from the Central Kalahari Game Reserve by the government to support tourism and conservation—an event that captured international media attention. Human rights organizations alleged that wildlife and park workers used physical force against San people who continued hunting on their former land. In her study of displaced San people, Sugawara (2002, 118) shows how displacement was materialized and experienced through food. As one man put it, "Food—those things which we used to eat on our sand . . . we cannot find such food. . . . All the food that was [available] at our land is absent here." In 2006, working together and collaborating with NGOs, the San were granted the right to return to their land to hunt and forage by the Botswana High Court (Solway 2009). Destruction of water supplies, intimidation by park ranchers, expansion of safari-style tourism, and the lack of hunting permits, have made resettlement difficult, pushing San into poverty, alcoholism, and illnesses such as tuberculosis and HIV/AIDS (Survival International 2019).

FIGURE 2.1 San hunters in the north Kalahari Desert. This illustration of "bushmen" was featured in a book titled *The Story of Africa and Its Explorers*, published in London in 1892 at the height of the British Empire. The author praised the abundance of game, including "antelopes and zebras and giraffes . . . to be had for the taking" and described "the natives" as "so skilled that they were able to run down a zebra on foot and spear the fleet animal as they came alongside of it."

Source: Brown (1892) https://www.flickr.com/photos/internetarchivebookimages/14597372348/.

This notion is backed by evidence that most hunter-gatherer societies did not establish permanent settlements, but instead migrated throughout the seasons according to food and water availability. However, we now know that foragers shared sophisticated knowledge of plants and animals, developed technology for hunting and domestic purposes, and manipulated the landscape. For example, hunter-gatherers, including Native Americans in the Northern Great Plains, harnessed fire for cooking, but also for warding off predators, hunting bison, managing ecosystems, and anchoring social gatherings (Roos, Zedeño, Hollenback, and Erlick 2018). Several hunter-gatherer tribes also domesticated animal species such as dogs and horses to help in their subsistence activities and occasionally provide food. Arctic hunter-gatherers, for example, have used wolf-dog hybrids like the Siberian husky for hunting, traveling, and companionship for centuries (Dumond 1987). Humans began breeding select plants thousands of years before agriculture, including tuberous species like sago palms, manioc, and taro that would complement foraged foods. For example, the Pumé and other nomadic tribes of northern South America have been cultivating manioc, which provides a reliable and rich source of carbohydrates and supports a diet based primarily on wild foods (Greaves and Kramer 2014). In fact, many societies have inhabited an in-between space between foraging and farming in which plants and animals are carefully managed but not fully cultivated or domesticated (Smith 2001).

Recognizing the coexistence of foraging and farming for thousands of years (and into the present) addresses another misconception about the hunter-gatherer food provisioning system: the idea that once farming is introduced, a swift and radical transition occurs leading to the abandon of foraging. While there is no doubt that exposure to farming has led to the transformation of hunter-gatherer societies around the world, often in dramatic and violent ways, there is also evidence that, in some cases, the introduction of cultivars might have helped sustain hunter-gatherer economies, rather than replaced them. By providing complementary food, cultivated plants allowed people to maintain and plan their foraging activities, which they often favored over farming. Some people may have farmed during temperate months and hunted during winter.

Indeed, the transition from foraging to agriculture has not been historically embraced by all societies and is likely the result of climatic stress, increasing population density, and declining access to the vast ecosystems on which people depended for food. While there is no consensus regarding what prompted the transition from gathering to producing food, political and economic forces constraining access to resources are likely to have played a central role in ushering change (see textbox 1.1 in chapter 1). This challenges the belief that agriculture was the result of a rational choice made by hunter-gatherers to adopt a presumably superior mode of food production. In fact, Indigenous people often attribute the transition to greed and the associated human desire to accumulate surpluses beyond what is necessary for survival by dominating nature and controlling land, labor, and capital. This sort of greed can be witnessed today in places like southern Africa (see textbox 2.1) and the Amazon, where hunter-gatherer livelihoods are being threatened by the privatization of land and its use for industrial farming, including ranching and feed crop production, by large national and international corporations.

Yet another misconception is the notion that foraging societies were organized by gender, with men hunting and women gathering plants. Such understanding is often used to justify contemporary gender divisions of labor as a natural evolution (see chapter 12). However, there is evidence that, at least in some societies, women participated in hunting and tasks were evenly shared by community members. Here again, it is important to note the diversity of social arrangements in foraging societies. While women and children participated in trapping, fishing, and small animal hunting, big game was typically reserved to men. Thus, the introduction of spears and new weapons that facilitated big game hunting prompted a new gendered division of labor that may not have existed before then.

In recent years, myths surrounding the hunter-gatherer diet have also been brought to the public's attention in the context of the rising popularity of the so-called "paleo" diet. Archaeological records indicate that, in times predating agriculture, most nutrients were obtained from hunted or fished animals whose lean meat, bone marrow, and internal organs were consumed. Dairy and carbohydrates were virtually absent from early diets since animals were not domesticated and grains not cultivated. This led some nutritionists to argue that "civilization diseases" like obesity and cancer could be prevented by adopting the diet of Paleolithic people. These claims, however, mostly ignore a host of other factors such as environmental conditions, food quality, and lifestyle that may have positively impacted the health of foragers. This push for meat-centric and high-protein diets, especially in the United States, has been criticized for its negative environmental and health effects.

Today, there is renewed attention to foraging as an alternative to mainstream agriculture. Chefs in high-end restaurants have turned to foraging to gather unadulterated ingredients. For instance, the French chef Marc Veyrat has repeatedly earned three Michelin stars—one of the highest awards in the Western culinary world—in part because of his "innovative" emphasis on native herbs and wildflowers, which he gathers in the Alpine landscape surrounding his restaurants. Similarly, René Redzepi—the executive chef at the world-renowned restaurant Noma in Copenhagen—has become a leader in promoting foraging to connect people to nature. His menu features seasonal "forest finds" as well as wild seafood and game. Redzepi even created an app—Vild Mad or wild food (2019)—to "teach you to read the landscape and to discover its culinary potential."

2.1.2. The Origins of Farming

Agriculture forever changed humans' relationship to nature and to each other. The domestication of plants and animals appears to have emerged about ten thousand years ago in the Fertile Crescent between the Tigris and Euphrates in what is today Iraq, Jordan, Syria, Lebanon, Israel, and Egypt (see figure 2.2). Archaeological evidence reveals the cultivation of wheat and barley and the rearing of sheep, pigs, goat, and cattle in that region. The recent discovery that figs were cultivated in the Jordan Valley some 11,400 years ago suggests that agriculture might have begun even earlier than previously thought (Joyce 2006). The physical geography of the region, including water, climate, and land resources, was conducive to agriculture and encouraged hunter-gatherers to settle and form villages. Overtime, some of these grew into cities and infrastructure such as irrigation, fortification walls, and roads developed. Political and social structures evolved to organize the allocation of resources and the division of labor. As the ability to generate an agricultural surplus grew, power relations became more complex. Societies grew more divided, with the emergence of a ruling class consisting of those who controlled the land and therefore the surplus of production. It was not until about 6000 BCE that agriculture would spread from the Near East to southeastern Europe.

Meanwhile, as illustrated in figure 2.2, agriculture emerged independently in other parts of the world, including East Asia, Central America, South America, North America, and Africa. New archaeological discoveries keep pushing the origins of agriculture in these regions to earlier dates. For instance, the oldest known rice paddy was discovered in Eastern China in 2007. It reveals evidence of flood and fire control as early as 7700 BCE (Owen 2007). Each of these regions had its own

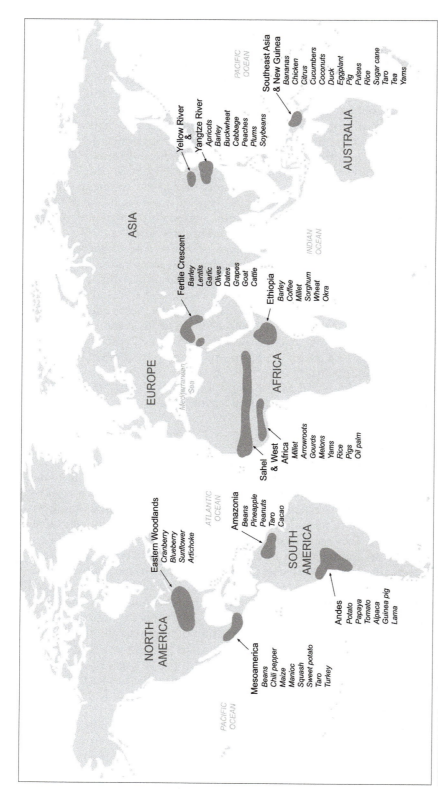

FIGURE 2.2 Map of early centers of agriculture.

Source: Author, adapted from Diamond (1997).

unique combination of crop cultivation and animal husbandry. For example, rice was dominant in East Asia, while maize and beans were the primary crops in Central and South America. Similarly, pigs, chickens, and buffalos were commonly raised in the former, while llama, alpaca and guinea pigs were found in the latter.

Much of the research on early agricultural societies has focused on their increasingly advanced and hierarchical social structures, which distinguish them from earlier and more communal forms of social organization. As with hunter-gatherer societies, however, these understandings are often steeped in misconception, particularly as it relates to the role of religion and violence. For instance, although many depict the Incans and Aztecs as religiously organized in ways that promoted violence and exploitation, recent scholars have challenged these ideas.

In addition to a more complex social organization, agricultural societies also differ from hunter-gatherers in their relationship to nature. The latter saw human life as contingent upon nature; the two are engaged in a reciprocal relationship in which the environment gives to and cares for people and vice versa. With the domestication of plants and animals came the notion that humans are separate from nature and could control it. The emergence of agriculture can thus be interpreted as the control of wildness for the purpose of generating services and products.

Agroecosystems 2.2.

Geographers have been particularly interested in the economic, social, and political characteristics of agriculture and the environment-society relationships in which various forms of agriculture are embedded. The concept of agroecosystem (sometimes called agriecosystem) refers to the socio-environmental characteristics of farming systems. Guy Robinson (2014) in *Geographies of Agriculture* identifies several key factors shaping agriculture in specific places (see figure 2.3), including human and physical inputs such as labor, machinery, water, and nutrients that are applied to the land to generate crops and produce livestock. As the land is cultivated, its physical characteristics are modified, leading to alterations of soil and water chemistry and impacting surrounding life. This, in turn, affects our ability to produce more food and support future human life. Social, economic, political, and environmental systems shape the relationship between inputs and outputs through a variety of interrelated processes, leading to very different forms of agriculture across the world. In other words, *agroecosystems* are socially produced ecosystems in which soil, plants, and animals are manipulated via labor- and capital-based technologies such as crop rotation, irrigation, seed selection, breeding practices, and the use of chemicals for the purpose of increasing biomass or output. They are dynamic and constantly changing as society and environment interact with each other.

2.2.1. Physical Systems

Physical systems' influence occurs primarily through climate, which affects temperature and precipitation; surface and ground water used for irrigation, watering animals, and aquaculture; landform, including slope and elevation; and soil quality typically measured in terms of depth, acidity, and nutrients. Together, these determine what plants can be grown and what animals can be raised in each region. Early regional geographers such as Chisholm (1889) identified world agricultural regions based on climate, landform, soil, vegetation, and/or surface water. Whittlesey (1936) added economic factors such as scale and destination of production to create a classification

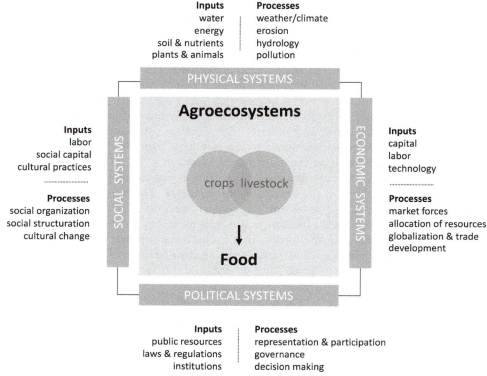

FIGURE 2.3 Conceptual framework for agroecosystems.

Source: Author, adapted from Robinson (2014, 2).

map showing eleven agricultural regions with clear distinctions between the Global South and the Global North. Descriptions and examples of these geographically specific types of farming are provided in section 2.3. below.

Soil is a key element of agroecosystems that embodies the dynamic relationship between environment and society to the extent that its quality is determined by physical and human processes that interact with each other. Soil is a living substance; it consists of minerals such as nitrogen, phosphorus, potassium and calcium, water, air, and organic matter including composting matter such as animal manure and food waste that increase its productivity. Industrialization, however, encourages monoculture, leading to a loss of soil nutrients associated with biodiversity and ecological cycling between plants and animals (see chapter 7). To address this issue, since the 1960s, many farmers have come to rely on synthetic fertilizers that increase yields in the short run but further degrade topsoil, leading to a vicious cycle of even greater use of fertilizer in the long run. These chemicals, including their production, distribution, and application, generate greenhouse gases such as carbon dioxide and nitrous oxide that contribute to climate change, exacerbating the loss of soil quality through rising temperatures and extreme weather events like droughts and/or floods.

Water is another crucial element of agroecosystems that, like soil, is affected by interrelated environmental and social dynamics. Because rainfall is very uneven and often unpredictable, farmers around the world have relied on irrigation for centuries, using wells, canals, pipes, pumps, sprinklers, and other types of infrastructure to redirect surface and ground water to fields. Today, it is estimated that 20 percent of the world's croplands are irrigated, including in desert areas, raising fears of aquifer depletion. For example, in Central Asia, the Aral Sea has been almost drained because of irrigated agricultural projects in Kazakhstan and Uzbekistan. In India, the irrigated land area has more than tripled in the past fifty years, allowing grain production in regions with seasonal or low rainfalls. However, there are serious concerns regarding the sustainability of this type of farming given the rate at which ground water is being depleted. Government-sponsored water projects have also led to conflict as people resent the transfer of water from one region to another. Not only is the heavy use of irrigation around the world reducing the supply of drinking water, but it is also changing entire ecosystems and having indirect effects on weather and climate. Furthermore, the rising use of chemicals such as fertilizers and pesticides has contaminated much of our water supply.

Clearly, the future of agriculture depends on our ability to maintain healthy soils and water supplies. In addition to improving resource management and conservation, this also requires that we address climate change. We return to these issues and the complex environmental impacts of food production in greater details in chapters 7 through 9.

2.2.2. Economic Systems

The economic system structures the relationship between factors of production, including land, labor, and capital, and their allocation to various uses. In the Global South, farming has traditionally been labor-intensive, with a very large share of the population involved in farming. The lack of *capital*, which means machinery, infrastructure, and technology, explains the relatively low level of productivity and high level of poverty among farmers who are primarily involved in *subsistence farming*. In that context, land ownership means power; it enhances the food security and resilience of smallholders. When *land* is unevenly distributed, however, it grants large owners, whether individual, corporate, or public, considerable control over landless farm workers. The economic consequences of land distribution make it a central element of agroecosystems.

Labor is another important component since farming has traditionally been a labor-intensive activity. Both today and in the past, farming has relied heavily on the exploitation of labor. Slavery—the most extreme form of labor abuse—has been used in various part of the world to control the supply of agricultural labor, especially in areas where land was plentiful but labor scarce. For example, millions of enslaved people from Africa were brought by force to work on colonial plantations in the United States and the Caribbean to grow sugar cane and other cash crops for the metropole. Indigenous people were also enslaved by colonizers across the Americas, including in the Spanish missions, where they were forced to farm the land and abandon their own food-provisioning practices.

The 2013 annual report on human trafficking prepared by the US State Department reported that there were approximately twenty-seven million trafficking victims in the world, including men, women and children forced to work in the food system. The report provides several examples, including children from Mali, Burkina Faso, and Benin forced to work on cocoa, coffee, and pineapple farms, ethnic minorities from India on Bangladeshi tea plantations, Brazilian men on Amazonian cattle ranches and sugar plantations, Central Asian and East European refugees in Southern Italian farms, and Southeast Asian migrants in the Thai commercial fishing and Indonesian palm oil industries, among others. Both today and in the past, indebtedness and restricted mobility are used to keep agricultural workers in bondage and servitude.

In the Global North, farming tends to be more capital-intensive (or industrial) and less dependent on labor. Capital owners, including agricultural corporations such as Monsanto, Bayer, Syngenta, and Cargill, have gained significant power over the process of production and the distribution of outputs. Labor demand remains important for certain crops but tends to be seasonal and poorly paid. This type of farm labor is overwhelmingly provided by immigrants whose legal status makes them particularly vulnerable to abuse. In the United States, most farm labor is provided by immigrants from Mexico, who have historically been recruited to fill lower-skilled temporary jobs. Development projects in the Global South, including the *Green Revolution* (see textbox 2.2), seek to increase agricultural productivity by injecting

---

Box 2.2   The Green Revolution

The Green Revolution was the idea, popularized in the late 1960s, that agricultural research and technology transfers would increase agricultural production in the Global South. New hybrid crops were engineered to be better adapted to local physical conditions and respond well to chemical inputs such as fertilizers and pesticides, with the assumption that yields would be significantly higher than with traditional crops. Extensive irrigation and mechanization were also important parts of this revolutionary approach to farming, leading to an increase in the average size of farms and a decline in the need for labor. This idea was so full of promise that, in 1970, the American agronomist Norman Borlaug—known as the "father of the Green Revolution" for having developed and introduced high-yielding varieties of wheat in Mexico, India, and Pakistan—received the Nobel Peace Prize.

India is one of the many countries where the Green Revolution transformed agriculture. Seen as a way to achieve food self-sufficiency and avoid the recurrence of chronic famine, the intensification of agriculture through the use of chemical inputs, mechanization, and irrigation was encouraged by the Indian government as early as the 1950s, with support from the World Bank, the US Agency for International Development, and the Ford and Rockefeller Foundations. In the 1960s, the cultivation of high-yielding varietals of wheat was encouraged by the Indian government's New Agricultural Strategy. Agricultural research centers were established in India and around the world to facilitate the development and adoption of new farming methods. The state of Punjab was one of the early adopters of the Green Revolution and became a leader in the cultivation of high-yielding wheat crops, with the state's wheat production increasing from 1.9 to 5.6 million tons between 1965 and 1972 and overall grain

production growing faster than any other states at an annual average of 6.4 percent between 1961–1962 and 1985–1986 (Singh and Kohli 1997).

Vandana Shiva (1991), one of the earliest and strongest critics of the Green Revolution, writes, "It has been a failure. It has led to reduced genetic diversity, increased vulnerability to pests, soil erosion, water shortages, reduced soil fertility, micronutrient deficiencies, soil contamination, reduced availability of nutritious food crops for the local population, the displacement of vast numbers of small farmers from their land, rural impoverishment and increased tensions and conflicts. The beneficiaries have been the agrochemical industry, large petrochemical companies, manufacturers of agricultural machinery, dam builders and large landowners" (p. 57). In other words, instead of creating self-sufficiency, the Green Revolution has increased dependence on the rest of the world by sabotaging traditional livelihoods, destroying environmental resources, and increasing reliance on expensive imported inputs. A major concern is the amount of water required for the engineered seeds to generate the promised yields.

Shiva equates the Green Revolution to a form of structural violence that has created both an ecological crisis and a cultural crisis causing the loss of lives in the Punjab. The first crisis refers to the environmental threat to life support systems. The second crisis describes "the erosion of social structures that make cultural diversity and plurality possible" and are now causing great ethnic conflict in the region. She attributes the latter to the high level of discontentment and indebtedness among farmers, in a context where the "Green Revolution state" controls agricultural policy, access to credit, and the price of inputs and commodities. In short, the high yields of the Green Revolution have come at very high costs for society and the environment.

capital through foreign aid and investment. Doing so often has serious implications for the environment and for labor, whose livelihood is no longer dependent on the subsistence crops they grow, but on wages paid by corporate farm owners. We will return to labor issues in chapter 4.

The intensity of capital and the industrialization of farming is tied to economies of scale; larger farms are more likely to be able to invest in expensive technology and benefit from it given the scale of their operations. As a result, land consolidation and monoculture have been on the rise in places like the United States, Argentina, Brazil, and Mexico where agriculture has become industrialized. This has an impact on the destination of outputs, which are no longer intended for local consumption but for commercial distribution and international trade. Together, these characteristics contribute to the definition of the modern agrifood system as corporate, global, commercial, and industrial (see chapter 3).

Economic concerns often ignore ecological factors to the extent that they do not represent a direct cost. Unlike land, labor, and capital, which farmers must pay for directly, environmental costs are externalities imposed on third parties and future generations. Growing attention is now given to these environmental costs and questions of sustainability (or lack thereof) in agriculture. The need to feed larger populations, combined with the urge to earn profits, have contributed to environmental degradation, which threatens future generations' ability to grow food. This topic will be covered in chapters 7 through 9. For now, it is important to remember that economic systems do not simply influence the organization of food production, but also the ecosystems in which it takes place, underlying a dynamic relationship between environment and society.

2.2.3. Political Systems

The political system interacts with the economic system in shaping agroecosystems. The state plays an important role in influencing access to resources such as land, labor, and capital and shaping markets through laws and policy. For *political economists*, political and economic forces operate in tandem to produce a system that is often geared toward the benefits of those in power. *Political ecologists* build on this tradition and emphasize how these power-laden processes interact with nature. In that perspective, nature is controlled and transformed by economic interests that are themselves supported by state interventions. These two related fields have contributed much to our understanding of political factors in shaping agroecosystems.

One of the primary ways the state shapes farming is through the enforcement of property rights. In the United States, the founding fathers, especially Thomas Jefferson, viewed land ownership and small-scale farming as the foundation of democracy. Yet, in newly independent America and in many other places around the world, land ownership has not been equally distributed and much of the land has been concentrated in the hands of a few powerful owners who today include a growing number of multinational corporations. This concentration of power underlies the exploitation of natural resources and labor and is a source of conflict in many regions. Landless peasants have limited resources to improve the land and increase productivity and face multiple barriers in escaping poverty, making it difficult to have a political voice.

For instance, in Latin America, 1 percent of landowners control more than half of the region's productive land while 80 percent occupy less than 13 percent of the land (Oxfam 2016). In Africa, there is a troubling trend of *land grabbing* by transnational investors—primarily Chinese companies and government agencies—in countries like Madagascar, Mozambique, and Zambia. According to a study by Arezki, Deininger, and Selod (2012), in 2009 alone 39.7 million hectares were being negotiated for sale to foreign investors, "more than the combined agriculturally cultivated areas of Belgium, Denmark, France, Germany, the Netherlands, and Switzerland" (p. 46).

Political factors are critical in explaining these land grabs. Evidence suggests that such deals are much more likely to take place in countries with high levels of corruption and weak land sector governance because it is easier for investors to bend the law and use their financial resources to enforce their own property rights. They are also more common in low-income countries where governments are struggling to attract investors and hoping that these land sales will lead to economic opportunities. For example, during the 2008 food crisis, which saw record increases in global food prices, the South Korean company Daewoo Logistics negotiated a deal with Madagascar's government to lease 1.3 million hectares—about half of the country's arable land—for ninety-nine years for free in order to grow corn and palm oil (Jung-a, Oliver, and Burgs 2008). The government assumed that this transaction would lead to employment opportunities, infrastructure investment, and technology transfer that would benefit the local population, but did not inform the people whose access to land would be severely compromised. A year later, popular protests led to the fall of President Ravalomanana and the deal's cancellation. Since then, Daewoo Logistics has moved on to other projects, including a large-scale palm oil operation in the politically unstable province of Papua in Indonesia (GRAIN 2018). This example illustrates the role of governments, as well as the importance of democratic forces and social movements, in shaping development trajectories and farming.

Popular movements in many countries have demanded *land reforms* to reduce economic inequality and alleviate poverty among landless farmers. Countries like China, India, Cuba, Vietnam, and Bolivia have redistributed land from large landowners, especially absentee owners, to landless peasants who had been cultivating it, often for generations. In many cases, ownership went from private landowners to the state or peasant collectives. These reforms have had mixed results in terms of poverty reduction, partly because of the incomplete nature of the redistribution and subsequent policies that continued to favor holders of large estates. Textbox 2.3 provides examples of land reform in Mexico and Tanzania.

Today, poor farm workers around the world continue to fight for their rights to access land, with little support from their government. The Landless Workers' Movement—*Movimento dos Trabalhadores Sem Terra* or MST—in Brazil is one of the largest concerted efforts to increase land access for low-income farmers. It does so by promoting the occupation of land that has been abandoned or is not being cultivated by its current owners, relying on a constitutional clause requiring that "property shall serve its social function." As of 2014, the movement won about 7.45 million hectares of land for 370,000 families, although these land occupations are often met with strong resistance by landowners and mitigated support from politicians.

Trade agreements and *structural adjustment programs* represent another important political factor shaping agroecosystems, particularly in the Global South. These topics will be covered in greater detail in chapter 3 on globalization, but it is important to emphasize their significance in promoting commercial and industrial farming and thereby transforming agroecosystems. Trade agreements encourage export-oriented growth based on cash-crops, leading to a decline in small-scale subsistence farming and a reorientation toward large-scale production of commodities like corn, soy, and coffee. While trade tends to favor industrialized economies that produce high-value-added goods, countries specializing in raw material and agricultural commodities are more vulnerable to world price fluctuations and often see their returns shrink in the face of growing global competition. In addition, trade increases dependency on imported goods, requiring Global South countries to continuously ramp up their production of cash crops to pay for these imports with severe threats to

Box 2.3 **Land Reform**

A famous example of land reform comes from Mexico where, beginning in the Spanish colonial period and leading up to the 1920 revolution, land had been concentrated in the hands of a few owners in the form of very large haciendas. The revolution temporarily halted the increasing concentration of land and put in motion a series of agrarian reforms through mobilization of the peasantry, including many Indigenous people who had been oppressed and marginalized by Spanish and Mexican landowners. The most sweeping reforms were put in place in the 1930s, under the presidency of Cardeñas, who passed the 1934 Agrarian Code and redistributed 450 million acres to *ejidos*—government-approved, communally-held cultivated plots of land. Since then, most elected officials have criticized the *ejidos* system for being economically inefficient and have adopted policies to encourage large commercial farming through subsidies and trade agreements. In 1992, President Salinas changed the Mexican constitution to allow the parcelization of *ejidos* and the sale of individual plots, opening the door for a new wave of land consolidation. Today, revolutionary figures such as Pancho Villa and Emiliano Zapata, whose statement that "the land belong to those who cultivate it" continue to inspire fights for the rights of Indigenous and landless people.

Another example of land reform comes from the Ujamaa program in Tanzania (Jennings 2017).

Ujamaa in Swahili means "familyhood" or "extended family" and symbolizes the communitarian version of socialism that influenced many African nations during the postcolonial era of the 1960s. In 1967, the Tanzanian government, led by its first president Julius Nyerere, adopted this program as a framework for its economic policy, which was to prioritize agriculture and self-reliance. A major part of this program was communal land ownership meant to engage the whole population in the project of postcolonial nation building through participation in agriculture at the village scale. Living and working together in interconnected villages would create a more egalitarian society. Implementation was met with some resistance, especially when households were forced to join newly created villages and the process became increasingly bureaucratic. By the 1990s, Ujamaa had been virtually abandoned in favor of market-oriented policies (Kaiser 1996). A major economic crisis in the 1980s forced Tanzania to adopt market reforms and austerity measures as part of a structural adjustment program negotiated with the World Bank and the International Monetary Fund in exchange for debt relief. For advocates of market reforms, Ujamaa was seen as the cause of Tanzania's financial troubles. Yet others argue that it promoted social cohesion in a multiethnic and multireligious society and fostered self-sufficiency.

the environment and the labor force. For example, following the North American Free Trade Agreement (NAFTA), corn production in Mexico became more industrial. Large farms growing genetically modified corn with modern technology progressively replaced small subsistence farms, reducing farmers' livelihood and raising domestic and international migration (see textbox 3.4).

Since the 1970s, structural adjustment programs have been implemented in many Global South countries to address severe fiscal and financial crises. These programs are part of conditionalities tied to loans made by the International Monetary Fund and other organizations to bail out indebted governments. Typically, they require governments to adopt a series of austerity measures, including a reduction in public expenses, the elimination of subsidies and trade barriers, the privatization of state-owned enterprises and assets, and deregulation of the economy. These measures reflect the neoliberal economic ideology that became dominant in the late 1970s in the United States government and at the International Monetary Fund, in which the United States has a powerful voice. Because all these institutions are in Washington, D.C., this became known as the "Washington Consensus."

Repercussions in the agricultural sector have been tremendous. *Privatization* meant that land and state farms had to be sold to private interests, including a growing number of multinational corporations. Consequently, many low-income farmers lost access to the land. Cuts in subsidies for seeds, fertilizers, and equipment, along with the removal of price floors, which guaranteed farmers a certain price for their crops, severely affected small farmers' ability to earn an income and stay in business. *Trade liberalization*—the removal of tariffs and quotas—led to increased global competition. As a result, world food prices dropped, making it difficult for local farmers to compete with imported goods, while at the same time reducing their potential to earn an income via exports. Under this new set of policies, foreign investments were promoted, leading to increases in the size and mechanization of farming. Farmers were encouraged to produce cash crops for export to generate revenue. This led to the spread of large-scale industrialized export-oriented monoculture throughout the Global South, as can be witnessed in the soybean fields of Argentina, the corn fields of Mexico, the cocoa plantations of Côte d'Ivoire, and the banana groves of Jamaica. These political changes partly explain why, today, almost half of the world's farmland produces just four crops: wheat, corn, soy, or rice.

Ironically, governments in the Global North support agriculture and farmers through a variety of policy programs, including subsidies and trade protections denied to the Global South. In the United States, the Farm Bill sets aside large sums of money to subsidize corn, wheat, soy, and other commodity crops (see textbox 3.2). In California, where there is a chronic drought, the state has subsidized irrigation by providing extensive infrastructure and pricing favorable to agriculture. Similarly, the European Union has a complex set of policies in support of agriculture, although many argue that such support has shrunk since the European economy has become more integrated and neoliberal.

In short, political systems shape the organization of farming by enforcing property rights, regulating the economy, and supporting specific industries. Political systems are both global and local; they follow global trends (see chapter 3), but they also reflect local institutional, economic, and cultural histories.

2.2.4. Social Systems

Farming is embedded in social and cultural relations that shape growing practices, divisions of labor, and relationships to nature. At the same time, agriculture has a profound impact on society as it becomes more concentrated, industrialized, and globalized. Broadly speaking, social systems relate to how agriculture is organized along the lines of class, race, ethnicity, gender, religion, and other social categories, with significant consequences for the well-being of farmers and the sustainability of farming. Influenced by rural sociology, many geographers adopted a political economic approach that emphasize class divisions in shaping agriculture. In the 1990s, however, the cultural turn reshaped the field of rural geography by drawing attention to the complexity of rural lives and places (Philo 1992).

As we learned in the previous sections, access to land and capital is an important factor in granting farmers the ability to control the process of production and its outputs. Because most farmers do not own land or capital, they must work for a wage or pay rent to cultivate someone else's land. Poverty is rampant among farm workers and tenants, both in the Global South and in the Global North. Ownership of land and capital is at the core of most definitions of *social class*, especially in Marxist thought. According to Marx, the working class comprises workers who do not own the means of production and sell their labor power for wages. This describes the position of most contemporary farm labor. Although peasants who own a small plot of land are technically not part of the working class, Marx argued that because they are dependent on capitalist creditors for the purchase of machinery and other inputs, they share the struggles of the working class (Katz 1992).

Race has also been historically important in shaping access to land and capital. People constructed and perceived by society as "other" because of their race (or ethnicity) have not enjoyed the same rights and protections as those in power. Racial discrimination negatively affects farmers in many parts of the world, where their ability to own land, obtain credit, and engage in market transactions, is restricted by their race or ethnicity. The history of African American, Asian, and Latino farm labor in the United States, from slavery to the contemporary exploitation of immigrants, illustrates this well. The 2017 Agricultural Census (USDA 2019a) reveals that 95 percent of farm producers (those who own or manage the farm) are white, 3.3 percent Hispanic, 1.7 percent Native American, 1.3 percent African American, and 0.6 percent Asian. Not only is this distribution not representative of the overall population, but it also differs drastically from the composition of the waged farm workforce, of which almost two-third are immigrants from Mexico and less than a third are white (USDA 2020).

In many societies, *religion* intersects with other social dimensions to create systemic barriers to opportunities, including the ability to sustain livelihoods through farming and fishing. There are few places where this pattern of social exclusion is as pronounced as in India, where the caste system continues to divide society and structure agricultural work and rural life despite having been officially abolished decades ago. Shudras are members of the lowest caste who provide most of the agricultural labor. Like sharecroppers, they are indebted to higher-caste landowners who grant them access to land in exchange for labor rents and political support. The most oppressed members of this caste, the Dalits or "untouchables" are excluded from most occupations and forced to perform tasks perceived too degrading for the rest of society, including scavenging as well as menial and seasonal agricultural labor.

Gender is another key social factor shaping agriculture. Throughout much of history, women have not had the right to own property and run a business. Farms—large or small—were typically transferred from fathers to sons, with wives and daughters having no legal claim to land or assets. This remains true in many parts of the world where women's rights are severely constrained. In the Global South, even as women's role in farming has increased, they experience an important gender gap. According to the FAO (2011a), women control less and lower-quality land than men, are less likely to use modern inputs, use less credit, have less education and access to extension services, and often do not have control over the income they generate. As a result of these disparities, women farmers achieve lower yields than men by an estimate of 20 to 30 percent. If resources were more equitably distributed, agricultural output would increase by as much as 4 percent, feeding millions of people and reducing food insecurity.

Focusing on *identity* and *difference*, feminist and cultural geographers have been writing about the "hegemonic masculinity" that infiltrates both farms and rural communities (Little 2017, Whatmore 1991). The perception of farm work as arduous and physically demanding associates farming activities with male labor and devalues women's contribution. In fact, women farmers are virtually absent from representations of farming in the Western world, except as "farmers' wives." The increasing role of technology in agriculture has done little to unsettle this bias to the extent that engineering is often viewed as a male profession. Rural lifestyles, particularly in farming communities of the Global North, are centered on the patriarchal institution of the family farm that supports a heteronormative version of masculinity. Some have argued that these gendered perspectives have prevented us from envisioning more caring relationships to nature and more sustainable approaches to farming.

Guided by this comprehensive framework for understanding agroecosystems, we now turn to various types of farming and describe their interrelated environmental, economic, political, and social characteristics.

2.3. Major Types of Farming

There is tremendous variety in the way societies have used natural resources to produce food around the world, with one of the major distinctions being between *commercial* and *subsistence* agriculture. Perhaps unsurprisingly, these different forms of cultivation have unique geographic patterns, which geographers have been keen to identify and map. Farming approaches tend to be divided between the Global South and the Global North, as we describe in the next subsections.

2.3.1. Nomadic Herding or Pastoral Nomadism

In regions where soil and climate characteristics make crop cultivation difficult, animal husbandry—herding livestock such as cattle, sheep, goats, and camels in search of grazing pasture and drinking water—has been an important form of farming. *Pastoral nomadism* retains some of the characteristics of hunter-gatherer modes of food production, although animals are domesticated and used for meat and dairy. It is common in arid areas like North Africa, the Middle East, Mongolia, and parts of China.

For instance, Bedouins are nomadic people who have herded goats and camels on the fringes of the Saharan, Syrian, and Arabian deserts stretching across North Africa and the Middle East. They lived on milk, yogurt, cheese, and meat from their animals, which they consumed and traded for other goods. However, national borders, political tensions, oil extraction, urbanization, and desertification have all contributed to the sedentarization of Bedouins in the past century.

2.3.2. Shifting Cultivation

In the tropical rainforests of central Africa, the Amazon, and Southeast Asia, where soil nutrient is low and vegetation is particularly dense, people have engaged in *shifting cultivation*, also known as "slash-and-burn" or "swidden-fallow." Here, farmers temporarily cultivate land that has been cleared of its vegetation, often through controlled fires that also help fertilize the soil. After a few years, the land is left to fallow and allowed to revert to its natural vegetation as farmers move on to other locations and repeat the process. The variety of plants cultivated through this process is usually limited to a few root or tuber crops (e.g., cassava), palms (e.g., açai), and/or fruits (e.g., guava). According to Stief (2019), between two hundred and five hundred million people use slash-and-burn agriculture today.

In recent years, the fate of tropical forests and their relationship to climate change has stimulated considerable interest in slash-and-burn practices, particularly in the Amazon basin of Brazil and Peru. Many conservation and development experts view these practices as backward, causing deforestation, and generating low yields (Henley 2011). However, a growing number of scholars argue that these accusations wrongly blame Indigenous agroforestry for problems caused by the growth of modern agriculture, transportation, and migration pressures. Indeed, some shifting agroecosystems, especially with proper agroforestry management, may hold lessons for sustainability to the extent that they do not rely on imported fossil energy for fertilizers, pesticides, and irrigation; represent far more biodiversity than alternative logging, ranching, and farming options; support diverse livelihoods for many people; and limit soil erosion and encourage carbon sequestration through fallow (Padoch and Sunderland 2013).

2.3.3. Intensive Subsistence (Wet Rice)

In humid regions of South and East Asia, paddies are formed for the cultivation of rice in irrigated deltas and floodplains and in rain-fed lowlands and upland terraces. This labor-intensive process has sustained large rural populations in countries like China, India, Vietnam, Burma, Indonesia, Laos, Thailand, and the Philippines where small-scale farms have dominated the landscape for millennia (see figure 2.4). Inputs come primarily in the form of labor from local households, animal power, and nature, including heavy rains. Outputs consist of rice combined with a few additional crops and subsistence livestock, poultry, or fish geared towards home consumption. Since the 1970s, however, the Green Revolution (see textbox 2.2) has transformed the way rice is cultivated in most of Asia, with the goal of increasing yield through genetic manipulations of cultivars. As a result, rice farming has become increasingly commercialized and centralized, particularly in irrigated areas that have a more consistent supply of water and are better suited for more capital-intensive large-scale farming than rain-fed areas.

For example, in northeastern Thailand, where rice paddies are found on rain-fed and drought-prone lowlands, smallholder farming remains dominant with 84 percent of farmland owned and operated by families (Rigg et al. 2016). Development economists often view this as problematic, because it presumably prevents economies of scale, modernization, and much-needed increases in productivity. Others, however, argue that smallholdings reflect the Thai identity and lifestyle and may be a key to sustaining rural livelihoods and eradicating extreme poverty

FIGURE 2.4 Farmer planting rice in an irrigated delta, Vietnam.

Source: https://pixabay.com/photos/rice-cultivation-rice-fields-4165415/.

(Rigg et al. 2016). In fact, evidence suggests that smallholders, especially family farmers, have greater incentives to care for the farm and improve productivity than hired waged labor on large-scale corporate farms. Perhaps because land ownership gives them some stability, smallholders have been particularly innovative at diversifying their activities, including relying on seasonal or temporary urban migration that generates remittances, adopting "micro-mechanization" like power tillers and motorbikes for transportation, and balancing multiple livelihoods both within and outside farming. Yet, climate change, the introduction of new cash crops like rubber and oil palm, global competition, and economic policies encouraging economies of scale continue to put pressure on smallholders to sell their land and abandon farming for urban livelihoods.

2.3.4. Intensive Subsistence (Other)

In dryer regions, where water needed to produce rice is not available, people turn to other crops that usually have a lower yield and are more prone to lasting droughts and occasional flooding. As with subsistence rice farming, this cultivation of staples is designed to feed growers themselves, rather than be commercialized through domestic or international trade. Because of the unreliable supply of water, famines are most common in these regions, including East Africa, Mexico, parts of China, and India.

For example, in Ethiopia, over three-fourths of the population is engaged in agriculture, which consists primarily of rain-fed subsistence farming of staple crops such as teff (see figure 2.5), wheat, barley, chickpeas and lentils—all of which are part of the traditional diet. Livestock and cash crops for exports, including coffee, are also important. The most fertile lands are in the highlands. However, deforestation, overgrazing, steep slopes, and prolonged droughts followed by heavy rains have caused severe soil erosion and reduced agricultural productivity, leading to repeated famines and chronic undernourishment.

FIGURE 2.5 Farmers harvesting teff near Axum in northern Ethiopia. Teff is a small staple grain traditionally used to make the flat bread called *injera*.

Source: Davey, A. https://www.flickr.com/photos/adavey/3131617016.

2.3.5. Plantation

This system was historically set up by Europeans through *colonialism* for the purpose of producing cash crops for exports. Although *plantations* are typically associated with slavery in places like the South of the United States and Latin America, they still exist today in areas where land is heavily concentrated in the hands of a few landlords and export-oriented growth is promoted.

As Sidney Mintz argues in *Sweetness and Power* (1986), sugar became a global commodity through the plantation system. Sugar cane roots were presumably brought to the Dominican Republic by Christopher Columbus and quickly became one of the most important commodities of the trans-Atlantic trade, forever changing the landscape of the Caribbean and Latin America. To satisfy Europeans' seemingly insatiable taste for sugar, large plantations—similar to feudal estates—were established in Northern Brazil, Cuba, Haiti, and other places with capital from the metropoles. Land was cleared for cane fields, local Indigenous populations were enslaved, and millions of additional enslaved people were brought in from Africa. The "white gold" boom enriched merchants and colonial elites, but ravaged the soil and impoverished the workers, setting in motion a vicious circle of *underdevelopment* that continues to affect the region today. When the slave trade ended and colonialism began to unravel, latifundia controlled by local elites replaced colonial plantations with little change to the organization of production and its dependence on foreign markets. Today, plantation-style operations, where commodities like coffee, bananas, cocoa, tea, and mangos are grown by landless peasants, are typically controlled by multinational corporations, such as Chiquita, Dole, and Del Monte (see textbox 5.1 in chapter 5). The orientation to foreign markets has been one of the central characteristics of plantations across time and space.

2.3.6. Mixed Commercial Crop and Livestock

In the Global North, popular representations of farming, from children's story books to artistic paintings, typically include a mix of animal husbandry and crop cultivation organized as a small-scale family-based enterprise. Although this type of farming can be found all over the world, it holds a special place in the popular imagination of North Americans and Europeans, for whom it is woven into visions of family, community, democracy, and independence (see textbox 2.4). The association of livestock with crops such as fruits, vegetables, and grains, promotes an ecological cycle of nutrients in which manure is used to fertilize plants and plants are used to feed animals. The mixing of trees and plants may also help manage soil and pests and minimize risk through diversification. The commercial nature of this type of farming implies that much of the output is sold to generate income and support both household consumption and reinvestment in the farm. Despite their powerful imagery, mixed commercial crop and livestock farms are threatened by monoculture and the economies of scale it promises.

2.3.7. Grain Farming

The cultivation of grains such as wheat, corn, soy, and barley for commercialization is the dominant form of farming in most high-income countries and has expanded rapidly in fast growing economies like Brazil, India, and Argentina. It is typically a capital-intensive process requiring large acreage of land and limited labor inputs. For example, in the United States, corn is grown in the Midwest's "Corn Belt," which includes the states of Iowa, Illinois, Indiana, and parts of Ohio, Michigan, Nebraska, Kansas, Minnesota, Missouri, North Dakota, and South Dakota. Together,

| Box 2.4 | **The Family Farm in North America** |

Thomas Jefferson, one of the Founding Fathers of the United States and its third president, envisioned an agrarian republic in which farmers were the most valuable citizens. In a letter to George Washington, he wrote: "Agriculture is our wisest pursuit, because it will in the end contribute most to real wealth, good morals, and happiness" (Jefferson 1787). For pastoralists like him, an economy built on family farms promotes hard-work ethics and the reproduction of family values, supports democracy, ensures America's self-sufficiency, and leads to a much better standard of living than the industrial and urban economies he had observed in Europe. The yeomen farmers who embodied this republican ideal were white settlers who lived on family farms they built themselves, raised their own animals, and grew their own food. Enslaved people were excluded from this vision of democracy. As the US territory expanded westward, hundreds of thousands of families acquired land to farm—a process facilitated by the 1862 Homestead Act which granted homesteaders ownership of "public land"—stolen from Native people—if they cultivated it.

Throughout US history, family farmers have been represented as embodying (mostly conservative) values of hard work, independence, family, and patriotism. Figure 2.6 illustrates the powerful association made between farming and patriotism in World War II propaganda. Today, although small family farms still represent the majority (89 percent) of farming operations in the United States, they only account for 25 percent of production (USDA 2019b). In contrast, the remaining 11 percent of farms, which include non-family farms and mid- to large-size family farms that rely on hired labor, generate 75 percent of output. Furthermore, due to consolidation, the total number of farms in the country has dropped from 6.8 million in 1935 (USDA 2005) to 2.2 million in 2017 (USDA 2019b) despite rapid population growth. In addition, over 40 percent of small farms' principal operators report a major occupation other than farming, suggesting that farming income alone is not sufficient (USDA 2018).

Yet small family farms remain a powerful source of inspiration in social and political thoughts. Author Wendell Berry in his famous essay *In Defense of the Family Farm* argues that "farms small enough to be farmed by families" have a special relationship to the land, in which caring is informed by intimate knowledge passed on through generations, and to their

FIGURE 2.6 Farming as a patriotic act. Poster created by the Office of War Information, 1943.

Source: https://commons.wikimedia.org/wiki/File:Get_Your _Farm_in_the_Fight.jpg.

neighborhoods on which they depend heavily. For him, farming as a family is "both a practical art and a spiritual discipline"–a much more meaningful form of labor than is possible in an industrial economy.

In a conversation with Berry, fellow Kentuckian and Black feminist author bell hooks (2008) points out the significance of family farms for generations of Black farmers, who are typically invisible from mainstream narratives. She reminisces on her childhood and tells stories of her grandparents and other Black farmers whose relationship to the land gave them physical and spiritual sustenance and supported a form of resistance to what she describes as the "imperialist white supremacist capitalist patriarchy." In a context strife with racial tensions, the Kentucky Hills offer the possibility of reframing racial relationships by creating a certain intimacy between Black and white people—a process she links to ideas of belonging and homemaking. Yet, today, just 1.3 percent of farmers are Black.

these states account for more than 85 percent of corn produced in the United States, which represents almost a third of global production (Statista 2020b). According to Pollan (2006), hybrid corn was key to the industrialization of corn farming in the 1950s and 1960s. With help from chemical fertilizers and pesticides developed in the 1940s, farmers of hybrid corn saw their yields increase dramatically in the post-WWII era. To fight the lower prices caused by rising supply, farmers expanded the scale of their operations and turned to mechanization. Animals and workers disappeared from the farm, along with other crops previously grown. Fields became highly productive "factories"—a far cry from the family farm ideal extolled by Berry (2009; see textbox 2.4). As one farmer told Pollan (2006, 40), "Growing corn is just riding tractors and spraying." Today, most corn grown in the United States is genetically modified to tolerate heavy applications of herbicides (such as Monsanto's Roundup). Like hybrids in the 1950s, GMOs are promising higher yields and further encouraging farmers to convert to monoculture.

With foreign investments and technology transfers, industrial farming of grain, particularly GMOs (Conrow 2018), has expanded around the world with China, India, Russia, Brazil, Argentina, Mexico, and Ukraine among the top ten producing countries, alongside the United States, the European Union, and Canada (FAO 2019a).

2.3.8. Livestock Ranching

This type of farming refers to sedentary or semi-sedentary animal rearing organized on a "ranch," which is defined as a large farm were animals are kept. It usually involves some level of capital investment in the form of feeding equipment, wells for water, paddocks, stables, fences, and so on. It is common in North and South America as well as in Northern Europe, where the soil is not ideal for the cultivation of most crops except forage. Common livestock includes cows, pigs, and chicken, which are now increasingly raised in concentrated animal feeding operations or CAFOs (see figure 2.7).

FIGURE 2.7 Concentrated animal feeding operation (CAFO) in Yuma, Arizona.

Source: Jeff Vanuga / Photo courtesy of USDA Natural Resources Conservation Service. https://commons.wikimedia.org/wiki/File:NRCSAZ02094_-_Arizona_(471)(NRCS_Photo_Gallery).jpg.

For instance, in Texas, where ranching has been historically the dominant agricultural industry, much of the production takes place on very large ranches. Of the 12.6 million cattle raised in the state in 2017, more than half were raised in farms with herd sizes above 500 (USDA 2019a). Among those, 3.8 million were raised on 203 farms with more than 5,000 head of cattle—that is an average of 18,688 cows. Thus, while many small ranching operations remain, most animals are kept on very large farms. Some of the largest facilities are owned by the Five Rivers Cattle Feeding Company that is controlled by the international beef processor JBS, reflecting both vertical and horizontal consolidation in the beef industry. The company operates twelve feedlots in the United States, including three in northern Texas, with an average capacity of 77,500 head of cattle (Food and Water Watch 2015).

Similar patterns have been observed for hogs and chickens. In the United States, according to the 2017 Census of Agriculture (USDA 2019a), 97 percent of hogs are raised in operations larger than one thousand head, with many large farms located in Iowa, Minnesota, and North Carolina. The average number of broiler chickens per farm is approximately thirty-eight thousand. Some of the largest farms, with over one hundred thousand chickens, are found in Ohio, Iowa, Pennsylvania, and California. As we will learn in chapter 8, these large animal feeding operations have significant environmental costs and raise ethical concerns regarding animal welfare.

2.3.9. Dairy

In many cases, animals are raised for dairy production, including butter, milk, cream, cheese, and yogurt. Most milk comes from cows (83 percent), buffalos (14 percent), goats (2 percent), sheep (1 percent), and camels (0.3 percent). Dairy farming has been common in Northern Europe, New Zealand, and the North of the United States (e.g., Wisconsin, Vermont). For instance, France has been producing milk and dairy products like cheese, which is an important part of the French diet, for centuries. Today, however, India is the leading producer of milk, followed by the United States, China, Pakistan, and Brazil. Most milk is processed in an industrial fashion to create a variety of products with a long and stable shelf life, including milk powder, dry whey products, and processed cheese (see textbox 5.3).

2.3.10. Mediterranean

This form of agriculture is often considered ideal because it combines various crops and livestock to create a diverse ecosystem adapted to the Mediterranean climate characterized by long dry and hot periods (see figure 2.8). This type of climate is found around the Mediterranean Sea (e.g., Spain, Morocco, Italy, Greece, Turkey), as well as in California, Chile, South Africa, and Southern Australia. It may be labor or capital intensive and may generate food for subsistence or for commercialization. However, because picking fruits and nuts tends to be more labor intensive than grain harvesting, labor has always been an important factor in Mediterranean agriculture. Common crops include permanent crops such as olives, citrus, grapes, and almonds. This is typically complemented by wheat, legumes (e.g., lentils, chickpeas), and livestock such as goat and sheep.

In the face of climate change and ongoing urbanization, water has become an increasingly scarce resource in Mediterranean climates. Adaptation to and mitigation of these new circumstances will be particularly important in sustaining this farming system.

FIGURE 2.8 Typical Mediterranean agricultural landscape. A combination of olive and almond trees, grains, and vegetable crops in Festos, Valley of Messara, Crete.

Source: Nat Pikozh, https://creativecommons.org/licenses/by/2.0.

2.3.11. Commercial Gardening

This system consists of farms focused on specialized crops of fruits and vegetables, including orchards and vineyards. These are typically found in proximity to urbanized areas, in places where physical characteristics make the cultivation of grain challenging. Although rarely described as gardening these days, the Central Valley of California illustrates this type of farming with its numerous orchards, vineyards, and large fields of fruits and vegetables. In cooler climates, greenhouses are often used to control temperature and humidity to facilitate growth. For example, the Netherlands is a major exporter of potatoes, onions, and other vegetables, most of which are produced in high-tech greenhouses (see figure 2.9), where yields are much higher than on outdoor land and the use of chemicals is significantly lower (Viviano 2017).

Conclusion 2.4.

Distinguishing between various agroecosystems draws attention to the importance of physical, social, and cultural geography in shaping food production. However, the classification of the world's agricultural regions into eleven distinct agroecosystems has been criticized for being rigid. One of the most common criticism pertains to its ethnocentric perspective in which farming patterns in the Global South are attributed to "backwardness" and/or unfavorable physical environments, ignoring the political and economic factors that have contributed to uneven farming patterns. Nevertheless, the classification highlights the tremendous variation in farming approaches across the globe and their relationships to physical environments. As Whittlesey

FIGURE 2.9 Greenhouses in Westland, the Netherlands. Greenhouses support the cultivation of crops that low temperature and overcast weather would otherwise make challenging.

Source: Creative Commons, https://commons.wikimedia.org/w/index.php?curid=130481.

(1936) who created one of the first and most well-known maps of global agroecosystems puts it, it provides valuable information by "clarifying man's comprehension of the limitations and opportunities which Nature poses as incentives and restraints to human ingenuity" (240). Much of the contemporary literature has moved away from such *geographic determinism* and is accounting for the political and economic forces underlying these uneven patterns. In addition, many researchers are now focusing their attention on how the physical characteristics of environments are evolving with human activity in ways that change agroecosystems and threaten the future of agriculture. According to many, contemporary agricultural systems "have lost their ecological foundation, as socioeconomic factors became the dominant driving forces in the food system" (Francis et al. 2003). Geographers have been instrumental in producing research that attends to the socioeconomic and environmental characteristics of agricultural systems and challenges nature-society binaries in ways that may help us design more sustainable systems.

Global Food Regimes 3

Outline

3.1. Globalization and Food Regimes
3.2. Early Globalization: The Spice Trade
3.3. Colonization of the New World and the Colonial Food Regime
3.4. Postcolonial Developmentalist Regimes
3.5. Contemporary Food Regimes
3.6. Homogenization and Differentiation
3.7. Conclusion

Learning Objectives

* Explore how globalization changed the way food is produced, distributed, and consumed.
* Distinguish between various stages of globalization, beginning with the spice trade and colonialism and leading to the contemporary food system.
* Use the concept of food regime to highlight the role of the state and corporate actors in shaping where, how, and for whom food is produced.
* Describe the main characteristics of today's industrial, capitalist, corporate, neoliberal, and global food regime, emphasizing the concentration of power.
* Examine the impact of globalization on the homogenization or differentiation of food cultures around the world.

Can you imagine Italy without tomato sauce, Texas without steaks, California without citrus, Belgium without chocolate, Hawaii without pineapples, Ireland without potatoes, or Thailand without chili peppers? You might be surprised that none of these iconic foods could be found in those places before Europe's colonization of the Americas and the vast global exchange of people, plants, animals, and microbes it unleashed.

The movement of food, going hand in hand with human migration, has been a central feature of history. Indeed, studying food reminds us that *globalization* is not new and forces us to think about it carefully, paying attention to the flows of food commodities, the evolution of food cultures and practices, and the political and economic forces and technologies shaping these global exchanges. Doing so reveals that, although the globalization of food is nothing new, its scope, organization, and impacts have changed dramatically over time, culminating in today's global, industrial, corporate, capitalist, and neoliberal food system.

In this chapter we explore how globalization changed the way food is produced, distributed, and consumed over time and space. We use the concept of *food regime* to highlight the role of state and corporate actors in creating and structuring the contemporary global food system and describe its main characteristics and impacts on society. After defining globalization, we describe how it relates to food and outline various stages in its development leading to the contemporary global food system. In our chronological examination of the early trade of luxury goods such as spices during medieval times, the colonization of the New World and the so-called Columbian exchange, the postcolonial regimes that followed World War II, and the neoliberal corporate regime that emerged in the 1970s, we emphasize the political and economic forces that shaped the nature and scope of globalization. We end the chapter with a discussion of homogenization and differentiation of food cultures around the globe.

3.1. Globalization and Food Regimes

Globalization has a bad reputation. The term has been used by many to represent everything that is bad about our economy and society. The globalization of food in particular has been blamed for problems such as world hunger, the obesity crisis, the ubiquity of McDonald's, rising inequality and poverty, environmental degradation, and climate change. Although there is certainly some truth in these claims, it may not be very informative to attribute all contemporary problems to globalization without defining clearly what we mean by it. If everything were explained by globalization, the concept would be entirely useless.

Globalization is often understood as the global expansion and integration of markets through trade and capital investments. However, globalization is not just economic; it also includes the dissemination of ideas and cultures, the movement of people, and the reorganization of political power and institutions. It is a spatial and temporal process in which human activity becomes increasingly transnational. Geographers often talk about *time-space compression* to describe how the world is "shrinking" at a rapid pace. This process can be elucidated and further refined by focusing on extensity, intensity, velocity, and impact (Held, McGrew, Goldblatt, and Perraton 1999). *Extensity* refers to the extent or coverage of interconnections over the earth's surface, reaching into increasingly remote and distant locations and involving a larger number of organizations. For instance, our taste for meat is prompting transnational corporations to clear more forest in the Amazon to create additional grazing land—a process that brings isolated places into global meat networks. *Intensity* draws attention to the volume of trade, the flow of ideas, and the consistency of global transactions. A growing share of our diet depends on imported products such as green beans from Kenya, grapes from Chile, and shrimp from Thailand, with such commodities traded in large volume along well-established supply chains (see chapter 5). *Velocity* indicates the increasing speed at which goods, services, and ideas travel. French cheese orders from restaurants in New York City can be met in a few hours thanks to information and transportation technologies. Finally, *impact* denotes the fact that events happening in one place have consequences in other places. For example, the so-called "Arab Spring"—the widespread anti-government protests that took place in Tunisia, Egypt, Yemen, Syria, and Libya in the spring of 2011—was partly triggered by the rising cost of bread caused by skyrocketing wheat prices linked to a drought in China, floods in Australia, and an export ban in Russia.

Scholars have cautioned us against accepting the determinist idea that globalization is an inevitable process that stretches through the world in uncontrollable ways. Indeed, globalization is the product of human activities that are sanctioned by political organizations that facilitate, regulate, or thwart such activities for economic purposes. The idea of *food regime*, which was originally formulated by Harriet Friedmann (1987), suggests that food production and consumption are organized by a "rule-governed structure" at the world scale. This structure consists of "sustained but nonetheless temporary constellations of interests and relationships" that are established by the most powerful actors, such as nation states, corporations, lobbies, and/or international organizations, to facilitate capital accumulation. Food regimes are both a reflection and an underpinning of the global economy and its hierarchies of power. They are therefore dynamic and evolve over time in response to economic crises and changing geopolitics.

Thus, it is important to view globalization as a political and economic system governed by rules, norms, and procedures set up by local, national, regional, or transnational institutions. These rules have changed over time and could conceivably be altered to produce a different regime. For example, as we will learn in the next section, international agencies like the World Bank, the International Monetary Fund, and the World Trade Organization have been instrumental in promoting the neoliberal global food regime that is in place today. Yet, under pressure from social movements around the world to address labor and environmental issues, a different type of food regime could be possible.

In addition to rejecting the deterministic views that dominate globalization narratives, many geographers have also questioned popular notions of "borderless" or "flat" worlds, which suggest that globalization removes barriers of entry into the world economy and levels the playing field. Borders still play a major role in dividing and shaping the world economy (Diener and Hagen 2009) and the world remains a very "rough landscape" in which economic, social, cultural, and environmental disparities continue to distinguish places (de Blij 2008). The study of food illustrates this unevenness. For example, the production of commodity crops continues to shift to the Global South, where the environmental impacts of monoculture are worsening (see chapter 7) and hunger remains a pressing issue affecting more than eight hundred million people (see chapter 6). Although globalization seems to have homogenized food around the world, with McDonald's, Pizza Hut, and Starbucks outlets popping up everywhere, there are clear signs that these trends have also spurred a revival of local cuisine and encouraged diversity. In short, there are qualitative differences in how globalization is experienced across space and time.

Early Globalization: The Spice Trade 3.2.

People have long traded basic staples like meat and grain products within their own societies. Until the fifteenth century, exchange with other societies over longer-distances was less prevalent and limited mostly to nonperishable luxury foods. The Romans, for example, imported products from various parts of their empire, including wine from Egypt and southern France and olive oil from Spain and Libya. They also imported spices, such as pepper, cinnamon, and cloves, from India.

The spice trade expanded dramatically during the Middle Ages and reached a high between the thirteenth and sixteenth centuries. Spices were used extensively by the rich to bring flavor and distinction to food and were valued for their medicinal properties (see textbox 3.1). Their appeal was caused in great part by their exoticism and association with the "Orient"—a mysterious and heavenly location to the East called India and the Indies. Spices were extremely expensive in Europe, and their trade was very lucrative, bringing wealth to merchants and cities such as Venice and Genoa.

The most common spices used in medieval cooking were pepper, cinnamon, ginger, saffron, nutmeg, clove, cardamom, coriander, mace, and sugar, which was not yet the mass-produced commodity it would become in the future and still considered

Box 3.1 **Spices in French Medieval Cuisine**

Guillaume Tirel—known as Taillevent—became royal chef in 1326, serving multiple generations of French queens and kings, including Charles V and Charles VI. His recipes are gathered in *Le Viandier*—one of the oldest and most famous French cookbooks. In this precious record of medieval cuisine, Taillevent describes loosely how to prepare dishes such as game hens with ginger and saffron; peacocks, swans, or pheasant in a rich sweet and sour sauce seasoned with vinegar, cinnamon, cloves, nutmeg, and "grains of paradise" (long pepper); and apple tart flavored with saffron, ginger, cinnamon, anise, and raisins. These dishes, served at weddings and other feasts, have more in common with Indian, Persian, or Moroccan cuisine than they do with contemporary French food. Instead of letting quality ingredients stand alone as modern chefs take pride in doing, medieval master cooks sought to create complex and flavorful dishes in which ingredients disappeared in spicy sauces.

According to historical records, medieval feasts consisted of three to four courses, each offering a handful of dishes. Ornate salt cellars and spice bowls were set on the table (see figure 3.1). Spiced red wine known as Hippocras—a name hinting at its potential medicinal properties—was served throughout the meal. Kitchen accounts reveal the importance of spices in the budget. For example, such records indicate that, in 1468, the Duke of Burgundy ordered 380 pounds of pepper for his wedding, confirming his status as one of the richest men in Europe.

FIGURE 3.1 *The Peacock Banquet*, anonymous painting, mid-fifteenth century. Elaborate dishes (such as Taillevent's peacock recipe, presented with its head and feathers) were a sign of status, as were salt cellars and spice bowls placed in front of guests of honor.

Source: https://commons.wikimedia.org/wiki/File:Banquet_du_paon.jpg.

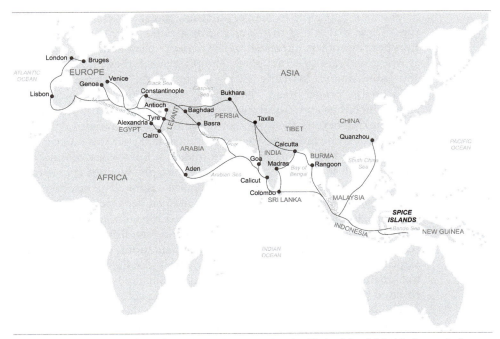

FIGURE 3.2 Map of medieval spice trade routes linking the "Spice Islands" to Venice and other European cities (eighth to fifteenth century).

Source: Author.

a spice. The vast majority came from Southeast Asia, including the Maluku or spice islands (Indonesia) and Ceylon (Sri Lanka), and transitioned through India. Although most of the trade went to China or stayed within the Indian subcontinent, a significant portion was destined for Europe, where demand was growing. Various trade routes were established by Arab and European merchants to link ports in the West coast of India to Venice and Genoa (see figure 3.2). These routes evolved over time according to wars, discoveries, and alliances. Shipments form "India"—as the region was often labeled—made their way to the Mediterranean through the Persian Gulf, transiting through Basra, Baghdad, and Constantinople, or the Red Sea, passing through Cairo and Alexandria. The western portion of the famous Silk Road was also used as an alternative land-based route, going through Tabriz and Samarkand in central Asia and reaching the Mediterranean through the Black Sea and Constantinople. Ultimately, the eastern Mediterranean cities that handled trade on these various routes came under the control of Islamic potentates, including the Ottomans, cutting into the profits of Christian spice traders. This prompted merchants to consider alternative routes that would bypass Islamic cities, eventually sending off Vasco de Gama, Christopher Columbus, and others on expeditions to reach India and the spice islands. Their discoveries would dramatically expand global trade beyond luxuries.

3.3. Colonization of the New World and the Colonial Food Regime

Christopher Columbus's 1492 arrival in the New World would change the global food system in dramatic and irreversible ways. As the Spanish, Portuguese, and eventually other Europeans settled in the Americas, they discovered a wide range of foods, from exotic pineapple to maize, potatoes, tomatoes, chili peppers, and sweet potatoes. These would find their way to Europe and other parts of the world. However, Europeans were less interested in native plants than in cultivating familiar crops such as wheat and raising cattle, goats, and pigs, which they introduced to the Caribbean and Latin America. As Earle (2012) argues, all the pineapple of the world could not make up for "proper bread." New World settlers sought to reproduce their European diets, which they considered superior in taste and health properties.

Thinking beyond their own consumption, Europeans introduced crops that could be shipped to the mainland to generate income. Sugar was one such crop. Originally from Southeast Asia, sugar had been traded in Europe for centuries and was in very high demand. While Europe's physical geography was not appropriate for the cultivation of sugar cane, the Caribbean had all the right elements, except for labor, given that Indigenous populations were small and had been decimated by the introduction of European diseases. Thus began the largest forced migration of human history, removing as many as fifteen million women and men from Africa over a four-hundred-year period to become slave labor in sugar plantations, gold mines, and other colonial extractive enterprises throughout the Americas (see figure 3.3). The early stage of the transatlantic slave trade, in the 1500s, was dominated by the Portuguese, who shipped enslaved people mostly to Brazil. Over time, the volume expanded and other nations such as Spain, France, the Netherlands, and England got involved in transporting enslaved people to the Caribbean, Mexico, and Columbia. Eventually, England would rise as the most important slave trader, fueling the economy of what would become the United States and engaging in lucrative transatlantic trade of commodities. Although trading enslaved people across continents was banned in 1808, enslaved people continued to be shipped from Africa to the Caribbean until the 1860s. By the mid-1700s, enslaved Africans working in plantations represented between 70 and 90 percent of the population in places like Cuba, Jamaica, and Haiti.

FIGURE 3.3 Enslaved Africans cutting sugarcane on the island of Antigua, painting by William Clark, 1832.

Source: The British Library (Public Domain) https://commons.wikimedia.org/wiki/File:Slaves_cutting_the_sugar_cane_-_Ten_Views_in_the_Island_of_Antigua_(1823),_plate_IV_-_BL.jpg.

The slave trade was extremely profitable, as was the trade of commodities such as sugar, molasses, rum, cotton, and tobacco that were produced by enslaved people for European consumers. The profits from this transatlantic trade financed the industrial revolution that took place in Europe, especially in England, in the first half of the eighteenth century, leading to a global division of labor in which industrial goods were produced in the Global North while raw materials, including food staples, were extracted or produced in the Global South. By the 1800s, the monoculture of crops for European markets had become common throughout the colonies, including cocoa in West Africa, bananas in Central America, sugar in the Caribbean, and coffee in Southeast Asia. Wheat and cattle, however, were concentrated in the Global North—mainly England, North America, and Australia, where they were produced on family farms and supported a standard meat, bread, and dairy diet. Built upon colonialism, the British-dominated food regime created a network of food exports and imports that spanned the world, increasing global dependence on traded food for consumption. Technological innovation in transportation, particularly steamships and railroads, hastened the pace of globalization in the second half of the nineteenth century.

Postcolonial Developmentalist Regimes 3.4.

Decolonization in the middle of the twentieth century, along with the Great Recession of the 1930s and World War II, destabilized British hegemony and led to the ascendance of the United States and the Soviet Union as the leading powers in the postcolonial food regime. Using its wheat surplus as a diplomatic tool, the United States granted food aid to war-torn Europe through the Marshall Plan and to newly independent countries through the Food for Peace program, presumably to promote development. The wheat surplus was the result of a domestic farm policy in which the government would buy farmers' wheat if they were unable to sell it at a guaranteed market price (see textbox 3.2). Combined with import restrictions, this farm policy protected American farmers and isolated the domestic wheat market from global pressures, keeping prices high. It also provided an effective means for the United States to increase dependence on US imports around the world. Rather than focusing on cheap food imports from colonial territories as England had, the United States' success rested primarily on creating markets abroad for its own food commodities, while continuing to import what it could not produce itself.

Meanwhile, postcolonial governments were eager to promote development and secure economic independence from former colonial rulers. Often described as "developmentalist states," they put in place programs to direct and promote economic growth, including subsidies and trade protections for agricultural sectors deemed important to the national economy. For example, in India, the "Grow More Food" program was designed to increase production and stop food imports that drained the public coffers. Nehru, who became the first prime minister of India in 1947, told the nation in a radio address: "If we do not produce enough food for our country, we become dependent upon other countries, and in a matter like food we cannot afford to be dependent" (quoted in Sherman 2013).

Yet farmers in formerly colonized countries struggled; competition with subsidized food from the United States forced them to keep prices and wages low. While some countries, such as India, attempted to increase self-sufficiency, others focused on export-oriented monoculture of commodities like sugar, pineapple, and coffee, to generate cash returns. Beginning in the 1960s, the Green Revolution encouraged these countries to modernize farming with seeds, chemicals, and equipment imported from the United States (see textbox 2.2). This technology, developed through research sponsored by government agencies and private foundations from the United States, would usher the transition to industrial agriculture. Farms grew larger and became more specialized, replacing subsistence production. However, as production of cash crops increased and yields rose, world prices fell and economic growth lagged, creating food insecurity in countries that had become dependent on imports to feed their people.

In addition to building a global market in which the United States became the "breadbasket" of the world, food aid also contributed to strengthening the geopolitical power of the United States. This was particularly important during the Cold War era (1947 to 1991) when the United States and the Soviet Union—the two superpowers and largest wheat producers in the world—were competing for allies in the so-called Third World through food and other types of assistance. While the Soviet Union supported Cuba, Vietnam, Laos, China, Cambodia, and other "Eastern Bloc" communist countries, the United States' influence extended over most of Latin America, the Philippines, and Thailand. Many African nations such as Angola, Ethiopia, Mozambique, Somalia, Namibia, and the Congo, were torn in proxy wars between the first and the second worlds, the scars of which can still be seen today. Political alignment with the United States would have a tremendous impact on economic and agricultural policy, including trade and foreign investment. Many aid recipients also became importers of US wheat and, by the early 1970s, the developing world was purchasing 40 percent of all US food exports. This is a major difference from earlier food regimes when most of the imports, whether spices, sugar, or tea, came from the Global South to the Global North. For many countries, the flow had now been reversed.

3.5. Contemporary Food Regimes

Catching up with economies that had received a head start on industrialization thanks to centuries of financial windfalls from colonialism proved to be extremely difficult, if not impossible for the Global South. By the 1980s, after two major global recessions triggered by rising oil prices (1972 and 1979), it became apparent that national economic programs were not meeting expectations of growth and development. To make matters worse, throughout the 1970s, the Global South experienced a major food crisis, with prices escalating and availability—including food aid—dwindling, making it very difficult for developing countries to import the food on which they had become dependent. The drop in grain supply was due to several factors including a drought in the Soviet Union, rising consumption of meat diverting grain to feed livestock, and a change in US agricultural policy that reduced the national stockpiles (see textbox 3.2). To address this immediate food (and oil) crisis and to continue to develop their economy, governments of low-income countries borrowed billions of dollars from international banks in the 1970s and accumulated large debts, which they became unable to repay. The 1980s are often described as a period of "debt crisis" because many Global South countries, especially in Latin America, defaulted on their loan obligations.

This debt crisis was dealt with by *structural adjustment programs* in which the International Monetary Fund (IMF) and the World Bank agreed to reschedule interest payments through new loans if indebted countries on the brink of financial collapse met specific conditions. Despite resistance from their people, government

| Box 3.2 | The US Farm Bill: A Tool for Controlling the Global Food System? |

Agricultural subsidies and trade protections have given the United States a significant advantage in the global food market, especially for cereals. Indeed, government support for staple crops has existed from the early days of the republic, by granting land to homesteaders and supporting agricultural research at land grant universities. Today, the so-called Farm Bill provides support for commodity crops in the form of farm subsidies, totaling more than $20 billion per year in 2018.

The Farm Bill came into existence during the Great Recession of the 1930s when food prices plummeted and many family farmers were at risk of losing their farm due to insufficient income and high debt. The 1933 Agricultural Adjustment Act was designed to alleviate farmer debt and increase revenue by providing loans to farmers to limit production and buying excess grain (that could be stored and released if needed). The act was renewed in 1938 when it became the first official Farm Bill. The surplus purchasing program would guarantee that prices would not fall below a minimum and that farmers would earn a decent income. In the post-WWII era, the surplus was used to support "developing nations and friendly countries."

In the 1970s, major reforms were enacted under the leadership of the United States Department of Agriculture (USDA)'s secretary Earl Butz. Rather than encouraging farmers to limit supply, Butz wanted to modernize agriculture and increase production—a strategy he famously summarized as "get big or get out." The 1973 Farm Bill set up target prices and deficiency payments, paying farmers for the difference between market and target prices on their crops instead of buying a surplus to guarantee a minimum price. This encouraged production and allowed market prices to drop while maintaining farmers' income, providing a huge benefit to corporations trading and processing commodities such as wheat, corn, and soybeans. Through consolidation, small farms disappeared, giving rise to megafarms that took advantage of new technology and economies of scale to produce cheap "inputs"

for food processing plants. Exports of US commodities were also encouraged through marketing assistance and specific loan programs.

The 1996 Farm Bill, known as the Freedom to Farm Act, provided farmers with direct payments independent from prices and production levels and created a subsidized crop insurance program. It also promoted exports. In subsequent versions of the bill, direct payments were curtailed and insurance programs expanded. Private companies provide insurance against revenue or yield shortfalls, with administrative cost and premium heavily subsidized by the government. Today, crop insurance has become the single largest farm subsidy program. Two other important programs include the Agricultural Risk Coverage and the Agricultural Loss Coverage that provide a similar type of insurance payment to farmers if revenues per acre or prices fall below set levels. Corn, soybean, and wheat farmers receive over 70 percent of these subsidies. Large farms are also benefiting disproportionately. For instance, 60 percent of subsidies go to the largest 10 percent of farms, and while three-fourths of farms with revenues above $100,000 receive federal subsidies, less than a third of those below that threshold do so (Edwards 2018). In fact, many small farmers who grow a variety of fruits and vegetables for local markets receive no subsidies at all.

Over the past century, the Farm Bill has shaped both domestic agriculture and the role of the United States in global food markets. Its primary objectives have shifted from supporting small farmers through loans and price support in the aftermath of the Great Recession to facilitating the United States' global influence through donation of surplus food after World War II and eventually encouraging industrialized production and trade of low-price commodities through direct payments and crop insurance programs. It appears that, through their lobbying efforts, large and powerful agrifood businesses have hijacked the Farm Bill, with dire consequences for small farmers around the world, the environment, and health (Imhoff 2019).

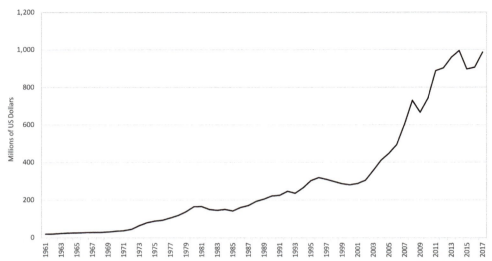

FIGURE 3.4 Value of global exports of food and animals, in millions of US dollars, 1961–2017.

Source: Author with data from FAO (2020a).

officials in these countries had little choice but to accept conditionalities, forcing them to liberalize trade, devalue their currency, deregulate the economy, cut public spending, and privatize nationalized industries. These policies would have tremendous consequences for food and agriculture. While Europe and the United States continued to subsidize their own agriculture, developing countries were asked by international organizations to eliminate such subsidies, further increasing the imbalance between the Global North and the Global South.

Monoculture of cash crops, undertaken in increasingly large and industrialized plantation-style farms, became even more widespread, especially in the Global South, where many of the remaining subsistence farms were displaced or enrolled into global production networks. Exports of commodity crops and livestock grew rapidly in the past fifty years as illustrated in figure 3.4.

Since the early 1970s, the composition of exports in terms of their value has also evolved (see figure 3.5), with cereals being replaced by fruits, vegetables, and nuts as the most important export category. Bananas, tomatoes, grapes, avocados, cashews, and almonds are high value crops, which can be more easily stored, processed, and transported today than in the past. Many are grown in Global South countries like Mexico (avocados, tomatoes), Vietnam (cashew nuts, mangos, tropical fruits), Chile

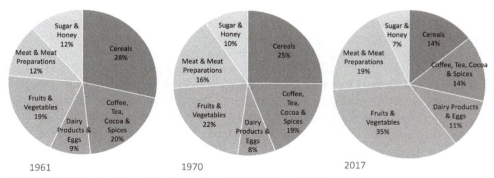

FIGURE 3.5 Composition of food exports, 1961, 1970, and 2017.

Source: Author with data from FAO (2020a).

(grapes, apples), Thailand (pineapples, tropical fruits, mangos), South Africa (oranges, grapes, apples), Peru (grapes, avocados, asparagus), and Ecuador (bananas). Cereals remain important, but as the United States' global dominance in the wheat market faded, maize and rice became more significant. Again, a growing share of these are now produced in the Global South, including maize in Brazil and Argentina and rice in India, Thailand, and Vietnam. Meat has also become traded globally in larger volumes, reflecting changes in diets around the world and raising fears of foodborne disease (see textbox 3.3). Although meat comes primarily from the Global North (i.e., the United States, Germany, Netherlands, Australia, Spain, Poland, and Canada), Brazil has become the second-largest meat-producing country in the world.

| Box 3.3 | **Global Meat Networks and Zoonotic Diseases** |

Meat production has quadrupled over the past fifty years. In 2017, more than eighty billion animals were slaughtered to satisfy expanding demand, with consumption reaching a global average of forty-three kilograms per person in 2014 (Ritchie and Roser 2020). In what is often described as a "livestock revolution," the size of production has increased dramatically in the past several decades. For example, today, it is not uncommon to find fifty thousand chickens in a single industrial-scale and vertically integrated factory farm in the United States, China, or Thailand. In addition, a growing share of animal production is traded internationally: about 20 percent of cattle and 12 percent of chickens and pigs—up from about 4 percent in 1960—are destined for exports either in the form of meat or as live animals. As a high-value commodity, meat represents almost a fifth of all exports (see figure 3.5). Global trade of live animals has also increased, particularly to countries with large Islamic populations in the Middle East, North Africa, and Southeast Asia, where halal slaughtering is preferred. In 2017 alone, according to data from the FAO (2020a), 1.8 billion chickens, 46 million pigs, 16 million sheep, 11 million cattle, 5 million goats, and almost half a million camels were traded live across national boundaries.

The transportation of live animals has raised concerns from animal rights and environmental advocates. Such transport imposes enormous suffering on animals, which are confined in tight containers for long journeys, and contributes to greenhouse gas emissions. Furthermore, these animals are more likely to be raised in large feedlots that are inhumane and impose significant environmental costs as we will discuss in chapter 8. In recent years, animal disease outbreaks have added to these concerns.

So-called *zoonotic diseases* are caused by germs, such as viruses, bacteria, parasites, and fungi, that spread between animals, including livestock as well as wildlife, and people. Such diseases can be contracted through direct contact with infected animals, indirect contact with contaminated areas like farms and rivers, vectors like mosquitos and ticks, or tainted food and water (Morse et al. 2012).

During the past decades, zoonotic disease risks have been heightened by deforestation that intensifies contact between wildlife, livestock, and people at the urban-rural interface; industrialization and concentration of meat production that provide ideal conditions for pathogens to survive, spread, and evolve and lead to greater soil and water contamination; increased transportation of live animals over long distances; and poor animal health infrastructure and monitoring of diseases. These causal mechanisms are entangled with economic globalization. International trade, in particular, contributes to the rapid spread of constantly mutating swine and avian flus and what Kimball (2015) calls "trade-related infections." Orientalist accounts blaming "bat-eating Chinese" for spreading diseases such as the avian flu and COVID-19 fail to acknowledge the economic networks underlying the geographic expansion and intensification of meat production and the associated global health threats. As geographer Rob Wallace shows in his book *Big Farms Make Big Flu* (2016), the emergence of highly virulent pathogens is directly tied to the rapid expansion of factory farming, itself a by-product of neoliberal policies and a major focus of transnational investment.

In contrast, coffee, tea, cocoa, and spices—quintessential colonial commodities—have become a smaller share of total exports, partly because of declining prices.

Today, China, the United States, India, and Brazil produce about half of the world's food and the top twenty producing countries generate 75 percent of the food supply (see figure 3.6). Over the last fifty years, however, the position of the United States has slipped, while that of China has grown drastically. Many European countries, including Germany, France, Spain, and Italy, also fell in the rankings. In contrast, as shown by the darker shade in figure 3.6, several Global South countries, such as India, Brazil, Indonesia, Mexico, Nigeria, Pakistan, Vietnam, Thailand, and Iran, raised their share of global food production. Genetically modified crops, engineered in the United States, began spreading around the world in the mid-1990s and helped these countries gain larger shares of global markets.

In many of the new agricultural economies, food commodities such as oil, soybeans, corn, and exotic produce are produced as part of an export-oriented development strategy. Several of these countries are *net exporters*, whose exports of food are significantly larger than their imports. When looking at the total value of net exports, China, the United States, Brazil, Argentina, Australia, New Zealand, France, and the Netherlands top the list. However, when looking at exports as a ratio of imports, other countries stand out as leaders, including Argentina (with $18 of export for every $1 of import), New Zealand, Paraguay, Côte d'Ivoire, Thailand, Uruguay, Brazil, Ecuador, and Vietnam.

Figure 3.7 shows the flows of food from major net exporting countries, as identified by the United Nations (2018), to various world regions. It ignores exports from smaller economies like Italy, South Africa, Israel, Colombia, and Côte d'Ivoire and countries that also import a lot of food such as China. Despite these limitations, several important trends are discernable in the map. First, it shows that food is traded across oceans and continents, connecting all regions of the world. Second, East Asia stands out as a major importer of food. This is primarily driven by China, whose growing middle-class population has fueled a global demand for food. Despite being the largest producer of food in the world, China must import large volumes of food, providing incentives for the Chinese government to engage in trade negotiations and direct investments in other countries, including "land grabs" in Africa (see chapter 2). Third, Brazil is shown as a major exporter with trade partners in every single continent and the lion's share of its exports going to China. As the largest Global South producer of food, Brazil is often called the "breadbasket of the world," robbing the United States of its nickname. Fourth, no single country in any regions of Africa or Western Asia (i.e., the Middle East) is considered a major net exporter, suggesting that they are either too small to be counted, self-sufficient, or import dependent, with the latter being the most common explanation. In fact, since the 1970s, Africa as a whole became a net food importer, despite large exports of cocoa from Côte d'Ivoire and Ghana, fruits, nuts, and vegetables from South Africa, Nigeria, and Kenya, and coffee from Ethiopia and Uganda. Finally, it is obvious from the clustering of arrows on the map that a very large share of trade occurs between countries within each regions. Regional trade agreements, such as NAFTA (see textbox 3.4), the European Union, and the Association of Southeast Asian Nations (ASEAN) Free Trade Area, increased these trade flows by removing barriers of trade between regional partners.

In many Global South countries, dependence on food exports is very high, making them extremely vulnerable to world price fluctuations. While some of the largest exporting countries in the world are in the Global North, they tend to have much more diversified economies, and agriculture only represents a small share of GDP. In contrast, in poorer countries of the Global South, agriculture is a larger component of GDP and food exports constitutes a much more significant share of all

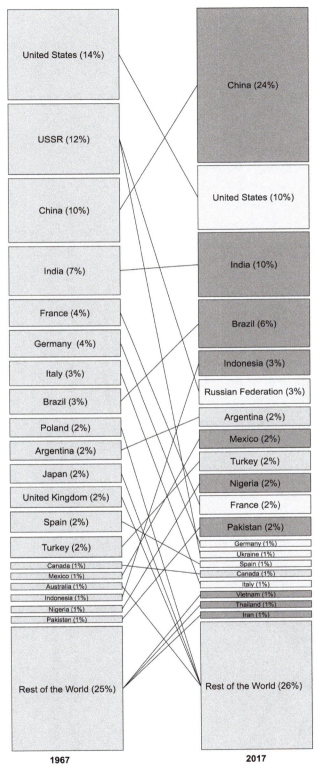

FIGURE 3.6 Leading food producing countries (fish excluded), ranked by share of global production, 1967 and 2017.

Source: Author with data from FAO (2020a).

1967 or no change   Increase   Decrease

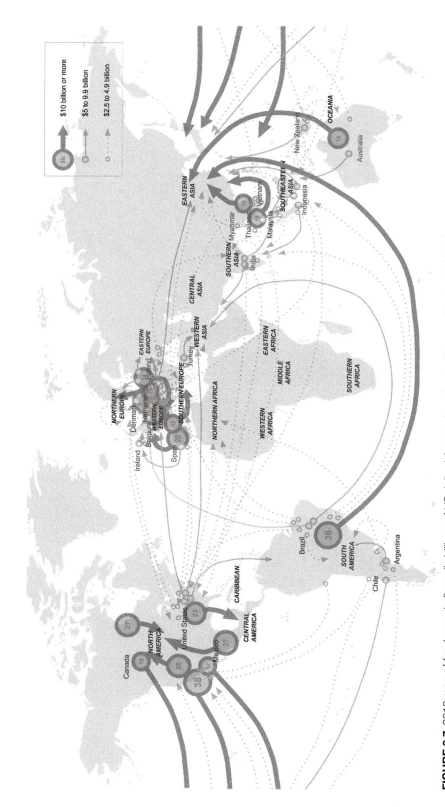

FIGURE 3.7 2018 map of food exports flows (in billions of US dollars) from major net exporting countries to world regions.

Source: Author with data from UNCTAD (2018).

| Box 3.4 | **NAFTA and Its Impact on Farming and Food in Mexico** |

Following a series of structural adjustment programs that began liberalizing and privatizing the Mexican economy in 1982, trade relations within North America were solidified in the North American Free Trade Agreement (NAFTA), which was signed by Canada, Mexico, and the United States and went into effect in 1994. NAFTA opened Mexican markets for US corporations to sell their products, invest their capital, and control production for global supply chains. Food, especially corn, was a central component of these transactions (Wise 2009).

Corn is believed to have originated in Mexico and has been cultivated for centuries in many parts of the country, where it is an essential component of the diet and is consumed daily in the form of tortillas and other preparations. Therefore, one would have expected corn to become one of Mexico's primary exports. Instead, heavily subsidized corn from the United States quickly flooded the market at prices against which small farmers could not compete—a strategy often described as "dumping." By 2000, 40 percent of the corn consumed in Mexico was imported from their northern neighbor. As a result, rural livelihoods were destroyed and small subsistence

farms dismantled. Land was consolidated into large industrial farms operated by transnational corporations, undoing the redistributive policies of the Mexican revolution and effectively prompting the "de-peasantization" of Mexico. In the two decades following the passage of NAFTA, over two million campesinos were displaced and forced to look for work elsewhere, fueling northward migration. By 2006, one out of ten Mexicans lived in the United States. People became increasingly dependent on the market to obtain food and the consumption of processed and imported products such as bread, pizza, and soda increased quickly. As Gálvez puts it (2018, 12), "the expansion of markets would be accomplished by the expansion of waistlines." In 2010, ethanol production drove up the global demand and prices of corn. By then, the Mexican corn industry had been mostly dismantled, making the dependency on US corn extremely costly.

In the context of NAFTA, Mexican exports of tomatoes, avocados, and mangos to the United States have increased. However, their revenues pale in comparison to the losses generated by corn, soybean, and wheat imports.

exports. This is the case in many African and Asian countries where more than a third of GDP comes from agriculture and food exports represent over three-fourths of merchandise exports. For example, in Gambia, agriculture accounted for 28 percent of GDP and about 90 percent of export earnings, with peanuts and peanut products representing about 70 percent of exports. Similarly, in Burundi, agriculture accounted for over 40 percent of GDP. Its primary exports are coffee and tea, which account for more than half of foreign exchange earnings. World price fluctuations of these commodities can wreak havoc in these economies, making them highly dependent on foreign aid and decreasing their sovereignty. In contrast, in the Netherlands, one of the world's largest net exporters, agriculture constitutes 1.6 percent of GDP, the same as the European Union average and slightly more than the United States, where it is below 1 percent.

Industrialization and specialization had been set in motion under previous food regimes, but what was new about the contemporary food regime was the growing role of transnational corporations and the neoliberal policies that enabled them to prosper. Structural adjustment programs and export-oriented development strategies were the expression of the *neoliberal* ideology that became popular in the 1980s—the idea that markets are better left alone and that government intervention should be limited to supporting these markets and encouraging free trade. In that context, transnational corporations grew more powerful and began replacing the United States as the leading economic power.

Structural adjustment programs, along with free trade agreements enforced first by the General Agreement on Tariffs and Trade (GATT) and, after 1995, by the World Trade Organization (WTO), opened doors for these corporate actors to expand their markets and acquire land, properties, and businesses at devalued prices, stretching their production and distribution networks globally. Foreign land acquisitions recorded by Land Matrix (2020) show (see table 3.1) that transnational corporations from China, India, the United States, Saudi Arabia, the United Arab Emirates, and other powerful countries have contracted large pieces of land for agricultural purposes, mostly in the Global South. In countries like Papua New Guinea, land changed hands between foreign investors so many times during the past two decades that the aggregate amount of contracted land is much larger than the amount of arable land. Some of the largest individual investors are corporations headquartered in Cyprus, Luxemburg, Switzerland, the United States, and the British Virgin Islands, where large banks managing hedge funds are located. This reflects the increased *financialization* of agriculture, in which institutional investors, such as private equity firm 3G Capital (see textbox 3.4), reallocate flows of capital to various places based on anticipated returns to shareholders.

After being dominated by the British empire and the United States, the global food regime is now ruled by *transnational agrifood corporations*, with headquarters in the Global North. As we will learn in chapter 5, corporate agrifood networks have become highly integrated and include all aspects of food from farm to plate, connecting plantation-style farms to food processing plants, major retailers, biotech companies and financial institutions. Among the world's most powerful corporations are several agrifood companies, including those that specialize in agrochemicals (e.g., Syngenta, Monsanto, Bayer, BASF, and DowDupont), grain distribution (Cargill, Archer Daniels Midland, Bunge, and Louis Dreyfus), livestock and meat (e.g., JBS, Smithfield, Tyson, BRF, Danish Crown, WH Group), fruits (e.g., Dole, Chiquita, Del Monte), food processing (e.g., Nestlé, Mars, Kraft Heinz, Coca-Cola, Pepsi-Cola, InBev and Unilever), retail (e.g., Walmart, Costco, Kroger, Lidl, Tesco), and restaurants (e.g., McDonald's, Starbucks, and Burger King). Today, in the United States, just ten corporations account for more than half of food products found in supermarkets.

The 153 agrifood companies on *Forbes*'s (2018) top-2,000 list had annual sales ranging between $1.4 and $500 billion, with the annual income of the largest fifty surpassing the GDP of seventy-seven countries. Only twenty-five countries had GDP larger than the revenue of the largest food retail corporation, Walmart. Figure 3.8 shows the largest agrifood corporations in the world by sector and reveals the importance of retailing and manufacturing. Food retailers—like Walmart—hold tremendous power in the global economy, squeezing the profit margins of farmers and food manufacturers and controlling consumers. Over the past two decades, *consolidation* has shrunk the numbers of supermarket chains. The global retail market in 2018 was worth just over $8 billion, with about 15 percent of sales by just ten companies. About half of the world's retailers consist of small independent stores, especially in the Global South. However, large companies have been expanding into Asia, where the grocery retail market is growing at a fast pace due to the sheer size of the population and rising income. Companies like Walmart, Tesco, and Carrefour have been investing in China, India, Thailand, and Indonesia for the past two decades now.

Today, processed foods hold a much larger share of trade than raw agricultural commodities like wheat and corn. This ranges from semi-processed food like roasted coffee, sugar, and fruit juices to manufactured food like sausages and cookies. By turning raw commodities into consumer-oriented food products that can be easily stored, transported, prepared, and marketed, food manufacturers and retailers are adding significant value and capturing a growing share of income. Additional value is added

TABLE 3.1 Top Target Countries for Large Land Acquisition Concluded 2000–2020

Target Country	Number of Contracts	Contracted Land (thousand hectares)	Transnational Share of Contracted Land %	Main Investors' Headquarter Countries	Arable Land (thousand hectares)	Ratio of Contracted Land to Arable Land
Russia	134	16,358	79	China, Japan, Switzerland, Cyprus, United States	123,122	0.13
Malaysia	38	5,419	8	India, China	882	6.14
Brazil	149	5,051	68	Netherlands, United States, Chile	80,976	0.06
Indonesia	166	4,623	65	China, Malaysia, Singapore	23,500	0.20
Papua New Guinea	46	4,112	94	United States, China, Malaysia, South Korea	300	13.71
Ukraine	272	3,984	83	Cyprus, Luxembourg, Saudi Arabia	32,776	0.12
Argentina	220	3,708	48	United States, Saudi Arabia	39,200	0.09
South Sudan	13	2,553	100	United Arab Emirates	–	–
Mozambique	133	2,454	90	Zimbabwe, Portugal, United States, Norway	5,650	0.43
Liberia	22	2,125	63	Italy, Malaysia, United Kingdom	500	4.25
Cambodia	179	1,743	46	China, Vietnam	3,800	0.46
Ethiopia	121	1,446	67	India, China, Saudi Arabia	15,119	0.10
Guyana	6	1,375	100	China, India	420	3.27
Myanmar	35	1,169	84	China	10,908	0.11
Uruguay	82	1,153	83	Argentina, Brazil, China, UK	2,411	0.48

Source: Computations by author based on data from Land Matrix (2020) and the World Bank (2020b)

Note: Excludes land intended for mining, oil and natural gas extraction, and forestry

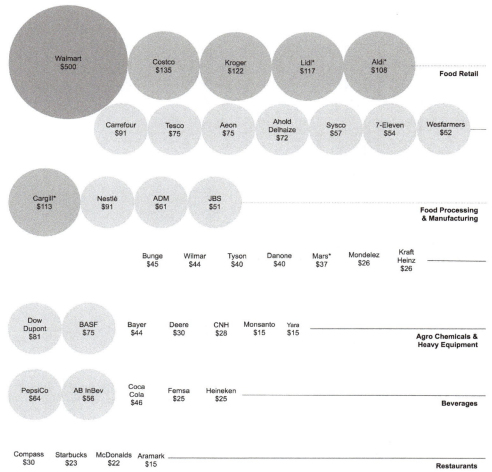

FIGURE 3.8 Large agrifood transnational corporations and sales in billions of US dollars, by sector. Note: Private companies (not traded on the stock market) are not included.

Source: Author with data from *Forbes* (2018).

when food is prepared and served in restaurants. In 2018, the global value of agriculture was $2.4 trillion according to the FAO, while food sales reached $7.5 trillion and food services totaled $3.4 trillion. Thus, much of the value is added at the tail end of the supply chain (see chapter 5), which itself is controlled by a few large corporations.

This vast corporate landscape is complicated by financing, mergers, and acquisitions. A handful of large companies own many related entities and brands, which they acquire to build linkages within the agrifood industry both *horizontally* (between companies supplying similar products) and *vertically* (between companies specializing in different components of a supply chains). Indeed, concentration has increased during the past decade as illustrated by the example of beer in textbox 3.5. With large advertising and lobbying budgets, these transnational corporations can exert huge influence on agricultural and trade policy nationally and internationally.

In summary, our contemporary global food system is defined by the following adjectives: corporate, neoliberal, capitalist, and industrial. To these, we could also add unsustainable and inequitable, as we will discover in upcoming chapters. Not only does it involve the trade of agricultural commodities and processed food, but it entails transnational investments in all aspects of food production and distribution.

Box 3.5 **Beer, Fast Food, and Transnational Capital**

A private equity firm with offices in New York and Brazil, 3G Capital, founded in 2004 by three of the richest men in Brazil, has led or participated in several major mergers and acquisitions in the food and beverage industry. Beginning in 2004, its founding partners, who already owned the Brazilian beer company Ambev, helped create the single largest beer company in the world by merging with Belgian Interbrew to form InBev, which in turn acquired Anheuser-Busch in the United States in 2008. Since then, the new company, AB InBev, has acquired Grupo Modelo in Mexico in 2013 and SABMiller in the United States in 2016. Today, the company controls 25 percent of global beer sales and 45 percent of profits, owning over six hundred brands of beer in about 150 countries.

Another major venture of 3G Capital was the creation of Restaurant Brands International through the acquisition of Burger King and its merger with large Canadian chain Tim Hortons and Popeyes Louisiana Kitchen. The most significant investment was made in 2013 in partnership with Berkshire Hathaway—Warren Buffet's conglomerate—to acquire the food company Heinz, which two years later acquired Kraft Foods Group. Despite an unsuccessful ambitious bid at acquiring competing food manufacturer Unilever in 2017, 3G Capital controls a major share of the food industry, with key positions in the beer, fast food, and processed food markets (see figure 3.9).

According to an article in the *Financial Times* (Daneshkhu, Whipp, Fontanella-Khan 2017), the equity firm is known for restructuring newly acquired companies through aggressive cost-cutting strategies, slashing jobs and closing low-performing plants, while raising profitability and shareholder returns.

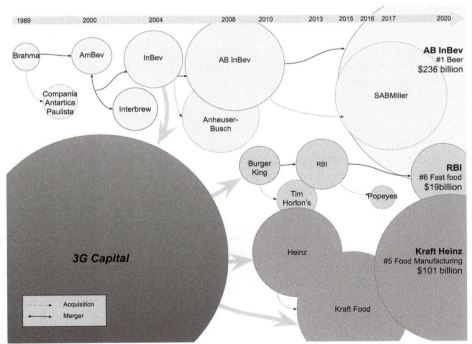

FIGURE 3.9 3G Capital's food industry assets and connections, 2020.

Source: Author with data from 3G Capital (2020), Daneshkhu, Whipp, Fontanella-Khan (2017) and the AgriFood Atlas (2017).

3.6. **Homogenization and Differentiation**

Resentment against globalization grew during the 1990s, resulting in regular protests at meetings of the International Monetary Fund and the World Bank, the World Trade Organization, and groups such as the G7, G8, and G20 that gather the world's wealthiest nations. One of the largest took place in Seattle in 1999 at the IMF/World Bank annual meeting, ending in violent clashes between demonstrators and the police and over six hundred arrests. Similar protests were organized in Genoa, Italy, Cancún, Mexico, and more recently Hamburg, Germany. In addition to concerns about growing inequality and poverty caused by structural adjustment programs and free trade agreements, protestors feared that globalization was robbing nations of their sovereignty and culture.

In popular discourse, globalization is often associated with *McDonaldization*—the idea that society is increasingly governed by the principles of efficiency, calculability, standardization, and control that characterize fast food production (Ritzer 2013). To many observers, this also suggests a *homogenization* of culture in which corporate convenience food is threatening unique local foodways, either directly as large food chains take over or destroy existing independent businesses or indirectly by influencing tastes and undermining culinary skills. At the same time, the presence of McDonald's in a country also reflects the spread of American values and global integration, which according to economist Thomas Friedman (2000), "makes warmongering risky and unpalatable to its people"—a controversial hypothesis known as the "golden arches theory of conflict prevention." Regardless of its impact, McDonald's has been one of the most enduring symbols of economic globalization, American corporate hegemony, and cultural imperialism.

The frozen fries and beef patties prepared at McDonalds' and other fast-food restaurants appear to come from nowhere. In other words, globalization has disembedded foods from the places where they are produced. Food production and consumption have become increasingly separated since we no longer need to rely on products grown locally or even domestically to prepare food. Indeed, most of us do not know where our food comes from. This is especially true of convenience foods that are highly processed and contain multiple ingredients. As we will see in chapter 5, this distance allows consumers to ignore the harsh realities of factory farming, including its impact on workers, animals, and nature.

Despite this disconnect, food and place remain deeply connected. In fact, globalization might reinvigorate geographic cultural differences and create new ones. This process is known as *glocalization*—the coexistence and interplay of globalizing (i.e., universalizing) and localizing (i.e., differentiating) processes. This has led to the adaptation of global food into local variations and to the spread of localism at the global scale. Rather than being devoid of cultural content or associated with hegemonic American or Western values, globally produced industrial foods often acquire meaning in the local places where they are prepared and consumed. The introduction of new foods within local settings often produces *hybridized* or *creolized* foods in which ingredients are mixed and dishes are adapted. In some cases, these global influences are strongly rejected in favor of local practices that are imagined as traditional, pure, and morally superior. Yet, the reinvigoration of traditional foodways often leads to the creation of new food, which might then be exported along with ideology surrounding localized food. The hamburger provides a good illustration of the cultural plasticity of food in the global context (see textbox 3.6). We will return to these cultural issues in chapter 10.

| Box 3.6 | The Hamburger: Global or Local? |

Few foods have received as much negative attention as the hamburger. Despite the German origin of the meat patty, it is perceived as a quintessential American dish and as such has come to represent what the French call "malbouffe" or junk food. It has been ridiculed for its simplicity, reviled for its costs to human health and the environment, and blamed for the loss of culinary skills and local traditions. Yet, it is consumed almost all over the world, at least in large cities with a significant middle class, thanks to companies like McDonald's that have revolutionized the way it is produced and opened franchises in almost fifty countries over the past four decades, including China, Indonesia, and India (Watson 1997).

The global popularity of the hamburger suggests the sort of cultural homogenization feared by globalization critics. The hamburger, however, is not a monolithic product. McDonald's menus vary from place to place, with vegan patties and chutney in India, panko-crusted filet-o-shrimp in Japan, sausage McNürnburgers in Germany, halal chicken McArabia in Saudi Arabia, and blue cheese and bacon burgers in France. Many Americans would not recognize these hybridized products that people in these places envision as American food. Over time, some of these international variations have inspired new recipes in the United States, reflecting the circular nature of globalization.

Burgers have also been adapted by chefs around the world who have reinvented the dish as gourmet versions using locally produced, sustainable, and/or rare ingredients to subvert globalization and reassert the significance of the local. Unlike the Big Mac, gourmet burgers have become very trendy in Paris, London, Buenos Aires, and other world cities, where affluent consumers line up at popular food trucks or spend money in highly rated restaurants for an elevated version of a meat patty in a bun, revealing that it is not just commodities that are circulated globally, but also knowledge and culture.

Conclusion 3.7.

Although global trading of food is not a new phenomenon, it has expanded and changed qualitatively over time. The concept of food regime is useful to understand the political and economic underpinnings of food production, distribution, and consumption and their evolution.

Early trade involved luxury goods such as spices, which made merchants wealthy. During the colonial era, commodities like sugar, cocoa, coffee, and bananas became mass produced on colonial plantations set up for the benefit of European trading companies and consumers. By the 1700s, England had come to dominate transatlantic and global trade, giving itself an enormous advantage in promoting industrialization at home. After World War II, the United States became the leading food producer, expanding its markets first through food assistance and then through imports of subsidized cereals. Despite efforts to become self-sufficient, newly independent Global South countries became increasingly dependent on food imports, contributing to their indebtedness in the 1970s and 1980s. Required to adopt market-oriented reforms by international lending organizations, these countries increased their exports of cash crops to generate currency and opened their economy to foreign goods and investors. The result is a highly integrated and industrialized global food economy controlled by transnational corporations.

The contemporary food regime exemplifies the growing extensity, intensity, velocity, and impact of globalization. However, these processes are highly uneven and unfold differently across space. The unique characteristics and histories of countries, regions, and localities shape their position in the global food regime, influencing the way they integrate global networks and how much power they have in controlling and benefiting from these networks. Thus, the global and the local are in constant interaction.

Labor Geographies of Food 4

Learning Objectives

* Appreciate the importance of food as a major source of livelihoods.
* Use the concept of labor geographies to examine the spatial organization of labor relations within the global food system.
* Distinguish between labor-intensive and capital-intensive food systems.
* Contextualize the rise of wage labor in agriculture.
* Recognize the exploitative nature of past and present labor arrangements in the agrifood economy and the significance of immigration, race, and gender in creating divisions of labor.
* Understand the role of space, place, and scale in shaping labor movements within the food system.

The food system, including farming and food processing, manufacturing, and services, is a major source of livelihood around the world. In low-income countries, it employs more people than any sector, mostly in the form of self-employed subsistence farming. Hired labor is more common in high-income countries, but is expanding everywhere as food is increasingly grown, processed, and sold in large operations and the locus of food labor is shifting from fields to factories.

Historically and to the present day, labor involved in the production and preparation of food has been associated with poverty, low earnings, long hours, and harsh working conditions compared to other economic sectors. Eighty percent of the world's poor people live in rural areas and rely primarily on subsistence agriculture to survive. Industrialization and land concentration, however, is leading to an increase in hired labor on and off farms. Around the world, the multiple sectors of the food system rely on some of the most vulnerable workers in society: impoverished rural populations, women, racial or ethnic minorities, and migrants. For example, subsistence farms depend heavily on the unpaid labor of women. Commercial farms often turn to migrants. Food processing and manufacturing plants, including slaughterhouses and meat packing plants, provide notoriously unsafe and low-paid employment, leading to high labor turn-over rates enabled by a large supply of migrant workers. Less physically demanding, but still tedious and repetitive, jobs such as cleaning, sorting, packaging, and labeling are typically performed by women. Closer to consumers, the rapidly growing food service economy—retail, restaurant, and catering—consists primarily of low-wage and *informal* jobs (i.e., jobs left un-protected in a context where regulations and protections have been put in place to ensure labor safety and minimize abuse).

Labor arrangements have changed over time and vary from place to place. Al-though exploitation appears to have decreased since the days of the transatlantic slave trade (see chapter 3), particularly in countries with stronger labor protections and social safety nets, many farm and food workers around the globe continue to work under slavery-like conditions. Globalization, which allows for the separation of pro-duction from consumption and makes labor invisible to consumers, has undoubtedly contributed to this process.

In this chapter, we introduce labor geographies as they relate to the global food system, including the farming, processing, manufacturing, and retailing stages. We distinguish between various institutional arrangements structuring work, noting vari-ations across time and space, and we explore the role of immigration, race, and gen-der in shaping labor processes.

4.1. Labor Geographies

Labor geographers approach the study of the economy and its spatial organization from the perspective of workers (Herod 2001, Castree, Coe, Ward, and Samers 2004). They pay particular attention to the role of *space* in controlling labor and shaping workers' lives, building on the notion that capitalism is a spatial system. As such, capitalism creates and transforms space in its search for profit by directing in-vestments to various places and altering their environment. This produces global, regional, and local economic landscapes in which employment and livelihood oppor-tunities are uneven. The general lack of *mobility* among workers limits their options and increases their likelihood of being exploited—a phenomenon described as "spa-tial entrapment." Three-fourths of the world's poorest people live in rural areas and are dependent on agriculture to survive. Even migrants, who on the surface seem to be highly mobile, are constrained by immigration policies and local labor mar-kets where they relocate. At the same time, corporate mobility dissuades collective bargaining and disciplines labor through threats of outsourcing jobs or moving else-where. In other words, the spatial context in which workers live and work matters in terms of their agency and well-being. Scholars within the subfield of labor geography also consider the role that the state plays in shaping the spatial organization of the economy by regulating the mobility of labor and capital.

Understanding labor organizing and its relationship to space and scale has been another major focus of the subdiscipline. Labor struggles are seen as spatial struggles and represent attempts to reshape the geographic structures and economic landscapes that are damaging to workers' lives. The ideal scale of labor mobilization, solidarity, and resistance has been debated among scholars who have investigated the success and failures of labor movements at the global, national, regional, industrial, urban, and workplace scales. Given the decline of labor unions since the 1970s, growing attention has been given to alternative labor geographies such as cooperative and collective projects as paths to increasing labor power.

One of the major premises of labor geography is that labor arrangements are em-bedded in locally distinctive social relations, political institutions, and cultural prac-tices that are nevertheless influenced by forces operating at larger scales (see chapter 3 on globalization). In that sense there are multiple labor geograph*ies*, emphasizing geographic differences in how labor is structured by industry, class, race, ethnicity, gender, etcetera and may be organized to improve work and resist exploitation.

Labor in the Food System 4.2.

Globally, food is the largest single source of employment (World Bank 2017), although work arrangements and conditions vary drastically across space. The exact number of jobs within the food system is difficult to assess for a variety of reasons. Agricultural employment is a common estimate, but it includes people working on non-food crops such as cotton, tobacco, and biofuels, which represent about 8 percent of harvested biomass globally (Carus and Dammer 2013) and excludes the many manufacturing and service jobs related to food processing, storage, transportation, manufacturing, distribution, retailing, restaurants, and waste management. Furthermore, in low-income countries, employment is often difficult to measure because many people, including children and women, work informally and without pay in *subsistence agriculture.*

In 2018, agriculture alone employed more than one billion workers or 27 percent of the global workforce. In very poor countries, especially in Africa, over two-thirds of the population works in agriculture. However, in the United States, Germany, and the United Kingdom, less than 1 percent does (see figure 4.1). As technological innovations and corporate takeovers have created larger and more heavily mechanized farms, the need for farm labor has decreased and owners of *capital-intensive* farming operations have turned to temporary hired workers to meet seasonal needs. Yet, even in these economies, labor remains critical for tasks that cannot be easily mechanized, including fruit and vegetable picking.

Over time, nonagricultural food system jobs have been growing, along with urbanization. Now that the majority of the world's population lives in cities, people rely on purchased, processed, and prepared food more often than in the past. In addition, the growth of international food trade (see chapter 3) requires more processing, logistics, and transportation services. In contrast with farming, food service (e.g., retail, restaurant, and catering) and manufacturing (e.g., processing, packaging) employment tends to increase with GDP, as illustrated in figure 4.2. Food manufacturing is especially important in middle-income countries where industrialization often begins with food processing, branching out of agriculture and creating vertical linkages. For example, in Brazil, meat processing has developed from its large livestock industry in the past two decades. Countries with large food exports also have a significant food service sector due to the importance of transportation, storage, and logistics.

An important distinction is made between *labor-intensive* and *capital-intensive* production (see figure 4.3). In the Global South, where labor is more readily available than capital, most aspects of food production, whether in farming, processing, or manufacturing, tend to be labor-intensive. The number of workers per hectare is much higher than in industrialized economies. The use of machinery such as tractors, artificial fertilizers, and other costly inputs is limited. In the Global North, where capital and technology are concentrated, production is primarily capital intensive, relying heavily on chemicals and machinery and requiring very few workers. As the capital-to-labor ratio rises, so does labor productivity, measured in output or yield per worker. Thus, economists typically view the process of development as one of *capital intensification*: as capital increases, it raises productivity and income, creating opportunities for further investments in capital. This view, however, pays little attention to the power dynamics described in chapter 3 and the fact that the economies of "developing" countries, including what they produce and their access to capital, have been shaped and controlled by more powerful actors in colonial and postcolonial global food regimes.

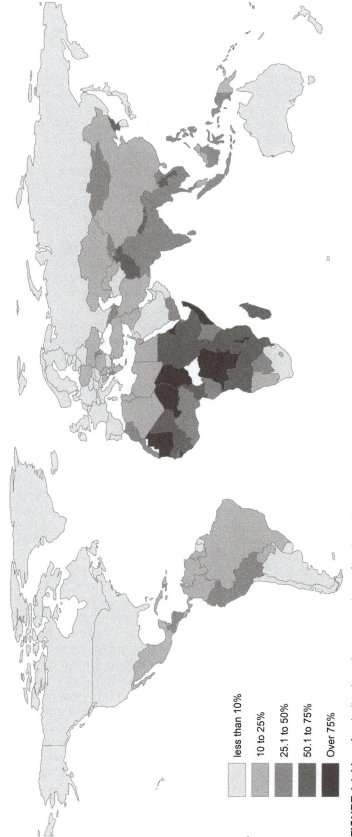

FIGURE 4.1 Map of agricultural employment as share of total employment, by country, 2018.

Source: Author with data from World Bank (2020a).

less than 10%

10 to 25%

25.1 to 50%

50.1 to 75%

Over 75%

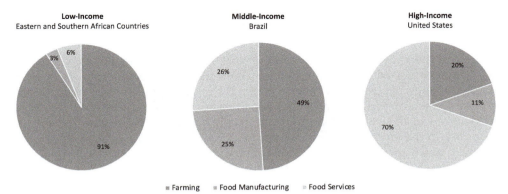

FIGURE 4.2 Agrifood system job composition: three examples by income level.

Source: Recreated from World Bank (2017), updated with data from the USDA (2020). Note: Food retailing and wholesale jobs excluded from these employment figures. Eastern and Southern Africa (ESA) include twenty-two countries in the Horn of Africa (e.g., Somalia), Eastern Africa (e.g., Tanzania), Southern Africa (e.g., Mozambique) and the Western Indian Ocean (e.g., Madagascar).

The notion of *agricultural technology treadmill*, which emerged as an early criticism of the Green Revolution, suggests that although the adoption of new technology may increase output, it also increases farmers' debt, decreases food prices because of overproduction, and depletes soils, which then require more fertilizer, pesticide, and irrigation. This leads to the adoption of newer technology in hope of increasing productivity to pay off debts and purchase imported inputs, repeating the cycle and increasing dependency. This idea explains why small farmers in countries like the United States earn very low incomes and are highly indebted, while agrifood corporations post record profits. It is also useful to understand how developing countries, which depend on wealthier nations for access to technology, are always on the treadmill, running behind countries that were able to adopt new technology earlier. In that context, there are growing calls to "get off the treadmill" and instead strengthen labor-intensive, small-scale, sustainable, and subsistence agriculture by attending to local livelihoods and ecologies, without relying on imported and expensive technology.

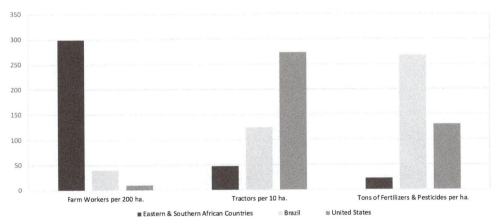

FIGURE 4.3 Selected indicators of labor and capital intensity in agriculture: three examples by income level.

Source: Author with FAO data (2020).

Note: Farm workers per hectare is computed using data from 2018, fertilizer and pesticide data are from 2017, and tractor data are from 2006. Tractor figure for Eastern and Southern African countries is based on six countries for which data were available.

4.3. The Rise of Wage Labor

Although subsistence farming remains important in many low-income countries where self-employed farmers and unpaid family members grow food for their own consumption, wage labor is quickly becoming the norm, especially as farms get larger and more mechanized. Globally, over half of farm workers are now hired workers, working on small farms or larger plantation-style operations (see figure 4.4). Wage labor is also the most common arrangement in food manufacturing and service industries, which are more industrialized and commoditized, meaning that outputs are typically produced to be sold for money rather than to be consumed directly.

Wage labor, however, is a relatively new invention that was brought about by the emergence of capitalism and the industrial revolution. Marxist scholars have written extensively about the so-called *agrarian question* which focuses on the *proletarization* of farm labor—its transition to wage labor—as capitalism expands. According to Marx, to understand labor relations, we need to look at how production is organized and who owns *factors of production* such as land and capital. Land has always been and remains essential in the production of food; without it, growing crops or raising livestock is impossible. Thus, those who own arable and fertile land have historically held the most power. In the past, large landowners have been able to control labor by granting peasants access to their land. Landless workers would keep a very small share of what they produce for their own subsistence and hand in the rest to the landowner. They were rarely paid in the form of money, but instead kept and exchanged food they grew. This type of arrangement was common in feudal Europe throughout the Middle Ages. It also describes *sharecropping*, which was widespread in the US South after slavery was abolished and remains common in many parts of the world, including South Asia.

Over time, capital has become an increasingly important input in the agrifood economy, allowing producers to increase their productivity through investments in

FIGURE 4.4 Workers harvesting green tea leaves on a large plantation in Vietnam. Many are women hired as casual day laborers, paid at piece-rate.

Source: https://www.pxfuel.com/en/free-photo-qoyte.

technology and economies of scale. This, however, required the monetization of economic transactions and the *commoditization* of labor. As landowners and merchants began earning money from trade, they invested in new technology, giving rise to the first factories. Land and labor became resources that could be exchanged for a price. Land was enclosed and peasants lost access unless they could buy it or rent it, forcing them to sell their labor for a wage. The factory model, whether in farming or food manufacturing, was built on the creation of wage labor through the separation of labor from both the means of production and the product of their labor. Capital owners—capitalists—pay workers a wage for their labor power and control the output, which they then sell to earn profits to be reinvested into additional capital, labor, or land. This process leads to *exploitation*, which Marxist scholars define as the ability to pay workers less than the true value of their labor. This dramatic transformation of the economy that began during the industrial revolution and continues to unfold around the world has stimulated the *commoditization* of food, which hired workers must purchase with their wages to feed themselves and their families. Thus, under capitalism, food production and consumption are clearly separated.

Labor arrangements reflect power relations between workers, who only own their labor, and capitalists, who own valuable inputs such as land and capital. As economies evolve in spatially differentiated ways, the value, intensity of use, and ownership patterns of these inputs change, altering labor relations. Feudal peasants and sharecroppers become waged workers, without much change to their socioeconomic status. Those who own land and capital seek to maintain and extend their power by controlling labor in ways that reflect economic, political, social, and cultural circumstances, leading to a variety of arrangements. Under capitalism, this means sustaining profits and capital accumulation through market expansion and cost cutting strategies, such as maintaining a large supply of cheap and disposable labor to work in farms, factories, retail shops, and restaurants.

Labor arrangements differ across the various sectors of the food system, with the main distinctions related to the formal or informal nature of the agreement (e.g., labor protected by a contract vs. unprotected casual labor), the length of employment (e.g., yearly, seasonal, temporary, daily), the mechanisms by which workers are paid (e.g., in-kind vs. cash, hourly wage, piece-rate, contract work), the level of regulation and their enforcement (e.g., labor and safety standards), and the strength of labor organizing and bargaining power (e.g., labor association and union representation).

Despite these important differences, there are similarities in working conditions. Tracing a tomato through North America, Barndt (2008) argues that work in fields, packing plants, trucks, warehouses, grocery stores, and fast-food restaurants operate along the same logic of maximizing profit and minimizing costs by relying on cheap labor from Mexico, often women. Similarly, in *Fast Food Nation*, Schlosser (2001) shows how the fate of fast-food workers is tied to that of cattle farm and meatpacking plant workers—all of whom work primarily for large corporations. Highlighting the linkages and similarities between all agrifood workers is useful to better understand labor processes and develop organizing strategies.

The distribution of labor in the food sector fits a pyramid shape, with the majority of jobs in agriculture, manufacturing, and services paying very low wages and offering limited upward mobility. Almost everywhere in the world, farm workers are at the bottom of the earning ladder, earning lower wages than workers in any other sector. Food-related middle-class jobs that offer greater protections and pay better wages are less common and consist primarily of managerial positions. There are even fewer high-paying jobs at the top, usually in corporate leadership or professional services such as finance, logistics, and scientific research. For example, in 2019, McDonald's chief executive compensation was almost two thousand times what the median worker earned. This ignores the workers who pick the tomatoes, slaughter the cows, or package the French fries on assembly lines and are often paid very low wages too. Such high CEO-to-worker compensation ratios are common throughout the US food service industry and higher than in any other sector (Ruetschlin 2014).

Globally, farm work often violates labor standards set by the International Labor Organization, including the freedom to choose where to work, the right to form unions and participate in decision-making, appropriate social protections, safe working conditions, decent wages, and equal treatment regardless of national origin, gender, race, ethnicity, and age (ILO 2003). Only 5 percent of agricultural workers worldwide benefit from workplace labor inspection and fewer than 10 percent of hired farm workers are organized and represented by labor unions. The lack of enforcement of work safety regulations is particularly damaging when it comes to exposure to toxic chemicals, which are widely used in agriculture and have been shown to cause death and disease among farm workers. According to the ILO (2003), at least one hundred seventy thousand agricultural workers were killed annually because of workplace accidents, including deaths from exposure to pesticides, and between three and four million people suffer from severe poisoning, cancer, reproductive abnormalities, and other chronic illnesses linked to hazardous chemicals such as DDT, which continues to be used illegally in many places.

Child labor is another common concern (see textbox 4.1), with 60 percent of the world's five- to seventeen-year-old workers employed in agriculture (ILO 2020). In the United States, labor laws on farm child labor that were passed in the 1930s are still in application today, in clear violation of ILO conventions. While those were meant to allow children to work on family farms, today they make it possible for farm operators to hire children of almost any age. Although some restrictions apply to children under twelve who must have their parents' consent to work, many of these are lifted for older children, such that any fourteen-year-old child is allowed to work on any farm using toxic chemicals and dangerous tools and machinery like tractors, chainsaws, high ladders, and sharp knives (Human Rights Watch 2010). A recent report to Congress (GAO 2018) shows that half of the 452 work-related fatalities killing children between 2003 and 2015 were in agriculture.

Work in the rest of the agrifood system tends to exhibit similar characteristics of low wages, high risk of injury, lack of enforcement of labor and safety regulations, and restrictions on freedom of association. Slaughterhouses and meat processing plants are notoriously dangerous, having changed little since Upton Sinclair wrote *The Jungle* in 1904. Many workers suffer from preventable stress injuries related to the sheer number of animals killed and the fast speed of assembly lines designed to meet increasing demand for meat and poultry. Financial compensation for such injuries is limited if not absent (Human Rights Watch 2004). Because a few firms control most of the industry, they have significant power in shaping government regulations and enforcement to their advantage.

Most food service jobs, including cooks and waiters, are also poorly paid and offer little security. In the United States, in 2019, there were more than twelve million people working in this industry. The average earning (including tips) for non-supervisors was about $14 per hour, with fast food workers earning less than $11 (CPS 2020). Average weekly hours of work hovered between twenty-four and twenty-five, leading to insufficient earnings to escape poverty. In fact, one in six restaurant employees lives below the official poverty—almost three times the rate experienced by other workers (Shierholz 2014). In addition, restaurant workers receive much lower benefits, such as employer-provided insurance and pensions, than workers in other industries. We will examine the nature of restaurant work in greater detail in chapters 11 and 12.

| Box 4.1 | **Child Labor in Low-Income Countries** |

Child labor is common around the world, especially in low-income countries where it is poorly regulated and culturally acceptable. For instance, in Africa, one in five children are engaged in *child labor*, which is defined as paid or unpaid child employment in a family, public, or commercial setting and excludes employees above the legal age or in permitted light work that does not interfere with schooling or harm children (ILO 2017). Child labor is especially widespread in the food system, including farming, fishing, food processing, and retail. Globally, the ILO (2020) estimates that there were at least ninety-eight million boys and girls engaged in child labor in agriculture in 2019, with almost two-thirds as unpaid family member and half between the ages of five and eleven. Several millions also work as seed sorters, street vendors, grocery baggers, and other related jobs.

Child labor often raises emotional outrage from people in the Global North whose notion of childhood is influenced by their own social, cultural, and economic context in which children are seen as innocent and vulnerable, lacking agency, and needing adult protection. However, not all child labor is exploitative and depriving children of their "childhood," causing heated debates among scholars and activists whose views on child labor can be categorized into three broad perspectives (Abebe and Bessell 2011).

In the first perspective, child labor should be banned entirely, granting children a universal right to play and learn in order to become productive adults. This *rights-based* approach is favored by international organizations like the UN and the ILO, which are heavily influenced by universalized Western understandings of children as adults in the making. For example, the United Nations' Sustainable Development Goals signed in 2015 has for a target to "take immediate measures . . . to secure the prohibition and elimination of the worst forms of child labor . . . and by 2025 end child labor in all its forms." Corporations such as Monsanto, eager to display corporate responsibility, have supported these initiatives.

Another perspective emphasizes the need to consider geography—social, economic, and cultural contexts—in determining what forms of child labor may be acceptable. They argue that schools are not the only places where children learn and that child labor might enhance learning. Children and young people's labor, such as helping on the farm or in a family enterprise, might contribute to their development, by providing them with valuable skills, experiences, and income. Such labor should not put children's health at risk or interfere with their schooling, but instead should enhance their learning and welfare. Universally preventing child labor without acknowledging geographic differences might actually be detrimental to some children. Jennings et al. (2006) showed that in Tijuana, Mexico, nine- to fourteen-year-old children working as "volunteer" grocery packers earned tips, demonstrated independence, acquired interpersonal skills, and were able to help their family, despite being unrecognized and unpaid by grocery stores.

A third perspective draws attention to the political and economic factors that have given rise to child labor and must be addressed to reduce its most deleterious effects. According to proponents of this view, universal rights and country-specific regulations will be ineffective as long as rural livelihoods are destroyed by economic globalization and policies like structural adjustment programs and trade agreements. By disrupting social reproduction, they increase the need for children to contribute to their households by helping in the fields, fetching water or wood, selling at local markets, or working for wages. Child labor results from the fact that children in poor countries are disproportionately affected by poverty, war, climate change, and epidemics caused or shaped by actors at broader scales. Katz (2004) shows in her book *Growing Up Global* how the everyday lives of children in Sudan have been deeply affected by economic restructuring that promoted export-oriented agriculture and led to an intensification and spatial expansion of child labor.

4.4. Migration and Race in Agrifood Labor

To ensure that labor is cheap, employers are eager to have an oversupply of workers at their disposal to be deployed in fields and factories as needed. One of the most common ways to produce this "reserve army" of labor has been to recruit migrant workers, either abroad or domestically, and keep them in misery so that they are willing to work at very low wages and on a temporary basis (Mitchell 2011). For example, in California, farmers have relied on labor from China and Japan in the 1800s, domestic southern Black migrants and dustbowl refugees in the 1930s, Filipino immigrants in the first half of the twentieth century, and Mexican workers throughout its history. As groups organized and gained power (see textbox 4.2), others were brought in through concerted efforts and changes in immigration policy.

The immiseration of farm labor, which guarantees its flexibility and affordability, is spatially produced through the regulation of *mobility* at various scales. The contradictory nature of immigration policy in many countries, which seem to oscillate between encouraging and restricting immigration, reflects this desire to control the labor supply. During times of labor shortages, governments tend to relax immigration restrictions in calculated ways, often under pressure from farm owners and their lobbies. Mobility across international borders is facilitated by recruiting schemes and guest worker visa programs, such as the Bracero program in the United States (see textbox 4.3).

Box 4.2	**Consumer Boycotts and the Farm Labor Movement: *Sí Se Puede!***

In the 1950s, the poor labor conditions on California farms, especially among immigrants, gave rise to a labor organizing movement. Larry Itliong, a Filipino American, became the leader of the Agricultural Worker Organizing Committee (AWOC) in 1957 and led a strike against grape growers in the Coachella Valley of Southern California in 1965, securing higher wages for the mostly Filipino workers. A few months later, AWOC led another strike against grape growers in Delano in the San Joaquin Valley. Cesar Chavez and Dolores Huerta, who were leading the predominantly Mexican National Farm Workers Association, joined the picket line, contributing to the success of the strike, which lasted five years. The two organizations combined forces and merged into the United Farm Workers (UFW), organizing non-violent protests, marches, consumer boycotts, and alliances with transportation workers. "*Sí se puede!*" (Yes we can!) became their motto. The movement spread across the state and the country, pressuring grocery stores and consumers to boycott grapes, wine, lettuce, and other produce unfairly grown. Today, UFW remains the largest farm workers union and continues to be active, especially in California.

Boycotts have become a common strategy to fight labor injustices in the food system. The Coalition of Immokalee Workers (CIW) has effectively enrolled consumers in their efforts to improve wages and working conditions for tomato farm workers through organized boycotts of retailers who refuse to participate in the Fair Food Program. The program demands that retailers pay a "fair price" for tomatoes, raising the cost by one cent per pound for tomatoes grown and picked according to certified standards set by workers, many of whom are Latinos and immigrants. While several companies, including Ahold, Whole Foods, McDonald's, and Subway, have signed the agreement, large grocery store chains Publix and Kroger and fast-food restaurant Wendy's have so far refused to do so and are now facing the threat of losing consumers, thanks to a well-organized boycott campaign led by mostly immigrant workers. The 2015 film *Food Chains: The Revolution in America's Fields* chronicles the Immokalee workers' strike against Publix. The Fair Food Program has expanded beyond Florida and tomato fields. For example, it supported the "Milk with Dignity" initiative in Vermont, leading to an agreement between dairy workers and the ice-cream company Ben & Jerry's (CIW 2020).

Box 4.3 **The Bracero and the H2-A Guest Worker Programs in the United States**

The Bracero program was first put in place in 1942, at a time when the United States feared that World War II would cause a shortage of farm labor. It was an agreement between the governments of Mexico and the United States to organize for the recruiting and employment of *braceros* or manual labor on a temporary basis. In its twenty-two years of existence, the program brought almost five million workers—two hundred thousand per year on average—from Mexico into the United States primarily to work in agriculture.

Under the agreement, braceros were to be adequately fed and housed, receive decent wages and proper health care, and become enrolled in a saving program in Mexico, which they could tap into after returning home. The recruitment and transportation of workers from Mexico and their employment in the United States was managed by the US government. However, there were many instances in which the terms of the agreement were broken: working and living conditions were not as expected, many farmers paid wages below the set minimum, some workers stayed beyond the terms of their agreement, and saving accounts were lost (to this day, despite several lawsuits and agreements, thousands of former Braceros or their survivors are still awaiting compensation from the vanished saving funds). Bracero workers, who were tied to a specific employer, were hesitant to request that their rights be upheld for fear of not being hired again.

Furthermore, braceros were often lent and borrowed by farmers to break labor strikes, weakening the bargaining power of non-bracero workers who typically earned even lower wages than those negotiated in the binational agreement. Resentment against braceros was high in many border states, especially in Texas, where violence was so common that Mexico banned the state from participating in the program for a few years. There were concerns that guest workers were taking jobs away from native-born farm workers and depressing wages, which eventually led to the end of the program in 1964 (Martin 2003, Mitchell 2011).

Since then, farm owners have been lobbying for new guest worker programs, arguing that an insufficient supply of labor will raise the cost of food. The H2-A program permits US employers to bring temporary foreign workers into the United States to perform seasonal agricultural work. Such employers must apply for the visa on behalf of workers by certifying that there are no US workers available to perform this work and that wages and working conditions meet basic standards. Workers are considered "non-immigrant," meaning that they must leave after the allowed period of stay. Like in the bracero program, workers are tied to specific employers, making them vulnerable to abuse and exploitation. Private recruiters make matters worse by taking advantage of vulnerable people who become indebted to them and forced to work (Farmworker Justice 2011).

The rigidity of guest worker programs has also contributed to large numbers of undocumented immigrants, who are even cheaper to hire and more easily disposable. Labor advocates have been arguing for the need for comprehensive immigration reform that would update guest worker programs and legalize undocumented immigrants who have been working in the United States for a long time (Farmworker Justice 2011). These reforms would help give all farm workers—domestic and foreign—bargaining power to improve wages, working conditions, health, and safety.

At the same time, migrant farm workers' mobility into other forms of employment is limited through restrictive immigration policy and race-based labor market segmentation. They have very little choice but to continue seeking employment in agriculture regardless of the working conditions. Some farm workers follow harvests and travel long distance across state boundaries in search of employment in seasonal jobs. The documentary film *La Cosecha* (*The Harvest*) provides an excellent illustration of the insecurity associated with seasonal farm work and its impact on children. Indeed, the term migrant farm worker is used by some to describe foreign workers who come temporarily to work in agriculture as well as workers whose job location changes with seasonal demand, requiring them to be shuttled to farms far from home or to move to follow crops.

The tightening and militarization of borders, as witnessed in the United States since the 1990s, contribute to the vulnerability of undocumented workers who live in fear of being deported and not being able to come back. Some employers, especially in large farms and meatpacking plants, use the threat of deportation as a way to control labor and limit individual complaints or collective organizing. While it has certainly weakened the bargaining power of immigrants, the restrictive immigration policies currently implemented in the United States do not seem to have reduced farm and business owners' reliance on immigrant labor.

In 2018, 73 percent of hired crop labor in the United States were immigrants, mostly from Mexico, and over 50 percent lacked work authorization (USDA 2020). Immigrants have also been historically overrepresented in food manufacturing and services industries. Research presented to the US Congress in 2015 showed that 20 percent of meat processing workers, restaurant cooks, and dishwashers were undocumented immigrants—well above their 5 percent share of the labor force (Passel 2015). Within the food manufacturing and service industries, immigrants and people of color tend to fill jobs with the lowest wages and employment security. For example, in restaurants, immigrants work as bussers, dishwashers, and cooking assistants. Rarely do they move up to chef, head waiter, or bartender positions that usually pay better and provide additional revenue in tips (see chapter 12).

These dynamics do not just apply to the United States. In Europe, migrants from Eastern Europe, including Poland, Ukraine, Romania, and Albania, provide a large share of farm labor. In Thailand and Indonesia, migrants from neighboring countries do so. The same occurs in the larger food producing countries of Latin America (e.g., Argentina, Brazil) and Africa (e.g., South Africa, Zimbabwe, Mozambique). In addition, many large farming operations also rely on domestic migrants coming from other parts of the country. For example, in Brazil, since the 1990s, many impoverished migrants have moved to the Cerrado savanna—a huge global bio-diversity hotspot at the edge of the Amazon that has become the agricultural frontier for soybean, cattle, sugar, and coffee production. They are often recruited by labor brokers—known as "gatos"—in urban slums and communities far from the farms, putting workers at greater risk of debt bondage and slavery (Bales 2012). As migrants arrived in greater numbers, subsistence farmers lost their access to land and livelihoods, forcing them to look for employment in the export-oriented and industrialized farms, producing for multinational trade corporations like Archer Daniels Midland, Cargill, and Louis Dreyfus Company (see chapter 3). In the Cerrado, as in most areas where industrial farming has recently expanded, the exploitation of labor parallels environmental destruction (see chapter 7).

Beyond selective immigration policy and recruitment strategies, scholars have argued that globalization and more specifically the structural adjustment programs enforced by lending institutions beginning in the 1980s (see chapter 3) have created a large supply of labor by undermining rural livelihoods in the Global South and encouraging urban and transnational migration. For example, much of the agricultural labor in California has been performed by Mexican immigrants who migrated in the 1980s and 1990s as the Mexican economy was being restructured and rural jobs eliminated.

Indebtedness is also a common mechanism to increase individual workers' dependence on employers. By providing transportation and housing to migrants, employers are able to coerce them to work until they pay off their debt. This type of labor arrangements, known in many places as "contract labor," has given agricultural companies an incentive to build labor camps where large numbers of workers can be

FIGURE 4.5 Lamesa farm workers community in Texas. The farm labor camp was built for migrant farm workers in 1941 and housed hundreds of Mexican migrant farm workers and their families for decades. They did not pay rent but were expected to contribute to the maintenance of the camp and a community fund. It closed in 1980 and was sold a few years later to a private owner who continued renting the dilapidated units to Latino farm workers.

Source: https://commons.wikimedia.org/wiki/File:Lamesa_Farm_Workers_Community_2018.jpg.

housed close to fields and factories but away from residential areas. These labor camps are notoriously insalubrious and overcrowded. In some cases, recognizing the importance of maintaining a sufficient supply of labor, government agencies subsidized the construction of such labor camps, as the US Farm Security Administration did in the 1940s (see figure 4.5). In the twenty-first century, 21 percent of farm workers in the United States are still housed in employer-provided housing, including insalubrious labor camps and motels, where their activities are regulated and surveilled, making it difficult to organize (California Rural Legal Assistance et al. 2010). The rapid spread of COVID-19 among farm workers in the central Valley of California in 2020 and 2021 illustrates the prevalence and risks of living in cramped quarters.

The tight connection between prisons and plantations provides an extreme illustration of the significance of controlling mobility in order to control labor (see textbox 4.4). At various points in the history of the United States, farm labor has been provided by people kept in captivity, including indentured servants, enslaved people, prisoners, or victims of human trafficking. Globally, there were approximately twenty-seven million trafficking victims in 2012, including men, women and children forced to work in the food system, especially in farming and fishing. A report by the US State Department (2013) provides several examples, including children from Mali, Burkina Faso, and Benin forced to work in cocoa, coffee, and pineapple farms; ethnic minorities from India in Bangladeshi tea plantations; Brazilian men in Amazonian cattle ranches and sugar plantations; Central Asian and East European refugees in Southern Italian farms; and Southeast Asian migrants in Thai commercial fishing and Indonesian palm oil industries, among others.

Most capitalist economies rely on racial hierarchies to differentiate access to mobility, segment labor markets, and ensure an oversupply of low-wage labor. Race, used to divide people and justify these divisions, has been at the core of spatial processes of immiseration and dispossession of agrifood labor. In addition to being migrants, many farm and food workers are people of color whose labor has historically been

Box 4.4   **Plantations and Prisons**

Slavery officially ended in the United States in 1865 with the signing of the Thirteenth Amendment, which states that "neither slavery nor involuntary servitude, *except as punishment for crime whereof the party shall have been duly convicted*, shall exist within the United States" (emphasis added). Thus, criminals were excluded from these new legal protections. Since then, forced labor has been commonly used as a form of punishment, especially in the South.

Because the Black Codes, followed by the Jim Crow laws, unfairly criminalized formerly enslaved people and Black people, incarceration increased after the Civil War. Concerned with the cost of keeping these convicts in custody, state governments in southern states, turned to former plantation owners to run penitentiaries. Under the practice of "convict leasing," prisoners were forced to work in the fields for no pay, generating large profits for those in charge. And thus began the profitable business of privatizing prisons in the United States.

According to Bauer (2019), prisons allowed former slaveowners to become even more economically successful than they had been under slavery. Examples include the Angola Plantation in Louisiana and the Parchman Farm in Mississippi (see figure 4.6). States also benefited from this practice. For example, between 1880 and 1904, 10 percent of Alabama's state budget came from state convict "leases," paid out of penitentiary profits. The Ku Klux Klan was involved in rounding up prisoners as well as managing penitentiaries, maintaining a racial hierarchy by ensuring that Black people remained in subordinate positions. Torture, including "watering," whipping, and flogging, was commonly used to control prisoners and caused numerous deaths. For example, about three thousand mostly Black Louisiana convicts died under the lease of a single owner between 1870 and 1903.

While the practice of convict leasing was phased out at the beginning of the twentieth century, prison labor did not end. For example, a documentary

FIGURE 4.6 Parchman Penal Farm, male prisoners hoeing in a field, 1911.

Source: Mississippi Department of Archives and History, Mississippi State Penitentiary Photo Collections.

produced by the Atlantic (Benns 2015), *Angola for Life: Rehabilitation and Reform Inside the Louisiana State Penitentiary*, shows through images resembling those of previous centuries how the largest maximum-security prison in the state of Louisiana continues to operate like a plantation. Because the relationship between the penitentiary and the inmates is not based on employment but on "rehabilitation" and "social benefit," labor protections typically afforded to employees are not enforced. As a result, prison workers work in harsh conditions and earn very low wages, which are frequently garnished to pay outstanding debts. Violence is

rampant and punishment, in the form of solitary confinement or revocation of family visitation, continues to be used to discipline labor.

For some scholars, prisons and plantations operate according to the same logic of *racial capitalism* (McKittrick 2013) in which the generation of profit involves the historical devaluation of Black bodies. This also relates to the concept of *necro-capitalism*—the idea that capital accumulation rests on "dispossession, death, torture, suicide, slavery, destruction of livelihoods and the general organization and management of violence" (Banerjee 2011, 324).

---

devalued because of their race or ethnicity. In the United States, the racialization of agrifood workers as brown, immigrant, temporary, and illegal "others" has been used to justify their poor treatment regardless of their place of birth and immigration status. In Brazil, where race relations are arguably very different, agricultural workers are primarily Afro-Brazilian migrants and internally displaced Indigenous men.

Women in Food Work 4.5.

Globally, women play a significant, yet mostly invisible, role in food production. According to the FAO (2011a), they represent 43 percent of the hired labor force. However, this statistic may be severely underestimated since unpaid subsistence farming work, in which many rural women engage, is not accurately counted in official metrics. Widely cited figures indicate that, in developing countries, between 60 and 80 percent of all farm labor is performed by women. And in the world's poorest countries, almost 80 percent of economically active women report agriculture as their main source of livelihood (FAO 2011a). Indeed, farm labor has become more feminized in recent decades, especially in Africa and Asia, challenging the long-held assumption that the modernization of farming would lead to a decline in female participation. As men migrate to urban areas in search of employment, women stay behind to cultivate the land and take care of children and the elderly. In many places, civil wars and the AIDS epidemics have also left many women in charge of subsistence agriculture.

Within particular agroecosystems (see chapter 2), men and women typically play different roles. Class, religion, ethnicity, and race intersects with gender in producing unique divisions of labor. The notions of "gendered crops" suggests that men and women cultivate different types of crops and are therefore differently vulnerable to shocks imposed by policies, global markets, technological change, environmental hazards, or climate change (Carr 2008, Sachs 2018). It also highlights the relationship between farming and the construction of gender, including ideas about femininity, masculinity, domestic responsibilities, and women's work. Although the gendering of agricultural work varies across cultural, political, and ecological contexts, in general, women's roles tend to be defined by their responsibility to feed and care for the family (see chapter 12). In many developing countries, women often fetch water, remove weeds, grow vegetables and other staple foods, and raise small animals like poultry near their homes. They often perform these labor-intensive tasks

FIGURE 4.7 Woman cleaning a field with a sleeping baby on her back, Ghana.

Source: https://commons.wikimedia.org/wiki/File:Women_in_farming.jpg.

while taking care of young children (see figure 4.7). Women also tend to be involved in the post-harvest manual processing of food, such as drying, chaffing, smoking, and grinding. This provides them with the ingredients they need to make meals for their household. In contrast, men are more likely to engage in cash-crops cultivation or cattle raising—a diversification strategy that allows household to earn money to purchase other types of food. Men's jobs tend to be slightly more capital-intensive. Because of this division of labor, when agricultural employment or income declines, men are more likely to be the ones moving to urban areas in search of other sources of income, while women stay home.

Gendered divisions of labor have also shaped family farms in the Global North, where women have traditionally been responsible for maintaining home gardens, milking animals, making butter and cheese, feeding chickens and rabbits, collecting eggs, and so on. These tasks are constructed as feminine because they take place in or near the home, allowing for interruption for childcare and other domestic duties, and are viewed as less physically demanding. During the twentieth century, the number of family farms in the Global North declined rapidly. Technological advances and specialization into single crops decreased the need for women's labor. Mostly male farm owners have come to rely on machinery and hired labor to prepare the ground, plant, fertilize, irrigate, and harvest crops. Home gardens, orchards, and chicken coops have virtually disappeared from industrialized farms.

Yet, although globally women provide most agricultural labor, men continue to hold the land, make decisions, and control women's labor. These gender gaps are reproduced through patriarchal social norms and gendered policies, such as rural extension, training, and credit, that operate under the assumption that men are the primary farmers despite the feminization of agriculture.

Globally, fewer than 15 percent of agricultural landholders are women. They mostly work on land they do not own and on plots that are smaller and less fertile. Because property and inheritance laws are biased against women, most land is owned by their husbands, fathers, or brothers, shaping their power in decision-making. Furthermore, women are rarely allowed to sign contracts on their own, excluding them from borrowing money, buying land, hiring workers, or participating in contract-farming arrangements in which they could obtain a relatively stable source of revenue for their crops. For example, in Kenya, where smallholders frequently engage in contract-farming of fruits and vegetables for exports, fewer than 10 percent of

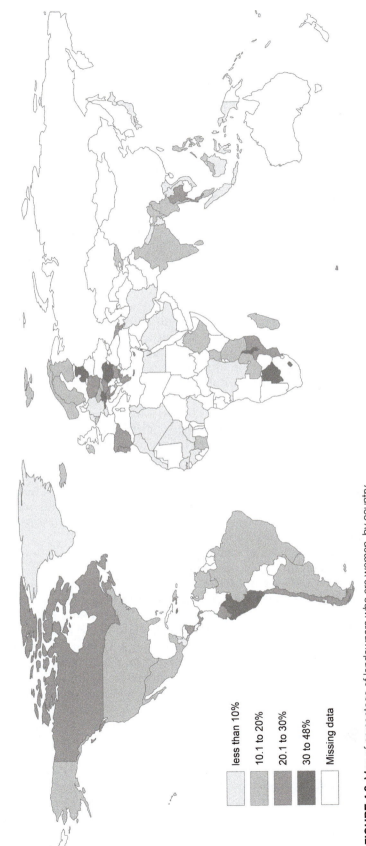

FIGURE 4.8 Map of percentage of landowners who are women, by country.

Source: Author with data from FAO (2011b).

contractors are women (Dolan 2001). This differential treatment of men and women in legal matters is particularly difficult for the growing number of single, widowed, divorced, or separated women, including those whose household has been split by the AIDS crisis or economic migration. It also severely impacts women's ability to access tools, seeds, fertilizers, pesticides, water, and credit, as well as information about technology, crops, management, and marketing. Women's access to finance is disproportionately low: they receive just 7 percent of total agricultural investment and 10 percent of aid (FAO 2011a). Similarly, in developing countries, extension services that provide technical assistance to farmers are only available to 5 percent of women working in agriculture. A major source of discrepancy is the lower level of education of women, who because of discrimination and social norms cannot attend school, especially beyond the primary level.

Women have also had a limited voice in decision-making at the household, community, national, and global scales where it is often assumed that they do not have the knowledge or expertise to make decisions regarding crop choice, soil management, marketing, or public policy. Legally, men are typically viewed as household heads, reinforcing the notion that women are just "helping," even though more than one in four households in Africa are headed by women. Such laws are embedded within patriarchal norms that devalue and subordinate women to roles of caretakers and govern their position within households and communities. In policy forums, women's voices are not equally represented. For example, between 2000 and 2010, only 30 percent of registered country delegates at UN Climate Change Summits, which deal extensively with agriculture, were women.

As a result of these gender gaps, women are less productive than men, generating 20 to 30 percent less food and earning lower incomes despite working more hours. Numerous studies show that gender differences in yields are not due to women's lower skills but to their limited access to inputs and knowledge. Addressing these gender gaps could increase women's productivity and income, thereby improving food security and the well-being of children (see textbox 4.5). According to the FAO (2011a), closing the land and input gap between men and women would raise total agricultural output by up to 4 percent. If this food were consumed locally, it could reduce the global toll of world hunger by 100 to 150 million people or 12 to 17 percent.

In the Global South, only a small proportion of women—seldom more than 5 percent—are engaged in waged agricultural jobs. When hired for wages on commercial farms, women's work tends to be more informal, temporary, and seasonal and therefore pay lower wages than men's work. In Africa, wages for casual farm labor are about twice as high for men as they are for women, who are more likely to be paid in kind and at piece rate. Because women's mobility into other forms of employment is often severely limited, farming offers the only option to feed their families. Women are also disproportionately represented in informal jobs related to the preparation and sale of food in public spaces and street markets (see figure 4.9).

In the United States, according to the latest US Census of Agriculture, in 2017, about 30 percent of farms were operated by women—up from 13 percent in 2012. The proportion of women making up agricultural hired workers has been relatively stable, around 26 percent (USDA 2020). There are significant differences between farm operators and farm employees, however. The recent increase of women as principal operators has been explained by changes in the agricultural census, allowing respondents to list multiple principal operators. It has also been tied to a recent increase in small organic farms, where women play an important role. Despite recent progress regarding gender, farm operators remain overwhelmingly white, with just 3 percent Hispanic, 1.7 percent Native American, 1.3 percent Black, and 0.6 percent Asian.

| Box 4.5 | The Social Costs of Gender Inequality in Malawi |

Malawi is a small and densely populated landlocked country in southeastern Africa. According to the FAO (2011a, 2015), it is one of the poorest countries in the world with 70 percent of the population living on less than $2 per day, 10 percent infected with HIV, and 85 percent dependent on agriculture for their livelihood. Major food crops include maize, groundnuts, sugar, potatoes, rice, cassava, and beans. The overwhelming majority of agricultural output (90 percent) comes from smallholder farms of less than one hectare of rain-fed land. What is produced on these small farms generates about 80 percent of Malawi's food and is essential to food security. Agricultural productivity is considered low, partly because of the high cost of imported inputs, poor irrigation, and a lack of infrastructure.

Fifty-two percent of agricultural labor is performed by women who have lower access to fertilizer, mechanical equipment, hired workers, credit, education, and training; hold smaller plots of land and fewer livestock assets; and experience gender discrimination in many areas of their lives (FAO 2015). Despite laws against it, child marriage is still common and half of girls are married before turning eighteen, often becoming unpaid farm laborers and domestic caretakers on their husbands' farms according to dominant patrilineal norms. Seventy percent of women farm managers are widowed, divorced, or separated, compared to just 3 percent of men, most of whom are married and rely on their wives for labor. As a result of these challenges, women's average earnings in rural areas are 35 percent lower than men's earnings and their productivity is significantly lower, making women more likely to turn to highly exploitative daily wage or casual labor, known as *ganyu* (Meijers et al. 2015).

Although the government of Malawi recognizes the importance of agriculture in promoting economic development, its commitment to reducing gender inequality has been weak and inconsistent. Too often gender issues are relegated to the background while economic development is prioritized under the assumption that it will benefit all equally. Indicators of the status of women in Malawi, whether based on education, life expectancy, economic outcomes, or political participation, score poorly. According to the FAO (2011a), the gender gap costs Malawi an estimated $100 million annually in lost productivity. Eliminating the gender gap could increase crop production by 7.3 percent, raise the GDP by 1.85 percent, and lift as many as 238,000 people out of poverty. Doing so might have positive environmental impacts since women tend to use less fertilizer, plant more diverse crops, and rely on agroecology. It would also reduce child labor, allowing young people to go back to school. Studies show that when women's income increases, household expenditure on food goes up disproportionately, improving nutrition and food security.

Malawi is heavily dependent on international aid from governments and NGOs. Many of these organizations seek to "empower women." Unfortunately, funding is highly unstable, and many projects are abandoned before being able to make a significant impact. Many initiatives, while well-meaning, are myopic in the sense that they do not tackle the deep causes of gender inequality such as disparities in control over land and labor, economic policies that reproduce gender inequality by disproportionately hurting women, and deeply entrenched patriarchal social norms.

For example, a short video on YouTube shows Malawian women using a fuel-efficient stove, known as *changu changu moto* (meaning "fast fast fire" in Chichewa), to cook a variety of dishes with food they have grown in their gardens (see https://www.youtube.com/watch?v=xA46sHliY74). The video highlights the disproportionate burden of reproductive activities or care work imposed on women and the significance of subsistence agriculture in structural everyday life in rural areas. The simple stove, promoted by American NGO "Ripple Africa" and built with local materials such as bricks and mud, frees up significant time for other tasks, potentially increasing women's agricultural productivity. Yet, it does not challenge gender roles.

FIGURE 4.9 Woman selling roasted corn on the side of a road in Midrand, South Africa.

Source: https://commons.wikimedia.org/wiki/File:Roasted_Mielies.jpg.

These figures contrast sharply with those of hired farm workers showing immigrants and people of color disproportionately represented. Among women working for wages in agriculture in the United States, the vast majority are racial and ethnic minorities employed in some of the lowest paid segments of the agrifood industry including picking crops such as tomatoes and strawberries. In addition to earning low wages, women in these occupations are disproportionately exposed to physical and sexual abuse. Indeed, 90 percent of female farm workers in the United States report that sexual violence is a major problem, revealing women working in fear of being raped and sexually harassed by their employers, supervisors, coworkers, or housing providers (California Rural Legal Assistance et al., 2010). Their precarious status, which itself is linked to their gender, poverty, race, and immigration status, makes it difficult to report these abuses. Similar reports of gender-based violence have been made by women employed in commercial farms in other countries.

Beyond agriculture, women around the world are also involved in many other aspects of food production. Like immigrants, they are over-represented in food service jobs, where their work is often devalued because cooking and serving food is viewed as unskilled labor that is a natural extension of female responsibilities (see chapter 12). For example, while men may be viewed as skilled and creative professional chefs, women are seen as traditional home cooks—a distinction that affects their employment opportunities and earnings. Women also tend to be overrepresented in food processing, including washing, packaging, and labeling produce for exports.

Gender relations are constantly being redefined. What is seen as male or female changes over time and space and evolves as women engage in negotiations within their household, village, and larger society. Global forces such as technological change, economic restructuring, and climate change also impact women's access to resources and economic opportunities. In the past twenty years, international organizations such as the FAO, the UN, and the World Bank, along with many nongovernmental organizations (NGOs), have begun to acknowledge the central role of women in food and agriculture (FAO 2011a, IAASTD 2009). For instance, the third of eight Millennium Development Goals established by the United Nations was to "promote gender equality and empower women," with the target of eliminating gender disparities in all levels of education by 2015. So far, efforts to reduce the gender gap have focused primarily on education, with the expectation that it would lead to changes in the realms of work, nutrition, health, political participation, and sustainability. Improving women's access to credit, including microlending, has been another important area of intervention for NGOs and community-based self-help organizations.

Conclusion 4.6.

The food economy has become much larger than agriculture. Although agriculture remains essential to livelihoods around the world, a growing number of jobs can be found in the processing of raw materials into increasingly complex food products and their storage, transportation, and marketing. Industrialization continues to shift jobs from fields to factories, supermarkets, and restaurants. As a result, both food and labor have become increasingly commoditized.

There are many types of labor arrangements within the agrifood system, including subsistence unpaid self-employment, contract labor, and formal and informal waged labor. The prevalence of these various forms of labor vary geographically according to economic conditions, regulations, cultural factors, and environmental characteristics. Despite differences across space and time, labor in the agrifood system is almost universally associated with poverty and exploitation, especially in comparison to employment in other industries. At the same time, it is becoming increasingly obvious that food and agriculture sustain billions of livelihoods. Supporting these livelihoods is not only important for the economic well-being and health of communities, including women, children, and people of color, it is critical to building a more sustainable food future.

Food Connections and Commodity Chains 5

Learning Objectives

- Investigate the networks and connections that organize the global circulation of food from farm to plate and the different ways that geographers have conceptualized them.
- Understand food commodities as socially produced objects of consumption that hide the production process, paying attention to geographic imagination.
- Locate power and agency within global food chains.
- Critically examine the transformative potential of fair trade and ethical consumerism.
- Explore efforts to relocalize the food economy and create alternative food networks.

In chapter 3, we learned about the globalization of food—the spatial processes associated with the increased extensity, intensity, velocity, and impact of transnational connections surrounding the production, distribution, and consumption of food. In this chapter, we focus specifically on the networks that underlie these connections between various components of the food system across space.

In recent years, the expression "farm to table" has become ubiquitous; it is used by activists to draw attention to the impacts of food consumption choices on farmers, by researchers to conceptualize connections between producers and consumers within food systems, and by businesses to market local foods. All raise important questions about where our food comes from. Today, it has become very difficult, if not impossible, to answer these questions. The industrialization, commoditization, and globalization of food, which we studied in the previous chapters, have created distance between producers and consumers, especially consumers from the Global North who tend to purchase heavily processed and beautifully packaged food from "nowhere." Since very few people in high-income regions grow their own food, they have limited knowledge of what the production process entails. The processing and packaging of food sold in supermarkets make it virtually impossible to trace the provenance of ingredients and hide potentially disturbing and exploitative aspects of their production under colorful labels and carefully produced images of pristine farms or exotic landscapes.

Yet, when we decide what to eat, we join networks of actors that have made it possible and desirable for us to consume specific foods. Depending on our circumstances and choices, these networks may be local and include just a few actors or they may span the globe and consist of hundreds of actors involved in bringing food to consumers through engineering, cultivation, packaging, marketing, financing, regulation, transportation, preparing, serving, and so on. Not only do material food products, money, and technology circulate through these networks, but knowledge is created and transferred between actors, including consumers who make practical and moral decisions about what food to buy, prepare, and eat.

In this chapter, we explore the connections that organize the circulation of food from farm to plate—or perhaps more accurately from laboratory to landfill. We examine how consumers relate to producers through commodities, via chains, networks, and circuits that influence global divisions of labor and the distribution of value. A better understanding of where power and agency reside within food chains points to the potential role of consumers in creating a more socially just and ecologically sound food system.

5.1. Commodity Fetishism and Global Chains

Rather than being consumed directly by those who farm it themselves, a larger and larger share of food is produced for exchange and sold for a monetary price. Price, becomes the primary indicator of value, making foods interchangeable and obscuring how they were produced and where they came from. For instance, for the average consumer, coffee is just coffee, regardless of whether it is grown in Ethiopia or Uganda, produced on a small family farm or large-scale exploitation, and harvested by household members or hired workers. What distinguishes one type of coffee from another is the price. Indeed, mainstream brands such as Folgers and Maxwell House, which are owned by Proctor and Gamble and Kraft Heinz corporations respectively, source their coffee from different places around the world depending on prices, with no or little noticeable impact on consumers. Industrialization, globalization, and new technologies, which have stretched the physical and social distance between the realms of production and consumption, have exacerbated this commoditization process and contributed to turning most foods into abstract and placeless commodities.

More than a hundred and fifty years ago, Marx described this phenomenon as *commodity fetishism*—the process of defining goods by their monetary value rather than the labor involved in producing them. This occurs when goods are produced for their "exchange value" rather than for their "use value." In Marx's words, "It is . . . precisely the finished form of the world of commodities—the money form—which conceals the social character of private labor and the social relations between individual workers, by making those relations appear as relations between material objects, instead of revealing them plainly" (Marx 1867, 335). Put it differently, commodity fetishism alienates producers from the product of their labor and allows consumers to ignore the realities of that labor, severing consumption from production. Geographer David Harvey illustrates this idea with the example of grapes that "sit upon the supermarket shelves mute" so that "we cannot see the fingerprints of exploitation upon them or tell immediately what part of the world they come from" (Harvey 1990, 422–23).

The food industry itself has sought to obscure the connections between farm and table, flooding supermarkets with plastic-wrapped products whose origin is hidden and replaced with a fabricated image meant to appease and entice consumers. For example, since 1958, Colombian coffee has been marketed in the United States via the fictional character of Juan Valdez—a happy coffee producer, wearing a poncho and wide brimmed hat, and flanked by his mule Conchita carrying sacks of coffee beans. Such cartoonish representations of producers mask the reality of agricultural labor (see chapter 4).

In addition to masking social relations of production, the commoditization of food also hides ecological relations in which land and natural resources are exploited. For Wendell Berry (2009, 228–29), "the industrial eater is one who does not know that eating is an agricultural act, who no longer knows or imagine the connections between eating and the land, and who is therefore necessarily passive and uncritical—in short a victim. When food, in the minds of eaters, is no longer associated with farming and with the land, then the eaters are suffering a kind of cultural amnesia that is misleading and dangerous." This quote reflects the common, yet contested, belief that consumers have become passive and uncritical actors at the mercy of corporations that manipulate their desires through marketing and advertising.

Beginning in the 1980s, the notion of commodity fetishism inspired scholars in sociology, geography, and political science to "lift the veil" hiding the social and ecological relations of production embodied in commodities. This emerging research was synthesized in the 1990s in the concept of *global commodity chain* (Gereffi and Korzeniewicz 1994), which incorporates three key aspects of the relations between producers and consumers.

First, global commodity chains encompass a sequence of activities that add value to a product as raw inputs are transformed into consumer outputs. Such chains, also known as *global value chains*, often begin with inputs production and farming and follow commodities through a series of steps related to sorting, processing, manufacturing, packaging, storing, and transporting, before it reaches consumers via marketing, advertising, and retailing in different types of outlets (see figure 5.1). Most of the value, which is reflected in the final price, is added in the later stages, giving food manufacturers and retailers a larger share of the revenue compared to producers. The highly unequal distribution of value within global commodity chains is illustrated in textbox 5.1 with the example of bananas destined to US consumers. While most scholars of commodity chains rely on this framework to offer a criticism of

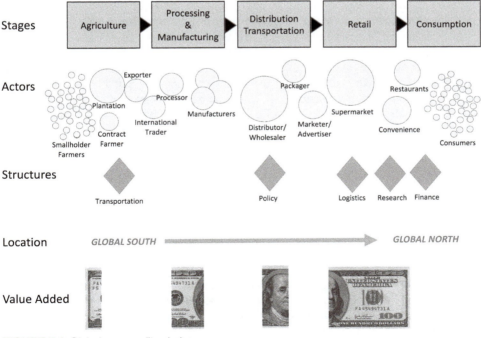

FIGURE 5.1 Global commodity chains.

Source: Author.

| Box 5.1 | **Global Commodity Chains of Bananas** |

The average US consumer eats over fifty bananas per year—more than any other fruit, including apples and oranges. Today, most of the bananas consumed in the United States come from Ecuador, Guatemala, and Costa Rica, where they are typically grown on large plantation-style farms that rely on hired labor. In Europe, consumption is even higher than in the United States, and bananas are imported from many more countries, including former European colonies in Africa and the Caribbean. These countries are highly dependent on the export of bananas for employment and income. To compete globally and keep prices low, plantation owners often pay workers very low wages on a piece rate or daily basis.

This is not a new arrangement; in the early 1900s, several Central American countries became known as "Banana Republics"—a term used today to describe countries whose political and economic system is dominated by a single commodity and controlled by powerful corporate interests. After the success of Jamaican bananas on the East Coast in the 1870s, American entrepreneurs began investing in banana plantations and transportation infrastructure in Central America, gaining tremendous power in countries like Honduras, Guatemala, and Costa Rica. Two gigantic corporations came to dominate the banana production for the US market: United Fruit Company and Standard Fruit Company, which would later become Chiquita and Dole respectively. They also held tremendous power over the state and were able to use their relationship with the US government to topple democratically elected leaders opposed to their ways of doing business, put dictators in charge, and influence economic policy that would allow them to maintain land ownership and control labor by resisting unionization (Koeppel 2008, Chapman 2014). For instance, as recently as the early 2000s, Chiquita allegedly funded a paramilitary organization in Colombia that targeted and killed union leaders, banana workers, and social activists, contributing to the country's long civil war (Patel 2012). According to Stone (2016), one of the reasons bananas are so cheap is the meddling of corporations in local politics and the constant involvement of the US government in supporting or protecting them.

If you ate a banana recently, you might have noticed the country of origin on the sticker. Perhaps it was from Guatemala, which exports about one hundred million forty-pound boxes of bananas to the United States each year, generating $1.6 billion in export revenue or 7 percent of its foreign currency. Most banana plantations in Guatemala are located along the Pacific and Caribbean coasts where marshes have been drained and forests cleared to make room for farming. The majority contract directly with large companies like Chiquita, Dole, or Del Monte. Although there are thousands of varietals, they almost exclusively grow the Cavendish banana (Koeppel 2008). Hired laborers work ten to twelve hours per day, six to seven days per week, earning about 100 quetzales ($12.90 US dollars) a day without any benefits to grow and harvest the bananas (Banana Link 2020a). These abysmal labor conditions are fostered by a price war between supermarkets and very limited unionization that is weakened by a lack of government protection, corporate interference, and violence against organizers including dozens of murders that have not been investigated. This is especially problematic in the southwestern part of the country along the Pacific Coast, where the lack of unionization gives producers an unfair advantage compared to companies along the Caribbean Coast that face more organized labor and have seen their share of the market shrink. This price war is threatening the progress made by unions such as SITRABI (Sindicato de Trabajadores Bananeros de Izabal) in raising wages and benefits (BananaLink 2020a).

In the United States, conventional bananas sold for an average of $.57 cents per pound in 2019—well below the price of apples or oranges. Most consumers are unaware that this low price comes at the expense of workers. Although it is very difficult to trace the physical travels of a particular fruit and to assess the value added at each step, researchers have attempted to do so in various contexts. These studies suggest that after being harvested, bananas are first washed, labeled, packaged, and boxed at or near the plantation, then they are shipped via trains or trucks to ports where they are loaded onto refrigerated containers and sent off to the United States on a large company-owned cargo ship (see figure 5.2). Upon arrival in places like Miami, Florida, or Long Beach, California, they are taken to temperature-controlled ripening facilities where they are bathed in ethylene gas and hydrocarbons for about a week until they are ready to be distributed to retailers throughout the country. There, they will be unpacked and displayed on produce shelves to be purchased by consumers.

FIGURE 5.2 Chiquita cargo ship loaded with refrigerated banana containers, Bocas del Toro, Panama.

Source: Karen Jain 2009 via Wikimedia Commons: https://commons.wikimedia.org/wiki/File:Chiquita_Schweiz_(ship,_1992)_001.jpg.

Value or profit is not distributed evenly along the commodity chain. Unfortunately, data from Guatemala are not readily available. However, data from BananaLink (2020b) suggest that for Latin American and Caribbean bananas destined to the European market, value is distributed as illustrated in figure 5.3, with workers receiving just about 8 percent of the total value generated along the chain. Much of the profit goes to retail corporations, who are powerful buyers and put tremendous pressure on lower levels of the commodity chain to ensure that bananas remain cheap, particularly since the price of bananas is apparently a common indicator used by price-conscious consumers to determine where to shop.

In recent years, this has been further complicated by the emergence of a new strain of the so-called Panama disease, also known as Tropical Race 4, that is threatening the Cavendish banana worldwide. Although, it has not yet reached Latin America, producers have been required to preventively spray their plants with expensive chemicals, raising their production costs, which they try to recoup by lowering labor costs.

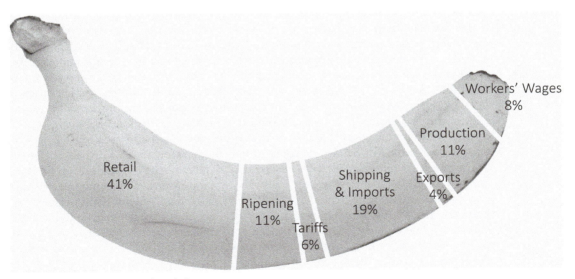

FIGURE 5.3 The banana value chain.

Source: Author based on data for Colombia, Ecuador, Costa Rica, and the Dominican Republic from BananaLink (2020b). Banana image from Evan-Amos via Wikimedia Commons: https://commons.wikimedia.org/wiki/File:Banana-Single.jpg.

globalization and its impact on inequality, others, including experts in development economics and business management, study how to strengthen, expand, and adapt supply chains to reduce transaction costs and stimulate economic growth (see for example McCullough, Pingali and Stamoulis 2008).

A second important and related aspect of global commodity chains is the power relations and governance structures underlying them. These chains did not emerge spontaneously and randomly out of the general expansion and deregulation of the global economy. Instead, they are organized through institutionalized arrangements resulting from strategic decisions made by powerful actors, including transnational corporations, international organizations, and national governments, to control access to resources in developing countries and to markets in more affluent economies. Thus, commodity chains represent layers of contractual transactions between public and private actors endowed with uneven amounts of power. One of the primary purposes of chains—or *filières* as they are called by French researchers who emphasize the institutional structure and regulation of commodity chains—is to minimize transaction costs (and maximize profit) by coordinating production and exchange. For instance, international bodies, such as the World Bank, the International Monetary Fund, and the World Trade Organization regulate and facilitate transactions by stimulating trade and foreign investment. In addition to being regulated by political bodies, chains are also maintained by financial and research organizations that govern the flow of money, technology, and knowledge.

Questions regarding the location and distribution of power within global commodity chains have animated scholars since the inception of the concept. Answers to these questions typically focus on the distribution of profit within the chain as the main predictor of power, which in a political economic framework stems from capital ownership.

A third element of global commodity chains is their spatiality and territoriality. Commodities are "spatial objects" that are created by moving through different places that become linked to each other. The secret nature of commodities is facilitated by distance and the technologies and institutions that make it possible for economic actors to extend their networks of production and "act at a distance" (Busch and Juska 1997). In its most basic formulation, it links firms from the Global South to the Global North. The latter (i.e., the *core*) is viewed as the locus of decision-making, where orders are placed, investments decided upon, technology designed, production flows engineered, and consumer markets concentrated. The former (i.e., the *periphery*) is where production takes place through the extraction of natural resources and the exploitation of labor, which are relatively abundant. This geographic division of labor relates to power to the extent that activities generating higher value added and capturing a larger share of profit tend to take place in the Global North, reproducing its ability to control actors further down the chain and creating barriers of entry to potential Global South competitors. Some interpret this division of labor as the result of *comparative advantages*, which determine each nations' economic strength (e.g., labor-intensive horticulture vs. capital-intensive food manufacturing) and its position along the chain as "lead" or "subordinate" agent. Such perspective, however, tends to ignore the historical political economic forces that underlie these comparative advantages, including processes of capital accumulation associated with colonialism that are emphasized by most commodity chain scholars working within the frameworks of political economy, food regimes, and world systems.

The linear connection between agents from the periphery to the core, however, is an oversimplification of the spatial organization of commodity chains, which can take on multiple forms. Over time, the spatiality of commodity chains evolves along with the type of commodities produced, steps required, actors involved, and regulatory frameworks in place. For instance, Fine and Leopold (1993), who prefer the term *systems of provision* to "commodity chain," argue that the structures connecting producers, distributors, and consumers differ depending on the type of commodities. These structures also change as new actors join the chain and others leave. For example, in farming regions, smallholders are being pulled into modern chains, not all of which are oriented toward exports. Observers have also noted the expansion of mid-level links and nodes in so-called emerging economies, where investments in food processing and manufacturing means that they are able to capture a larger share of value. In some cases, chains are shortened or intermediaries removed as new technologies allow retailers to bypass so-called "middle-men" and purchase directly from producers. In addition, there are differences between "Fordist" *producer-led* integrated chains which are ruled by large manufacturing corporations to minimize costs of production for products such as cars and computers and *demand-driven* "post-Fordist" chains for highly differentiated retail goods like fashion, toys, and food that require more just-in-time flexibility and rest on outsourcing and subcontracting with numerous competing producers to respond quickly to consumer preferences and corporate retailers' demands. Such distinctions raise questions about where chains begin and end and point to their potential *circularity*.

Although extremely useful in conceptualizing the spatiality of social relations underlying the production of food commodities, the idea of global commodity chains has been criticized for being too linear, putting too much emphasis on production, ignoring consumption, culture, and nature, and being too deterministic by suppressing human agency. These criticisms led scholars to pay more attention to the realm of consumption and to rethink food connections in terms of networks, circuits, and assemblages in which not only money and material goods circulate, but also knowledge and values.

Bringing in Consumption: 5.2.
Networks and Circuits

So far in this book, we have paid relatively little attention to consumption. Instead, we have focused primarily on the political and economic forces, including global food regimes, labor arrangements, and commodity chains, that organize the production of food across space and time. Perhaps, this reflects the long-standing emphasis that engineers, agronomists, economists, geographers, and other scientists and practitioners place on the material production of food, prioritizing quantity of output to stimulate economic development and address food insecurity. In that framework, food consumption, especially its cultural and symbolic aspects, has often been seen as frivolous and feminine—the world of housewives—and therefore less worthy of inquiry.

Yet, consumption is as integral to the food system as production. After all, food is produced to be consumed and therefore the way consumers relate to food is deeply intertwined with how and where it is produced. While production determines what is available for consumers in particular places, geographically contingent incomes, tastes, cultures, values, and social identities of consumers influence the food commodities they desire, purchase, and ingest. Mundane and taken-for-granted activities such as grabbing lunch at a fast-food restaurant, stopping at the supermarket after work, going out for a date in a fancy dining establishment, or grilling meat on the barbecue connect people and places and pull consumers into a web of social relations. Thus, it is important to intersect consumption with production to gain a better understanding of food systems and their geographies.

Since the 1990s, consumption has received significantly more attention from scholars, including geographers interested in understanding relationships and experiences surrounding food. Gill Valentine and David Bell's *Consuming Geographies*, first published in 1997, exemplifies this shift from production to consumption by emphasizing activities such as shopping, cooking, and eating, and paying attention to social and place identities, spatial connections, and cultural meanings of food. The move away from production-centric accounts of the economy and the emphasis on consumption as critical in driving economic growth, shaping cultural identities, and organizing social relations was part of a bigger shift in human geography, known as "the cultural turn"—the reorientation of scholarly work toward culture as integral to the creation of meaning and value.

This rapprochement between culture and economy has led many human geographers to explore the "social lives" and "hidden geographies" of food commodities in an attempt to understand how consumers relate to them and how this impacts the people and places associated with these particular commodities. This was influenced by anthropologist Arjun Appadurai's work on "the social lives of things" (1986) where he argues that commodities gain value through exchange as they become social objects that carry meaning for consumers. For example, scholars have examined the "social biographies" of chickens (Watts 2004), papayas (Cook 2004), and green beans (Friedberg 2003). Instead of starting with agriculture or food production as in the classic commodity chain model, they begin their analysis by focusing on eating and typically work their way through the material and discursive processes that transformed food into a meaningful and valuable commodity.

Focusing on consumption draws attention to the narratives surrounding commodities—in this case, the stories and imageries that accompany food and shape its consumption. For geographers, these meanings are intensely geographic and tied to the places, proximate and distant, real and imagined, where they are bestowed onto commodities (see textbox 5.2). This occurs as much in production as in consumption—when consumers shop, desire, imagine, learn, touch, cook, taste, and interact with food physically and emotionally. Understanding food consumption as a cultural phenomenon that connects things, people, and places, beyond the economistic relations of global commodity chains, seems to be particularly relevant in affluent places where consumers are faced with an abundance of food options that allows them to express their identities through lifestyles and consumption choices. It is therefore not surprising that much of the research on this topic focuses on the lives of globally traded exotic commodities (e.g., coffee, bananas, chocolate) and the social, cultural, political, and economic connections they weave across the earth.

While some scholars have adapted the idea of commodity chain to study how these meanings and knowledges emerge and travel between actors and places, many prefer more fluid concepts such as networks and circuits. Both ideas share much in common and the terms are often used interchangeably. *Networks* emphasize the complex and multidirectional webs of interdependence that connect many individual and collective actors through flows of materials and information that support the production and consumption of particular commodities (Whatmore and Thorne 1997, Busch and Joska 1997). The related notion of *circuit* posits that these connections operate in a nonlinear fashion, without a clear beginning and ending point (Leslie and Reimer 1999, Jackson 2002). This approach does not privilege production or consumption, but instead recognizes the multiple and nonsequential points of connection between consumers and producers.

| Box 5.2 | **Food and the Geographic Imagination** |

All human experiences "take place" somewhere; they are embedded in and shaped by particular places. Our everyday activities, including eating and obtaining food, cannot easily be abstracted from the geographic contexts where they occur. This embeddedness is what makes each experience unique and different and helps define cultures. Yet, as work on food commodities suggests, consumers have limited contact with the places where their food is grown, harvested, processed, and stored before it meets them in a market or restaurant. This is where *geographic imagination* comes into play; it affords consumers ways of envisioning the kind of places where food comes from, giving it unique meaning. Such imagination, which can be thought of as a form of knowledge and spatial consciousness, can also be powerful in making empathetic connections across space and fostering social justice. For instance, urban consumers might feel connected to farmers when they "imagine" the places where their food is grown, which might in turn motivate them to support rural communities.

At the same time, the spatial images we hold in our minds can easily be manipulated and distorted, preventing us from grasping material reality. Advertisers know that all too well, as their work consists of stimulating positive geographic imaginaries attached to food, whether pristine rural landscapes, trendy urban neighborhoods, or exotic places. When referring to these unconscious and subjective perceptions of place, geographers often use the term *imaginaries*.

The idea of geographic imagination is useful for understanding how consumers attach meaning and symbolic value to particular commodities. It also points to a specific form of *commodity fetishism* in which imagined places contribute to hiding the social relations of production. For example, the advertising for New Zealand beef and lamb in figure 5.4 reveals a "pure" and "natural" landscape that masks the industrial nature of meat production and the animal abuse associated with it.

FIGURE 5.4 Advertising banner for Beef & Lamb New Zealand's "Taste Pure Nature" branding campaign in California.

Source: https://beeflambnz.com/news-views/blnz-celebrates-successful-year-taste-pure-nature.

These conceptualizations have been influenced by *Actor-Network Theory* (ANT) and the work of Bruno Latour, John Law, and Michael Callon. ANT's main premise is that knowledge is the result of networked interactions between human and nonhuman actors and "actants," such as technology, documents, buildings, and living organisms that have agency within these networks to the extent that they enable or trigger action. While the original focus of ANT was to understand scientific knowledge, the idea of networks has now been used to understand all sorts of social phenomenon and knowledges, including ideas constitutive of food. In that framework, an object—such as a particular food—does not have any inherent properties, its value and character emerge out of the network of relationships in which it circulates as a commodity. There is no single independent social force, such as capital, that produces value and meaning. Instead, the value of a specific commodity is in the networks themselves; it originates in the situated interactions between actors and actants that give it both material value and symbolic meaning. This is not to say that power does not exist. Some actants clearly have more power in enrolling others and, as a result, may influence the durability of networks and their ability to achieve desired outcomes (Bosco 2006). Yet, this power is contingent, relational, and unstable (Lockie and Kitto 2000).

ANT also posits that such networks are constantly in a state of flux, enrolling new actants, generating different knowledge, and shaping the value of commodities. Cook and Crang (1996) use the idea of *displacement* to show how particular food cultures emerge by replacing, transforming, and pushing aside certain knowledges as commodities circulate from one site to another. For instance, Hollander (2003) discusses how different and competing narratives about sugar (e.g., as an exotic, unhealthy, luxurious, pure, or colonial commodity) emerge, stick, and influence consumers, while others are displaced. This process of "meaning making" parallels a process of identity formation and negotiation for the various actors who engage with the commodity.

Another important characteristic of networks, from an ANT perspective, is their spatiality. Networks "touch down" in specific places or nodes where actants come together to contribute in one way or another to the lives of commodities. Although the physical distance between these nodes may be stretched and networks lengthened, nodes are always localized in places that represent "the intersection of many trails" (Latour 2005, 204). This notion has much in common with contemporary geographers' conception of place as relational (see chapter 1). Thus, one way to understand networks is to explore key places within them. For example, a street market, a home kitchen, a farm, a factory, or a warehouse could become a window to explore connections between actants and their contribution to creating food commodities.

The role of cultural geographers is to "get inside the network, go with the flows, and look to connect" (Cloke, Crang, and Goodwin 2005, 49). In other words, the primary objective of commodity analysis is not to remove the veil of fetishism to reveal the true (capitalist) relations of production, but to understand how the processes of fetishization and commoditization occur as things, people, and places interact and produce new and changing geographic knowledges, images, and symbols that define commodities. Many have adopted this approach in studying various food commodities, such as coffee (Whatmore and Thorne 1997), sugar (Hollander 2003), beef (Stassart and Whatmore 2003), imitation crab (Mansfield 2003), and hot pepper sauce (Cook and Harrison 2007).

For example, Cook (2004) "followed" a papaya as it traveled long distances, stopped in multiple places, and went through many hands before becoming the exotic fruit showcased in the produce aisle of a London's supermarket and catching the attention of an affluent and educated shopper with an adventurous palate and a penchant for exoticism and authenticity bolstered by her artistic background, numerous trips abroad, and interest in cooking shows, cookbooks, and lifestyle magazines. For cultural geographers, these magazines or the layout of produce aisles are as relevant as seeds and fertilizers in the creation of commodities. The papaya would not be what it is without the seed, the packaging, the label, the warehouse, the farmer, the foreman, the importer, the computer software tracking the price of exotic fruits, the supermarket shelves, the produce stacker, the health reports, the magazine, and the many other people, things, and places that created it materially, emotionally, and discursively. Dupuis's study of milk as America's "perfect food" illustrates a similar approach to telling the story of commodities (see textbox 5.3).

Box 5.3 **Dairy Networks and the Story of Milk**

Milk is a hybrid commodity; it is produced by cows for their calves and literally extracted from their udders for human consumption. This fact alone ought to turn off many consumers, but there is more: milk is also prone to carry dangerous and deadly disease; it is expensive to produce and store; it has a very large carbon footprint; and it causes discomfort and allergic reactions in many people. Yet, milk is often viewed as "the perfect food," especially in the United States. How did this story emerge?

In her book *Nature's Perfect Food: How Milk Became America's Drink*, sociologist Melanie Dupuis (2002) tells the story of milk and shows how it took hold of American society. She argues that until the nineteenth century, drinking fluid milk was not common. Instead, people consumed smaller quantities of dairy in the form of cheese, butter,

and yogurt. By 1940, however, the average American was consuming over a pint of milk per day—a figure that continued to increase until the 1970s. Dupuis sought to answer the question "why do we drink milk?" by investigating how it became a universalized symbol of health.

Although Dupuis does not rely explicitly on the concept of network or mention actor network theory in her book, her research can be interpreted through that lens. She painstakingly shows how various human and nonhuman actors came together to produce milk as a healthy and quintessential American staple. Some of these actors were "boosters" who promoted milk because they economically benefited from doing so. For a long time, the dairy industry has actively advertised milk as a healthy food, as illustrated in figure 5.5 and in the

(continued)

FIGURE 5.5 Milk advertising in the *Daily Oregonian* newspaper, February 11, 1920.

Source: Oregon Dairyman's League and Portland Milk Distributions (1920), via Wikimedia Commons: https://commons.wikimedia.org/wiki/File:1920_Oregonian_milk_ad.jpeg.

Box 5.3   *Continued*

more recent "Got Milk?" campaign in the United States. Other actors are enrolled in this project through a wide-cast net, including refrigerators in urban homes, romanticized notions of a rural past when the family cow provided sustenance, doctors and health professionals, government regulations and food safety standards, large dairy corporations and small dairy farms, dietary guidelines, religion, pasteurization technology, children's bodies, women and mothers' identities, education, school lunch programs, advertising, and the media. The story of milk is shaped as much by the material aspects of its production as it is by the discourses, symbols, and debates surrounding it. Similarly, milk production as an economic pursuit is shaped by the notion of milk as perfect food. Thus, production

and consumption are deeply interconnected. These social, cultural, economic, and political relations underlying milk are imbued with power. As Dupuis shows, power shapes the story of milk in privileging white northern Europeans' ideas of purity, health, and perfect living, while suppressing other views, including those of people of color who are disproportionately lactose intolerant.

In the past two decades, the story of milk as perfect food has been reexamined by scholars, activists, and consumers. By foregrounding this story in the networks of actors that underlie it, they have shed light on hidden processes, telling a different and darker story that is likely related to the sustained decline in milk consumption in the United States.

Criticism of consumption-oriented network approaches often focus on the difficulty of identifying entry points or targets of intervention within complex networks for addressing pressing issues such as social inequality and environmental degradation associated with consumerism (see Goss 2004 and Hartwick 2000). Some also lament the lack of theory in these approaches, which tend to "describe" social phenomenon through their particular and unique networks rather than "explain" why certain types of networks emerge and produce undesirable outcomes. By putting so much emphasis on cultures of consumption, they argue, we might lose track of the relations of production that underlie commodities and reproduce a sort of fetishism. As Hartwick (2000, 1179) puts it, "in a world where objects have autonomous powers beyond comprehension, the still-existing connections with production are lost in imagistic reverie." She worries that researchers spend too much time focusing on consumer identities and not enough on the social and environmental conditions of those who produce consumer goods. There are also concerns that emphasizing the cultural and symbolic significance of commodities might inadvertently support a consumerist status quo. These related criticisms do not negate the importance of consumption in the lives of commodities but rather ask for correctives that would reconnect the symbolic to the material and the consumers to producers.

5.3. Ethical Consumerism and Fair Trade

The focus on consumption and culture came at a time when many were decrying the spread and homogenization of consumerism and raising a series of moral questions regarding the impact of food consumption on workers, animals, environment, and society. Some scholars argue that this has prompted a "quality turn" in the way consumers relate to food, with growing interest in knowing more about their food and its various qualities, such as taste, nutrition, health, origin, authenticity, fairness, and sustainability. Ethical consumerism grew out of these concerns and the belief that revealing the real life of commodities—or de-fetishizing them—would spark a new kind of activism led by better-informed consumers who would put pressure on producers by "voting with their fork."

On the surface, making the geographic life of commodities more transparent to encourage more ethical methods of production seems to align with the goals of those working to shed light on global commodity chains and lift the veil of commodity fetishism. Questions arise, however, regarding the source and nature of that geographical knowledge, the kind of activism it might enable, its primary actors, and its potential effects. Answering these questions requires that we look beyond traditional commodity chains and consider the realm of consumption, including how particular commodities become known as ethical.

There are two primary mechanisms to inform consumers about the quality of their products: certification labels that provide guarantees regarding certain aspects of production and shorter networks that put consumers directly in contact with producers. Both have grown exponentially in recent years and consist of creating new or alternative markets that would elude the negative effects of mainstream markets.

Various descriptors, such as *fair trade, humanely raised, organic, sustainable*, and *locally grown*, increasingly appear on the packaging of foods, advertising their superior quality to anxious consumers. Labels advertising products as fair trade came first, quickly followed by ones that designate them as organic. Others followed suit and continue to proliferate as consumers' sensitivities evolve. The idea of fair trade emerged in the early 1980s from the recognition that producers in the Global South were receiving a disproportionately small share of profits compared to manufacturers, distributors, and retailers in the Global North, reflecting the then emerging body of research on world systems and global commodity chains. Over time, it evolved to address concerns beyond farmers' livelihoods, such as environmental sustainability and gender equity.

Fair trade began first with nonprofit stores that sold products from a network of vetted producers and countries, such as coffee from a Mexican growers' cooperative or chocolate made with cocoa from a Côte d'Ivoire association of small farmers. This strategy reached only a small number of consumers who were able and willing to spend the time and money to shop in these specialized stores. Fair trade eventually broke through conventional distribution channels by creating distinctive labels that would be affixed to products distributed through regular intermediaries. Since then, sale volumes, premium received, certified producer organizations, workers involved, participating retailers, and consumers of fair trade products have all increased rapidly (see figure 5.6).

This expansion has been facilitated by the creation of Fairtrade Labelling Organization International (FLO) in Germany in 1997. Now known as Fairtrade

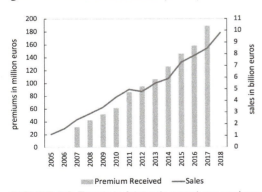
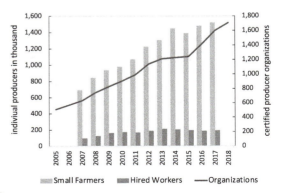

FIGURE 5.6 Trends in fair trade premiums and producers.

Source: Author with data from Fairtrade International (2011, 2013, 2014, 2020).

International, FLO brought together third-party organizations from Europe, North America, Japan, New Zealand, and Australia to coordinate and standardize the certification process by which qualified producers are selected to receive premiums above market prices and distributors are granted the right to use the Fairtrade label to market their products. While Fairtrade International remains the leading certifying organization globally, Fair Trade USA has become an important player in recent years. The latter split from the former in 2012 and distinguishes itself by its willingness to work with all kinds of producer organizations, rather than prioritizing small farmers. This includes plantation-style farms that rely on hired labor to produce large quantities of coffee or other commodities for the growing fair trade market in the United States.

Although there are small differences between certifying organizations, they follow a very similar set of general criteria to define fair trade, which requires producers, whether cooperative associations of small growers or larger companies relying on hired or contract labor, to agree to the following:

- Promote democratic organization of production that allows producers to participate in the decision-making process
- Allow employees of larger companies to organize collectively in trade unions and associations that can advocate on their behalf for better wages, working conditions, health benefits, and other protections
- Not rely on child and forced labor
- Work toward developing more sustainable production methods by minimizing negative environmental impacts and adopting conservation strategies

Buyers who want to use the label to market products in the Global North also agree to:

- Purchase directly from producer organizations
- Pay a premium that would guarantee minimum wages and the creation of a social fund to support community projects such as building schools, clinics, storage facilities, wells, or other infrastructure.
- Make advance payments to help finance the purchase of inputs without forcing producers to acquire large debts
- Sustain long-term contracts that ensure a steady flow of income and facilitate planning

In 2017, according to Fairtrade International (2020), there were 1,599 Fairtrade certified producer organizations in seventy-five countries—up by 32 percent from 1,210 just five years prior—representing 1.6 million farmers and hired workers, who received $212 million in premiums. Meanwhile, Fair Trade USA (2019) reports working with 1,250 business partners in forty-six countries, involving 1.6 million workers and generating $105 million in premiums for farmers and their communities. Because many producers are certified by both organizations, adding these figures would lead to an overestimation of the size of the fair trade economy. Thus, I only used data from Fairtrade International to create figures 5.6 to 5.8.

However impressive the Fairtrade International figures look in terms of growth, it is important to note that the $10 billion in sale pale in comparison to the $7.5 trillion of global food sales, representing just 0.13 percent of all sales. Even if sales from Fair Trade USA and other smaller organizations were included, certified fair trade commerce would still capture less than 1 percent of all food sold worldwide. Similarly, the number of individual producers (3.2 million for both leading organizations combined) represent less than 3 percent of the global agricultural workforce.

Premiums paid to producer organizations averaged $219,000 in 2017, representing about $122 per farmer. Most premiums were used for education and training, housing development, production capacity building, infrastructure projects, and direct payments to farmers. According to Fairtrade International, this makes fair trade an effective means to address many of the United Nations' Sustainable Development Goals (SDGS), including reducing poverty, eliminating hunger, promoting health and education, supporting decent work and economic growth, and creating sustainable communities.

The location of production and consumption clearly aligns with the divide between Global South and Global North (see figure 5.7). Most producer organizations are located in Latin America (803), followed by Africa (535) and Asia (261), with Peru, Côte d'Ivoire, Colombia, India, and Kenya having the most certified organizations. The biggest consumer markets in total volume of sales are found in the United States, the United Kingdom, and Germany. However, Switzerland, Sweden, Ireland, Finland, Austria, and the United Kingdom are the only countries where more than 1 percent of all food sales are fair trade certified. The United States is one of the largest and fastest expanding markets, yet fair trade represents only 0.1 percent of food sales, suggesting that there is room to grow. Very few countries, except Brazil, South Africa, India, and Kenya, which have a growing economy and middle-class population, engage in fair trade as both consumers and producers.

A handful of commodities dominate fair trade (figure 5.8). Coffee is perhaps the quintessential fair trade commodity and one of the first to become part of such conventions. Bananas have become the most important commodity in terms of production volume and premium earned. However, because it is less labor intensive than

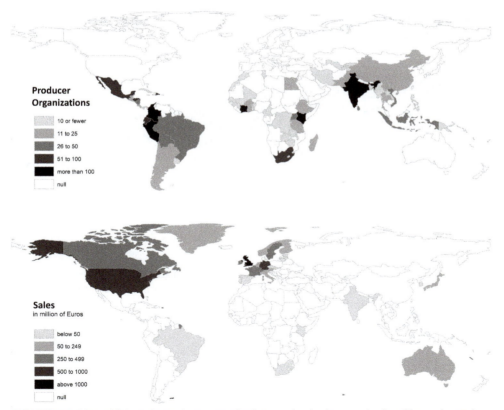

FIGURE 5.7 Map of fair trade producer organizations and sales by country (in millions of euros), 2018.

Source: Author with data from Fairtrade International (2020).

Production Volume (in Metric Tonnes)

Farmers & Workers

Premium Earned (in Euros)

Area of Cultivation (in Hectares)

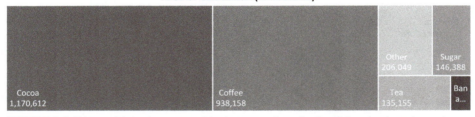

FIGURE 5.8 Primary fair trade commodities in terms of production, labor, land, and premiums.

Source: Author with data from Fairtrade International (2020).

crops such as coffee, cocoa, and tea, it requires fewer workers and does not have the same impact in supporting livelihoods.

Researchers have investigated the impact of fair trade since its early days and raised several concerns regarding its potential to address social inequality and environmental degradation. A first concern regards barriers of entry that limit participation. Many small food producers in the Global South are attracted to the benefit of fair trade but find it difficult to obtain certification. The application and inspection processes are bureaucratic and expensive. In some cases, it requires growers to invest in new technologies to meet sustainability criteria, putting an additional burden on them. These barriers of entry can be more easily navigated by large companies with deeper pockets and corporate connections upstream in the supply chain than by small independent farmers who may lack the organizational capacity.

Another concern relates to the use of premiums, which do not always fund projects that meet the most urgent needs of communities, whose members often have limited say in the allocation of resources because workers associations and trade unions are created and controlled by special interests. Instead, corporate buyers might promote projects that enhance perceptions of corporate responsibility that appeal to Global North consumers. For example, funds may be allocated toward a processing facility that will benefit the exporter more than the farmers or toward a school building without addressing the economic obstacles that keep children out of school (e.g., the cost of books, supplies, uniforms, and transportation, and the opportunity cost associated with not working).

Fair trade is limited to very few products and is virtually nonexistent for cereals, vegetables, and most fruits. Some argue that the top commodities might be reaching a market saturation point, where it is unlikely that demand will continue to increase. Wine grapes is one of the few new products that is gaining market share. This raises a more fundamental critique of fair trade: the fact that it rests on the logic of market expansion and the need to convince more consumers to pay premium prices. This is problematic for both producers and consumers and hampers the transformative potential of fair trade.

For producers, this means that they remain at the mercy of world market prices and must find outlet for their commodities. For example, in 2008, the world price of coffee plummeted and, although growers certified as fair trade were protected by a price floor that included a premium, it made it harder for them to compete against conventional coffee. The documentary film *Black Gold* poignantly illustrates this issue by depicting the struggle of a fair trade-certified Ethiopian cooperative to market its coffee in Europe and the United States. One of the reasons for this intense competition is the fact that many mainstream distributors have learned from fair trade and begun marketing their products differently, emphasizing places of origin and showcasing growers to appeal to consumers' moral values.

On the consumer side, it appears that ethical consumerism might be more about giving buyers a clear conscience than about changing the food system. Because it helps assuage the guilt of consumers, fair trade might actually encourage consumerism in all its wasteful and destructive aspects. As Žižek (2010, 117) puts it: "For the price of a couple of cappuccinos, you can continue in your ignorant and pleasurable life, not only not feeling any guilt, but even feeling good for having participated in the struggle against suffering!" The de-fetishization of commodities prompted by fair trade's transparency and accountability might indeed lead to re-fetishization, giving commodities a progressive meaning that hides remaining problems and suggests that capitalism can be reformed. To attract consumers, fair trade distributors must brand, market, and advertise their products as ethical. This often occurs by invoking geographic imaginaries that hint at the connections between consumers and producers, but nevertheless reify and essentialize them in the service of profit (Evans and Joassart-Marcelli 2018).

Figure 5.9 illustrates this process of re-fetishization with an advertising for fair trade coffee. The text that accompanies the ad lets coffee farmers identified by name explain what fair trade does for them. For Ivania in Nicaragua, "Fairtrade has changed [her] life and the lives of people in [her] community." For Zinabua in Ethiopia, who is portrayed in the photograph, it gave her an opportunity to take a leadership role, making her "very proud" and giving her the desire "to encourage other female farmers to participate in the economic and social development of community life." She also views fair trade NGOs as filling a gap in infrastructure investment—"meaning change is happening faster than it would if we just relied on the government." For Luis Augusto Blandón in Colombia, fair trade gives him great pleasure "knowing that [his] coffee is going to many places" giving unknown consumers "a great product." For Mr. Wesi in Ethiopia, it made his hard work pay off, "improving [his] livelihood" and allowing him to dream about "educat[ing] [his grandchildren] properly

FIGURE 5.9 Fair trade coffee advertisement.

Source: Fairtrade Canada, Fact Sheets: https://promo.fairtrade.ca/download/taste-the-good-in-your-coffee-poster/.

so they can be part of the country's social and enterprising future." Such narratives produce images of happy, proud, and dignified workers in pristine environments, with whom consumers can connect through their purchase. Notice how women are identified by their first name while men are called by their full name or title, suggesting that the narratives of fair trade might also be gendered. Indeed, as in figure 5.6, the fair trade representations developed to attract ethical consumers disproportionately include images of women happily going about their daily work.

If political economists are correct that capitalism is inherently exploitative of people and nature, then such re-fetishization will not address the root causes of social inequality, labor exploitation, animal abuse, and environmental degradation. Because it seeks to expand rather than overhaul or replace markets, fair trade arguably exemplifies *neoliberalism*—the belief that markets are better left alone and that government intervention should be limited to supporting these markets and encouraging free trade (see chapter 3). This idea underlies the expansion of social entrepreneurship and ethical consumerism, which shift the responsibility for improving human lives and preserving nature onto individual consumers and producers who can do good by participating in new markets. Redistributing resources, altering systems of production, and changing international trade regimes through political means are not part of the fair trade model or any neoliberal agenda.

5.4. Localism and Alternative Food Networks

The difficulty of enforcing and monitoring fair trade at a global scale has led many "shoppers who care" to look for options closer to home. In the past two decades, so-called alternative networks have emerged to reconnect consumers and producers at the local scale. This movement has its origin in the recognition of the deleterious effects of food globalization on people and nature, as illustrated by the ubiquitous phrase and unofficial motto of the movement: "Think globally, eat locally." If globalization means alienation, destruction, and exploitation, then *(re)localization* could mean the opposite. Removing physical distance between consumers and producers would presumably increase transparency and accountability on both sides. Arguably, closer ties between eaters and farmers, institutionalized in local food networks, foster community, increase participatory decision-making, support livelihoods, and encourage sustainability.

The most common alternative exchange platforms include farmers markets, community-supported agriculture, and cooperative stores. Although street food markets have been common features in cities around the world for centuries, they have experienced a recent revival. In the United States, the number of farmers markets grew from 1,765 in 1994 to 8,807 in 2020—an almost 400 percent increase. These markets typically prioritize certified organic producers that are considered "local" (e.g., operate within a certain distance radius or regional boundary).

Community-supported agriculture (CSA) is a model in which a farm relies on membership contributions to fund (at least part of) its operations. This annual or monthly investment generates a return in the form of regular seasonal produce deliveries. In some cases, to provide members with greater diversity and consistency, small farmers join forces and create a distribution hub. In coastal areas, the CSA model has also been used to support fishing communities. Many CSA farms host membership meetings and volunteer workdays that help create a sense of community and partnership between producers and consumers. Cooperative stores are typically owned by consumers who invest in the store through their membership fees and/or by farmers and other food producers who use the store as a distribution hub. They have been popular in areas ignored or abandoned by mainstream retailers, including low-income urban neighborhoods and rural towns.

A more radical approach to food provisioning consists of people growing their own food. In recent years, gardening has experienced a resurgence in the Global North, particularly in cities where *urban agriculture* is seen as a means of achieving food justice, self-sufficiency, and security (see chapter 11). In addition to private spaces, it can take place in shared spaces on vacant lots, school yards, rooftops, abandoned buildings, walls, alleys, and public easements. Often organized as a co-op, nonprofit, or land trust, these spaces can be divided into individual allotments or plots or run collectively as a farm. In 2016, there were an estimated eighteen thousand community gardens in the United States, providing space for over three million gardeners to grow food—a small but growing number compared to the thirty-seven million people who garden at home (National Gardening Association 2014).

One such garden is Glenwood Green Acres Community Garden in North Philadelphia (see figure 5.6). The garden was first established in the mid-1980s on the site of an abandoned factory by the busy railroad tracks linking Philadelphia to New York. Located in one of the poorest neighborhoods of the city, it provided a space where residents, mostly Black, could grow food to feed their families and neighbors. At 3.5 acres, it is one of the largest of over four hundred gardens in Philadelphia. It is also one of very few that are protected by a land trust—a legal mechanism in which the land was acquired by a philanthropic organization to be used and managed as a garden by the community. Indeed, securing access to land with clean and healthy soil is one of the most serious challenges facing urban growers.

What motivates consumers to purchase or grow food through these various alternative channels is not always clear and can even be contradictory, underscoring the

FIGURE 5.10 Glenwood Green Acres Community Garden, Philadelphia.

Source: Tony, Glenwood Green Acres, https://commons.wikimedia.org/wiki/File:2009_Glenwood_community_garden_Philadelphia_3496476276.jpg.

idea that the alternative food movement consists of not one but multiple movements. Some are motivated by "quality" typically understood as seasonal, fresh, organic, healthy, and wholesome. These motives are the most individualistic to the extent that they primarily enhance the health and pleasure of those who purchase and ingest the food. Others want to support local economies and keep money circulating in the community, out of the hands of large and transnational corporations. Many believe that alternative networks underlie a more sustainable approach to farming by reducing "food miles" and making growers accountable. A large number also argue that alternative food practices bring about *food justice* by increasing participation of marginalized groups, reducing dependency on corporate and industrial food providers, and promoting food security.

These various motives, however, may conflict with each other. For instance, those who desire quality food are willing to pay a premium for locally grown lettuce, grass-fed beef, free-range chicken, organic strawberries, bio-dynamic wine, and artisanal cheese, creating an elite market for these "gourmet" foods that are unaffordable to the majority of low- and middle-income consumers. High prices are justified by the moral argument that if people really care about good food, especially when it comes to their own health, they would be willing to pay its real cost. In that context, the burden of responsibility is put on individuals and those who cannot pay are viewed as unaware, uncaring, undisciplined, and/or irresponsible—in short, immoral. The market-based consumption politics that determine the value and meaning of locally grown food go against the goals of food justice by tying the ability to participate and affect change to consumer income, thereby excluding lower-income consumers. Even among those trying to bypass the market and grow their own food, the availability of resources such as time, land, knowledge, and tools influences participation by favoring more affluent gardeners for whom gardening is a hobby rather than a necessity. In other words, conflating the local with fair or sustainable is problematic if the structure of alternative food networks remains embedded in social, political, economic, and cultural relations in which some groups are denied access or agency.

A growing number of scholars have noted the mainstreaming of alternative food networks, suggesting that even though consumer products may be different, the production and distribution mechanisms remain profit driven and shaped by class, race, and gender relations. For instance, critical observers have described farmers markets and food co-ops as "white space" catering to the needs of affluent and primarily white consumers while ignoring larger structural issues such as systemic racism, gender inequality, and class (Allen 2010, Slocum 2007, Guthman 2008). New research is focusing on the connection between alternative food spaces and gentrification, suggesting that the development of farmers markets, community gardens, and food co-ops in urban neighborhoods might attract educated, affluent, and primarily white residents, displacing poorer longtime residents, including immigrants and people of color (see chapter 11).

Increasingly, alternative foods, including their spaces and practices, have been absorbed by many large corporations. For instance, locally grown produce can now be found on many supermarket shelves, without any clear indication of where it actually comes from and how local is defined. Farmers markets have been turned into tourist destinations and entertainment zones. Like fair trade, alternative foods have the potential to subvert exploitative relations of production by creating new markets. However, because they need to appeal to consumers to succeed, ethical commodities are susceptible to a troublesome re-fetishization, which, after uncovering certain aspects of the relations of production, replaces that knowledge with morally benevolent meanings, giving commodities "magical" powers beyond what Marx might have envisioned in the mid-1800s.

Conclusion 5.5.

Our food system is woven out of countless networks that connect producers, distributors, retailers, and consumers over space. Commodities may be conceptualized as socially produced by these networks, which have become the object of scholarly scrutiny since globalization intensified in the 1970s.

The concept of global commodity chain first came to the fore, emphasizing the political and economic relationships between Global North consumers and Global South producers. The focus was primarily on uncovering relations of production and their power-laden nature, which are presumably hidden from consumers by the commoditization and fetishization of food. The idea of global commodity chain—and its variations such as supply and value chains—have been criticized for being too rigid, putting too much emphasis on production, and ignoring the cultural and symbolic meanings of food that influence consumption.

In response, scholars turned to networks and circuits to theorize the role of consumption and draw attention to the discursive aspects of food. According to some, this shift of emphasis from production to consumption might have gone too far in stressing the individual, subjective, and symbolic—gutting explanations of power and politics. These two broad perspectives, however, ought not to be mutually exclusive; consumption and production are constituted simultaneously and intertwined in complex networks in which the material and informational are blurred. Paying attention to the culture of commodities does not require that we lose sight of the political economic aspects of their production.

Recognizing the significance of production-consumption relations has spurred new forms of food activism aimed at reforming food production via consumer-led interventions, including fair trade and alternative food networks. By voting with their forks, consumers presumably (re)connect with producers in mutually beneficial ways. The rapid expansion of ethical food consumerism, however, has raised some concerns regarding the nature of the consumer-producer relationship and its transformative potential.

Global Food Crises
Hunger and Malnutrition

6

Outline

6.1. Definitions and Trends
6.2. Explanations of Hunger and Food Insecurity
6.3. Beyond Hunger: Malnutrition
6.4. Eradicating Hunger and Enhancing Food Security
6.5. Conclusion

Learning Objectives

- Outline geographic disparities in the prevalence, nature, and experience of hunger and malnutrition.
- Understand conceptual and measurement differences between the notions of food insecurity, hunger, famine, and malnutrition.
- Compare explanations of hunger and malnutrition, including food scarcity, entitlements, and political ecology.
- Think about hunger as a form of violence.
- Articulate the connections between food insecurity and obesity.
- Contrast different approaches to enhancing food security.
- Introduce the concepts of right to food and food sovereignty.

In 2021, amid the COVID-19 pandemic, 957 million people were estimated to be hungry—more than one in every eight. If we add those with uncertain access to nutritious and sufficient food, the ratio of people who suffer from food deprivation or nutrient deficiency goes up to more than one in four (FAO 2019b). In Nigeria, Venezuela, Haiti, Yemen, Somalia, India, the United States, and many other places, millions of people experience the deleterious effects of not having enough to eat.

Anyone with any empathy finds hunger reprehensible. In a "world of plenty," the thought that fifteen thousand children under five years of age die each day for lack of nutrients provokes sadness, shame, and/or outrage. Hunger, even in the most affluent societies, is a poignant symbol of our collective incompetence at meeting the most basic of human needs. Yet, politicians around the world have made promises to eradicate hunger and pledged to uphold the *right to food*. The latter was first mentioned in the 1948 Universal Declaration of Human Rights as part of the right to an adequate standard of living and recognized specifically in the 1966 International Covenant on Economic, Social and Cultural Rights. In 2000, the United Nations declared the "eradication of hunger" a priority in its Millennium Development Goals. The same year, it established the position of Special Rapporteur on the Right to Food, with the mandate to develop an integrated approach to promoting the full realization of the right to food defined by food *availability* (through production or purchase), *accessibility* (economic and physical access), and *adequacy* (capacity to meet dietary needs). Such a definition goes beyond "freedom from hunger" by requiring that food be nutritious, safe, diverse, and provided with dignity. The right to food was reiterated in 2016 in the "zero hunger" pledge of the Sustainable Development Goals. Since then, the prevalence of undernourishment has declined in Asia and Latin America, but progress in Africa and the Middle East has been very slow and in the past five years global rates have increased slightly. The COVID pandemic, which started in 2020, has reversed most of the recent progress. All this raises serious questions about our ability to feed the world and the political will to uphold basic human rights.

Such questions have animated policy makers and scientists for centuries. Thomas Malthus wrote: "The power of population is indefinitely greater than the power in the earth to produce subsistence for man" (1798, 44). According to Malthus, population tended to grow exponentially, while the earth's resources and its ability to generate food were finite. Any agricultural productivity increase that could potentially raise people out of poverty would be canceled by unbridled population growth, "subject[ing] the lower classes of the society to distress and . . . prevent[ing] any great permanent amelioration of their condition" (19). Thus, for Malthus and the neo-Malthusians who are still inspired by his theory today (even though population is now six times larger and the incidence of premature death from hunger is lower than in the late 1700s), population growth must be stopped to eliminate hunger. This argument remains popular today, particularly in media and political outlets prone to fearmongering, and profoundly influences how we think about hunger.

Less pessimistic perspectives suggest that hunger and food insecurity can be curtailed if we take steps to address the unique circumstances that give rise to hunger in specific places and the system forces that underlie increasing pressure on the earth's finite resources and their uneven allocation. While population growth is certainly a factor in increasing demand for food, other dynamics are highlighted in these explanations. For instance, researchers have emphasized the role of conflict, natural disasters, and economic downturns in triggering and reproducing food crises. Others argue that structural inequalities and social marginalization make certain groups more vulnerable to hunger. Finally, those working within the political economic tradition posit that the neoliberal corporate food regime itself, with its emphasis on industrialization, specialization, and trade, has undermined people's ability to control their own food and be self-sufficient. At stake is not so much whether we can produce enough food, but how we can do so to ensure that everybody is fed and receives the nutrients they need to lead healthy lives.

In this chapter, we examine the extent and causes of global hunger and the larger malnutrition crisis that has contributed to the seemingly paradoxical rise of obesity alongside undernutrition. We begin with a series of important definitions and draw attention to geographic disparities in the prevalence, nature, and experience of hunger and malnutrition. Attending to these differences point to the complexity of global hunger and its embeddedness in local economies, politics, cultures, and physical environments—characteristics that ought to influence proposed solutions.

6.1. Definitions and Trends

Hunger typically refers to the physical sensation of discomfort, pain, or weakness that is caused by a lack of food and triggers a desire to eat. Most of us have experienced hunger and might occasionally claim to be "starving" when we skip a meal. For scientists, however, hunger is a loose and ambiguous term that means different things, such as undernutrition, malnutrition, food insecurity, and famine. Understanding hunger in all its complexity requires that we define these terms.

6.1.1. Undernutrition

Chronic *undernutrition*—the insufficient intake of macronutrients and micronutrients sustained over an extended period of time—is one of the most common aspects of hunger. Undernutrition manifests itself in child *stunting* (low height for age), child *wasting* (low weight for height), maternal underweight (low weight among women of reproductive age), and micronutrient deficiencies (lack of vitamins and minerals)—all of which can be measured with relatively simple tests. These conditions are particularly damaging to the cognitive and physical development of young children, raising their susceptibility to infectious disease and likelihood of staying in poverty.

In 2018, there were 820 million chronically undernourished people in the world (FAO 2019b), with 144 million children under five stunted and 47 million wasted (UNICEF 2020). Most undernourished people live in developing countries, as highlighted in figure 6.1. Asia is home to the largest number of undernourished people—over half a billion in total, with 194 million in India alone and 122 million in China. However, the Asian continent, especially Southeast Asia (e.g., Vietnam, Indonesia, Laos, Myanmar), made significant progress in reducing their prevalence rate in the past two decades, which is now well below Africa's (11 compared to 20 percent). Indeed, the situation in Africa has worsened since 2015, with hunger rising in all subregions, especially in Middle (e.g., Central African Republic, Chad, Democratic Republic of the Congo) and Eastern Africa (e.g., Zimbabwe, Zambia, Rwanda, Uganda). The absolute number of undernourished people in Latin America and the Caribbean has dropped by 34 percent since 2000, lowering the average prevalence rate to 6.5 percent. The most significant progress took place in South American countries like Peru, Colombia, Ecuador, and Bolivia, while improvements have been slower in Central America and the Caribbean, particularly in Haiti, where approximately half of the population is undernourished.

6.1.2. Malnutrition

There is growing recognition that hunger leads to overlapping deficiencies and manifests itself in different ways, including overweight and obesity, which had previously been thought of as the result of affluence. Although it may seem counterintuitive, a chronic lack of nutritious food means that poor people are more likely to turn to cheaper, less nutritious, and more caloric foods. As Raj Patel (2012) puts it, the world is both "stuffed and starved." The inclusive notion of *malnutrition* refers to deficiencies, excesses, or imbalances in a person's intake of energy and/or nutrients. Thus, in addition to undernutrition, this concept also includes micronutrient-related deficiencies (e.g., anemia), overweight, obesity, and diet-related noncommunicable diseases (e.g., diabetes, heart disease).

In 2018, there were an estimated 2.1 billion people who suffered from *micro-nutritional deficiencies* globally (FAO 2019b). The most common deficiencies relate to iron, vitamin A, iodine, and zinc. These are particularly damaging for pregnant women and young children who have higher physiological requirements, which is why data are typically reported for these groups. Micronutrient deficiency is described as *hidden hunger* because its effects are less visible than those of undernutrition, but nevertheless severely impact vulnerability to disease, cognitive development, general well-being, and productivity. For instance, anemia, which is caused by a lack of iron or vitamin B_{12}, is the most widespread nutritional disorder in the world, affecting 33 percent of women of reproductive age (fifteen to forty-nine years old) and causing up to 20 percent of maternal deaths (Ritchie and Roser 2018). Like most nutritional deficiencies, it is particularly common in poor countries like India, Yemen, Mozambique, and Myanmar, where diets lack diversity.

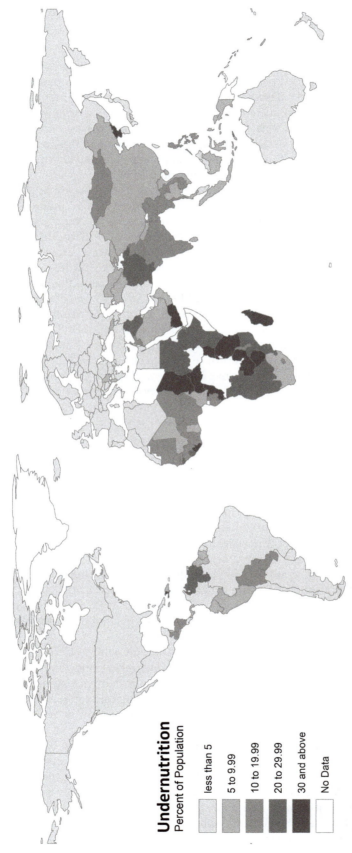

FIGURE 6.1 Map of prevalence of undernourishment by country, 2016–2018 average.

Source: Author with data from FAO (2020b).

Obesity and overweight—another aspect of malnutrition—have been on the rise globally (see chapter 13). Obesity alone was responsible for 4.7 million or 8 percent of premature deaths in 2017. In 2016, 13 percent of the world's adult population was considered obese and 39 percent was overweight. The body mass index (BMI), which is computed using a ratio of weight to height, is one of the most common measures used to determine obesity (BMI over 30) and overweight (BMI over 25). While these concerns predominantly affect people in rich countries, especially the United States where 70 percent of the population is considered overweight, it has become a global issue (see figure 6.2), with countries as different as Brazil, Iran, Turkey, Tunisia, Afghanistan, Botswana, Mozambique, Ethiopia, Thailand, and Indonesia seeing the percentage of overweight adults in their populations double or even triple in the past fifty years. In China, it rose from less than a tenth of the population in 1975 to over a third in 2016. Pacific islands such as Samoa, Tonga, and Fiji have the highest rates of obesity in the world, raising questions about what geographic factors underlie this concentration pattern (see textbox 6.5 below).

6.1.3. Food Insecurity

The term *food security* emerged in the 1990s and is now widely used to describe a condition that "exists when all people, at all times, have physical and economic access to sufficient safe and nutritious food that meets their dietary needs and food preferences for an active and healthy life." It was developed as a supplement to indicators of undernutrition to emphasize multiple dimensions of *food insecurity* related to the four pillars of food security: availability, access, utilization, and stability.

Hunger is an extreme form of food insecurity, which can also manifest itself in other conditions such as micronutrient deficiency and overweight that are associated with malnutrition. In its 2019 *State of Food Security and Nutrition* report, the FAO introduced the idea of moderate food insecurity in which people might have access to food but face uncertainty about how long it will last and may need to reduce quantity and/or quality to get by.

In 2019, there were about 800 million people globally who suffered from severe food insecurity—a number close to that of undernourished people—and another 1.2 billion with moderate food insecurity, bringing the total of food insecure people to approximately 2 billion people. Geographic disparities in severe food insecurities parallel those illustrated in figure 6.1. However, the patterns of moderate food insecurity differ and reveals that a much broader range of countries suffer from uncertain access to nutritious food. For instance, although just 1 percent of North America and Europe's population was severely food insecure in 2019, the figure rose to 8 percent (or 89 million people) after adding moderately food insecure people. This figure aligns with survey data collected by the United States Department of Agriculture, indicating that 11.1 percent of households were food insecure and 4.3 percent had very low food security in 2018.

6.1.4. Famine

Famine refers to widespread and systemic hunger that causes elevated mortality over a specific period of time. It is the result of cumulative failures of food systems that require urgent food and nutrition assistance to avoid deaths by *starvation*. Famines are closely associated with the concept of *acute malnutrition*. In contrast to the slow violence of chronic malnutrition which robs people of the physical and cognitive strengths conducive to living a flourishing life, acute malnutrition is an immediate life-threatening condition.

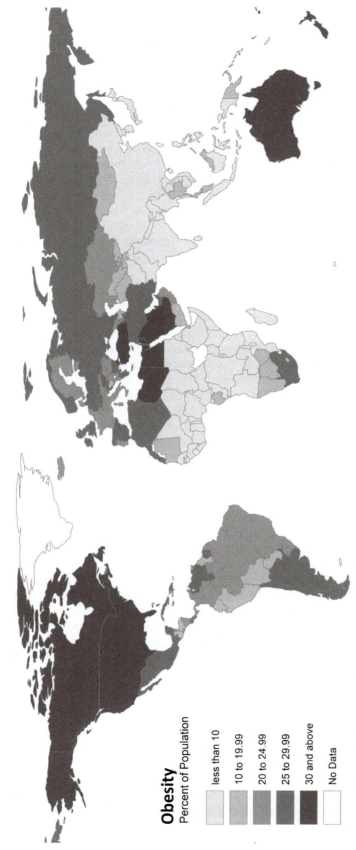

Obesity
Percent of Population

- less than 10
- 10 to 19.99
- 20 to 24.99
- 25 to 29.99
- 30 and above
- No Data

FIGURE 6.2 Map of percentage of adults who are obese by country, 2016.

Source: Author with data from WHO (2020).

One of the most infamous famines is the Irish potato famine that began in 1845 and killed approximately one million people through starvation over the following five years as a result of a phytophthora infestation that destroyed about three-fourths of the potato crop. During the twentieth century, an estimated seventy million people died of famines. The largest famines occurred in China (1927, 1929, 1943, 1958–1961), Russia (1921–1922, 1941–1944), Ukraine (1932–1934, 1946–1947), Iran (1917–1919, 1942–1943), North Korea (1995–1999), India (1943), Cambodia (1975–1979), and Nigeria (1967–1970). The number and severity of famines have steadily declined since the 1970s. However, war, conflict, and climate change appear to be reversing this trend with major crises currently taking place in Syria and Venezuela, among others.

Today, international organizations monitor the globe for signs of impending food crises taking into account climate shocks, natural disasters, conflict, political insecurity, population displacements, and economic factors that could affect the supply and demand of food. For instance, the FAO's Global Information and Early Warning System (GIEWS) was created by the United Nations in 1975 to help predict and hopefully circumvent impending famines in all countries of the world. In 1985, the United States Agency for International Development (USAID), created its own Famine Early Warning System's Network (FEWS-NET) which monitors food insecurity in twenty-eight countries with close connections to the United States. These and other organizations use the Integrated Food Insecurity Phase Classification (IPC) to determine the severity of crises and warn the international community of the need for assistance (see textbox 6.1).

Countries that have experienced major food crises in 2018 include Afghanistan, Yemen, the Democratic Republic of Congo, Zimbabwe, Sudan, South Sudan, Nigeria, and Madagascar. The COVID-19 pandemic, which began in early 2020, has raised concerns linked to unemployment and lack of income, falling sources of foreign currency, restrictions on trade and the movement of food, and possible surges in food and oil prices that are being closely monitored.

6.1.5. Measurement Issues

Various concepts of food deprivation rely on different measurement and estimation methodologies that are limited by available and imperfect data and might change over time, creating confusion about the actual number of people suffering from hunger. Undernutrition and malnutrition are typically estimated from direct observation, including weight, height, age, and biological tests, complemented by surveys of food expenditure and caloric intake. Food insecurity is measured indirectly through a series of survey questions related to people's experiences and behaviors. These distinctions are important not just for determining how many people are hungry or lack access to food, but also in shaping policy responses.

For instance, famines trigger international emergency responses in the form of humanitarian food aid. However, the criteria to define such crisis situations are arbitrary. For example, famine, which corresponds to the Phase 5 of the IPC, is defined by three specific criteria: at least 20 percent of households facing an extreme lack of food, at least 30 percent of children under five suffering from acute malnutrition, and at least two people out of every ten thousand dying each day (IPC 2019). Each of these criteria must be measured rigorously before declaring a famine.

Some experts argue that technical changes in how the FAO measures hunger are politically motivated and underestimate the full extent of the problem. For instance, as Holt-Giménez (2019) discusses, in order to meet its millennial development target of cutting world hunger by half between 2000 and 2015, the FAO changed its

Box 6.1   **Classifying Food Insecurity**

The Integrated Food Security Phase Classification (IPC 2008) is a standardized scale developed to assess the severity of food crises and determine the appropriate level of response by relying on existing data and information from a variety of sources. It was developed by international (e.g., WFP, FAO, USAID) and nongovernmental (e.g., Care International, Oxfam, Save the Children, World Vision) organizations that use it in their decision-making process. As outlined in table 6.1., criteria for each phase rely on measurable indicators of mortality, nutrition, and livelihood considered to be "objective indicators" of food insecurity. In addition, it recognizes that food crises are dynamic and therefore, for each phase, also assesses the risk of worsening. The IPC outlines appropriate responses for each phase classification, ranging from livelihood support to food distribution and management of displaced people.

TABLE 6.1   **Integrated Food Security Phase Classification**

Phase	Description	Response	
1 Generally Food Secure	Households are meeting their basic food and nonfood needs without unsustainable coping strategies that depletes assets. Stable and adequate calorie intake. Low mortality, acute malnutrition, and stunting.	Assistance to limited pockets of food insecure groups. Agricultural livelihood development programs. Prevention by removing structural barriers to food security	LIVELIHOOD SUPPORT AND DEVELOPMENT
2 Moderately/ Borderline Food Insecure	Household food consumption is minimally adequate (minimal calory, unstable and lacking diversity). At least 1 out of 5 households reduces nonfood expenditure or engages in unsustainable coping strategies. Low mortality, but evidence of some malnutrition and stunting.	Provision of safety nets. Protection and enhancement of livelihoods. Contingency plan and close monitoring	
3 Acute Food & Livelihood Crisis	Household food consumption below daily requirement, reduction in nonfood expenditure, crisis coping strategies that deplete existing livelihood assets. Disease and epidemics. Higher mortality, malnutrition, and stunting. Civic instability	Strategic intervention to increase food access and protect livelihoods. Implementation of contingency plan. Close monitoring. Support health and sanitation sector.	URGENT HUMANITARIAN ACTION REQUIRED
4 Humanitarian Emergency	Acute deficit in food consumption. Mortality, acute malnutrition, and stunting. Pandemics. Civic tension and conflict. Irreversible destruction of livelihood assets. Population displacement.	Urgent provision of food aid. Protection of vulnerable groups. Provision of shelter, health, and sanitation. Close monitoring.	
5 Famine/ Humanitarian Catastrophe	Acute deficit in food consumption. High mortality, acute malnutrition, and stunting. Pandemics. Intense conflict and widespread civic insecurity. Collapse of livelihood assets. Massive displacement of population.	Critically urgent protection of human life. Comprehensive basic needs assistance. Management of conflict and population displacement.	

Source: Based on IPC Global Partners (2008)

benchmark estimate of hungry people several times and switched the base period to 1990. Without these methodological adjustments, the number of hungry people today may be closer to 1.5 billion.

Another critical methodological consideration is the scale of measurement. For example, food insecurity might refer to global, national, regional, household, or individual concerns. At larger scales, food insecurity typically means that there is insufficient food to feed people within a geographic area, leading scholars and policy makers to focus on average per capita food production, dietary energy, or protein supply. At the household or individual scale, food insecurity tends to imply a lack of income and resources as well as individual characteristics and behaviors that make it difficult for specific individuals to obtain nutritious food. At these scales, scholars pay attention to things like gender, livelihood, and the functioning of markets. According to Jarosz (2011), the scalar shift in conceptualizing and measuring food insecurity at the individual scale has been paralleled by a policy shift away from promoting national self-sufficiency toward enhancing individual acquisition of food in the global market, reflecting an increasingly neoliberal agenda.

Explanations of Hunger and Food Security 6.2.

What causes food insecurity? Answering this question typically points to poverty. However, behind the almost obvious connection between poverty and hunger lies several important social dynamics that influence who is hungry and where hunger takes place.

Before diving into various explanations, it is important to remember what we learned in the previous section regarding the different ways of measuring hunger and food insecurity. For example, chronic hunger and famines are not the same, yet the two are often conflated, confusing explanations. Because famines are extreme and catastrophic events, they have received significant attention from researchers in disciplines such as economics, political science, and geography that emphasize specific triggers leading to sudden disruptions in food supplies, including war, conflict, natural disasters, and economic downturns. Chronic hunger, however, is a more permanent condition that may be exacerbated by the same triggers but is generally produced by more systematic structural forces that deprive people of access to food over extended periods of time. Focusing on triggers such as natural disasters may normalize hunger as unavoidable, particularly in regions prone to extreme weather events and conditions, and hide the fact that hunger is socially produced.

Similarly, undernutrition and malnutrition are two distinct aspects of food insecurity that may be explained by different processes. Recent work on malnutrition, including the rise in obesity around the world, points to the fact that food insecurity may not be entirely caused by a lack of food but rather by too much food lacking in nutrients. This means that we must look beyond agricultural productivity and consider why our current food regime prioritizes the production of food that does not meet human needs.

Looking beyond poverty, we can identify three types of explanations for the prevalence of hunger in the world. The first focuses on the scarcity of food, which is due to insufficient supply caused by low productivity and underdeveloped markets. The second emphasizes individual capacity to obtain food and the factors, such as gender, class, and ethnicity, that influence it. The third takes a political economic approach and looks at inequality within the food regime for explanations of hunger that highlight the role of capitalism and neoliberalism in creating unequal access to nutritious food. Although these explanations emphasize different causes of hunger, food insecurity is a complex problem caused by overlapping social, economic, and political factors that create the conditions making people food insecure in a particular time and place. Famines and acute food insecurity typically occur in places that already suffer from chronic food insecurity and are brought over the edge by triggers and exceptional events that are not themselves the direct cause of hunger. De Waal (2018) makes a similar distinction between *structural* and *proximate* factors, emphasizing the need to understand deeply rooted causes of hunger beyond the most immediate catalysts.

6.2.1. Food Scarcity

As noted in the introduction to this chapter, there is a common belief rooted in Malthusian theory that we do not produce enough food on earth to feed the growing global population, which is predicted to approach ten billion by 2050. Although this argument has now been mostly debunked, it became the dominant explanation for chronic hunger and famines in the late 1960s when so-called neo-Malthusians raised alarms about impending mass starvation.

An important implication of Malthus's theory is the distinction he makes between, on the one hand, wealthy and industrious people who can control their impulse to reproduce, invest in the future, and provide for themselves and, on the other hand, poor people whose sexual passions and carelessness keep them in misery. This distinction between "us" and "them" is reproduced in international policy that dictate the fate of developing countries without questioning the consumption habits of those in wealthier nations, thereby blaming the victims of hunger and shifting the responsibility on them.

The assumption that chronic hunger in the Global South is caused by low agricultural productivity combined with high fertility rates has led international organizations to engage in a two-prong approach of raising yields and reducing fertility as a long-term strategy to eliminate hunger. The first objective is to be achieved through economic development strategies including investment in technology, specialization, and international trade. The Green Revolution, which became popular in India in the 1960s in response to multiple devastating famines and severe food insecurity in the region, illustrates this focus on raising productivity with a large influx of capital (see textbox 2.2). The second objective of reducing fertility could be achieved through education, public health campaigns, and more direct controls such as forced sterilization and abortions. In China, the infamous one-child policy was implemented in 1980 to curb population growth, which was viewed as outpacing the country's ability to produce food. It ended in 2016.

In recent years, the recognition that the earth's climate is changing and that its natural resources are being threatened by overexploitation have reinvigorated this line of argument. Several environmentalists, concerned that the earth's *carrying capacity*—its ability to support or carry its existing population—has been stretched to a dangerous tipping point, are advocating for population control and green technology.

Despite their popularity, these ideas have been criticized by scholars from a variety of perspectives who argue that the world generates enough food to feed ten billion people and that most of this food is actually produced by the very people who suffer from chronic hunger. If so, why are these people hungry and why do international organizations continue advocating for significant increases in food production by 2050? Answering these questions requires that we look beyond food scarcity.

6.2.2. Entitlement Theory, Vulnerability, and Resilience

In 1998, Amartya Sen received the Nobel Prize in Economics for his comprehensive work on famines. His main contribution was to show that famines and chronic hunger are not caused by a lack of food, but by people's inability to access it, which itself is linked to a lack of what he calls *entitlements* (Sen 1982 and 1984). Rather than being caused by an insufficient supply of food, hunger is attributed to demand-side factors that make it difficult if not impossible for certain people to purchase or obtain food. Sen (1984, 497) defines entitlement as "the set of alternative commodity bundles that a person can command in a society using the totality of rights and opportunities that he or she faces." This refers to the various means through which a person may legally obtain food, either by growing it, buying it, working for it, or receiving it from others. Livelihood indicators in the IPC phase classification (see textbox 6.1) aim at assessing entitlements by measuring five types of "assets" or "capitals" that people can utilize to obtain food, including human, social, financial, cultural, and physical. The loss or depletion of these assets are indicators of the severity of food insecurity.

It should come as no surprise that some people have more entitlements than others because they own more assets to obtain food, have more rights to exchange these assets for food, and/or have better access to markets allowing them to do so. Women in developing countries typically have lower entitlements because they own fewer assets, have limited rights, and face numerous social and cultural barriers to employment and participation in markets (see chapter 4). This restricts their ability to grow, sell, or purchase food. Sen (1992) estimated that, in the early 1990s, there were about one hundred million "missing women" in the developing world who died prematurely because of gender biases causing malnutrition and neglect (Sen 1992). Such biases are reproduced at different scales, in households, schools, labor markets, and policy-making circles.

The notion that malnutrition is the result of low entitlements draws attention to *vulnerability* and *resilience*, which have become important concepts in the development and environmental change literature (Adger 2006). Specifically, scholars and policy makers have sought to identify individual factors, such as gender, race, ethnicity, age, marital status, sexuality, etcetera, that explain low levels of entitlements, translating into increased vulnerability to external shocks such as food price increases, climate change, natural disaster, or conflict. While vulnerability is related to poverty, it is a broader concept that emphasizes individual inability to control and use assets to secure food. Poor people's vulnerability may differ depending on their circumstances and the types of safety nets in place.

Within this broad framework, hunger is not caused by a lack of food, but by the vulnerability of certain individuals or groups who have low entitlements and are unable to withstand or adapt to economic, political, and/or environmental stressors. For Sen and other economists, this is primarily attributed to poorly functioning markets that weaken people's ability to exchange their assets for food. For example, famines have been attributed to food hording, price fixing, speculation, underdeveloped credit markets, and trade barriers. More recently, the 2008 global food crisis was blamed on the combined and sudden increases in the prices of oil and food, which severely weakened people's exchange entitlements, making it difficult for them to obtain food with the resources they had at their disposal. Focusing on economic and institutional factors underlying hunger has made great strides in challenging supply-side theories. Yet, for many scholars, it did not go far enough in showing how food insecurity (and famines more specifically) are politically produced by human actions.

6.2.3. Political Ecology of Hunger

If we produce enough food to sustain the current population, as per capita food production indicators suggest, then the fact that some have food and others do not is a political economic question related to social inequality in food distribution. This sort of inequality cannot be explained by vulnerability alone. Instead, we must consider the structural and political economic underpinnings of vulnerability—that is the way entitlements are distributed, won, or lost.

Michael Watts (1983) describes chronic food insecurity in the Sahel, including oil-rich Nigeria, as "a massive political failure." Alex de Waal (1997, 7) echoes this idea when he describes famines as "catastrophic breakdown in government capacity or willingness to do what [is] known to be necessary to prevent famine." The history of famines over the past centuries suggest that their incidence and severity coincide with *colonialism*. Mike Davis (2002) provides evidence to this argument in his book titled *The Late Victorian Holocausts*, in which he documents how British imperial policies caused the deaths of millions in India. Although severe El Niño events triggered crop failures, recurring famines were exacerbated by colonial economic policy that removed food from starving areas for commercial purposes (including exports to England) and dismantled traditional protections against starvation in the case of natural disasters by commodifying food. Historical records indicate that there were thirty-one severe famines in India during the 120 years of British rule, compared to seventeen recorded in the previous two millennia.

The extended World Wars also caused several famines, including the German siege of Leningrad in Russia in the winter of 1944. Likewise, the political turmoil that characterized the postcolonial context after World War II, including the rise of totalitarian regimes and civil wars, led to large-scale starvation such as the Biafran famine in Nigeria between 1967 and 1970 (two million deaths), the Khmer Rouge famine in Cambodia between 1975 and 1979 (1.5 to 2 million deaths), and the Second Congo War from 1998 to 2004 (2.7 million deaths).

For these scholars, chronic hunger and famine are on a continuum, going from the "silent violence" of malnutrition to the death toll of massive starvation—all caused by human actions. Many authors, including de Waal, emphasize the role of conflict in creating the conditions for hunger and starvation. Today, most countries closely monitored by the FAO's Global Information and Early Warning System (GIEWS) are involved in international or domestic conflict, including Syria, Yemen, Afghanistan, the Democratic Republic of Congo, Sudan, South Sudan, and Venezuela. In conflict conditions, food is often used as a weapon. For instance, Israel has been using food to pressure and control Palestine, causing widespread hunger (see textbox 6.2). The displacement of people generated by conflict presents a tremendous risk of hunger and famine because it uproots people from their communities, livelihoods, and support systems, as illustrated by the ongoing cases of Syria and Venezuela.

Looking at famines through the lens of political failures also points to the role of the international community in failing to recognize crises and act appropriately. For example, today, Yemen is experiencing one of the worst humanitarian crises of this century, with the United Nations classifying 14.3 million people as being in acute need of food assistance and 3.2 million requiring treatment for acute malnutrition. Over 3 million people live in refugee camps, having been displaced by the conflict between Houthi rebel militias supported by Iran and government forces backed by Saudi Arabia, a major ally of the United States and NATO. Yet, as I write this book, there has been very little effort by the international community to commit resources toward addressing this crisis, partly because of the ongoing conflict and the lack of political will to get involved.

Box 6.2   **Hunger and Conflict in Palestine**

Food insecurity is widespread in Palestine, where one out of four people lives in poverty. The situation is particularly dire in Gaza where 53 percent of the population are poor, 68.5 percent are food insecure, and 49 percent are unemployed (WFP 2020). Children are severely impacted, with 15 percent of children under five experiencing stunted growth, 50 percent being anemic, and 27 percent suffering from food-related parasitic infections (Radi, El-Sayed, Nofal, and Abdeen 2013).

Gaza is a densely populated twenty-five-mile-long territory that borders the Mediterranean, Egypt, and Israel and is physically separated from other Palestinian territories. The majority of Gaza's two million residents are Palestinian refugees. Growing tension between Israel and Gaza have led to further control and restriction by Israel, which the United Nations considers to be occupying Gaza. In 2006, Gaza residents elected leaders from Hamas—a party that the United States and Israel consider a terrorist organization. Since then, an Israeli blockade (or siege) has restricted the movement of goods and people in and out of Gaza, which has been nick-named "the world's largest prison." Specifically, the blockade restricts the entry of money, construction materials, medical supplies, food, and "dual-use items" that could become potential weapons, including fertilizers, irrigation pipes, seeds, fuel, and other supplies necessary for people to grow food.

Pasta, rice, garlic, cooking oil, chickpeas, lentils, etcetera have been blocked at some entry points. The food allowed to be imported from Israel is expensive and unaffordable to most, given the high rate of unemployment and much lower purchasing power in Gaza. In addition to controlling land access, Israel also controls the coastal zone, restricting fishing activities and food aid shipments (WFP 2020).

Whenever demonstrations along the border fence flare up, as they did in 2018, Israel responds by further tightening the blockade, resulting in complete control over the livelihoods and well-being of Gazan residents and resulting in what the United Nations calls "de-development" (WFP 2020). Under the Trump administration, the United-States, which has historically backed Israel, canceled more than $200 million in aid to Palestinians in Gaza and the West Bank, severely weakening the international capacity to alleviate hunger.

In short, food insecurity is caused by a lack of economic development, including limited capacity to grow food and restrictions on imports, both of which are the result of economic warfare led by Israel. Indeed, United Nations experts have repetitively recognized the siege of Gaza as a form of "collective punishment," which is a violation of international laws to the extent that it affects innocent civilians. Food insecurity is unlikely to be alleviated without addressing its political roots.

Although conflict is a cause of food insecurity, it might also be a result of food insecurity, as illustrated in the adage "hungry people are angry people." Hunger and conflict are both coproduced by larger political, economic, and environmental dynamics. Those working within a political ecology framework emphasize the role of politics (or power relations) in shaping the human-environment connections that define the long-term ability of people to feed themselves. For political ecologists, hunger and famine are caused by the fatal interaction of environmental degradation, global economy, and political failure in which individuals and groups are robbed of their livelihoods and ability to control food. They argue that government failures are inseparable from the expansion of global capitalism which weakens political institutions through corruption and greater dependency on foreign powers. Similarly, environmental degradation and climate change are not just natural disasters, but human-made disasters accelerated by the commoditization of land, labor, food, and nature. In other words, economic downturns, natural disasters, and conflicts are not "external" triggers, but symptoms of deep economic, social, and environmental crises inherent to our global economic system. Since the 1980s, structural adjustment programs and neoliberal policies meant to encourage free trade (see chapter 3) have had particularly disastrous effects in many countries where they destabilized traditional food livelihoods and encouraged environmentally harmful agricultural practices that have resulted in increased food insecurity.

For instance, enough rice had been produced in Haiti to feed its population until the IMF lent the government money in 1986 after the departure of the dictator Jean-Claude Duvalier. Conditions for the loan required Haiti to remove tariffs on rice imports and reduce assistance to farmers to encourage free trade. According to Gros (2010), it made it impossible for Haitian farmers to compete with heavily subsidized or donated American rice. What became known locally as "Miami rice" flooded the country and, within a couple of years, many farmers stopped working the land. Imports of rice increased, making consumers more vulnerable to price hikes. Another IMF loan in 1994 required further trade liberalization, this time focusing on the sugar market. Haiti, once the largest exporter in the world, became a net importer of sugar. The chicken industry suffered the same fate, with imports of less desirable pieces, such as wings, legs, gizzards, necks, and feet, from the United States replacing domestic production. Agriculture collapsed, farmers lost their jobs, moved to Port-au-Prince, and joined a growing number of impoverished, unemployed, and informal workers in increasingly unmanageable and violent slums. When world food prices skyrocketed in 2008 and a major earthquake hit in 2010, Haiti was structurally ill-prepared to withstand the shocks: its economy had been hollowed out and its government stripped of its ability to intervene (Gros 2010). Widespread hunger, disease, and social tension ensued. The economic policies underlying the conditions of poverty, vulnerability, and food insecurity in Haiti are described by Farmer (2011) as a form of *structural violence*—less visible than political conflict but similarly deadly.

Watts's (1983) research in Nigeria illustrates a similar process of "silent violence" by dispossession, linking hunger to the country's incorporation into the global economy following the discovery of oil. The large and sudden revenues generated by oil led to political instability, cronyism, corruption, waste and inefficiency, agricultural decline, increased dependence of imports, losses of livelihoods, rising inequality, rampant poverty, ethnic tensions, a growing Islamic insurgency, and environmental devastation, including pollution in the oil-rich delta and soil degradation in the arid northern region. In short, integration into the world economy has increased the severity of shocks, such as rising food prices or environmental disasters, while weakening the capacity of the social system to withstand them.

These examples, which illustrate the political ecology perspective, seriously challenge the idea that the development of markets and the advent of capitalism into the Global South reduce would the incidence of famines and raise food security.

6.3. Beyond Hunger: Malnutrition

Malnutrition is increasingly being recognized as a phenomenon as devastating as hunger, even if its effects are slower and less noticeable. As noted above, globally, about one billion people suffer from hunger, more than two billion struggle with micronutrient deficiencies, and an even larger number are considered overweight. Combined, these various aspects of malnutrition are causing millions of premature deaths each year. It is only recently, however, that researchers began looking at hunger and obesity together.

There are multiple explanations for the so-called global obesity epidemic, including demand-side explanations that emphasize individual biological factors, changes in lifestyles, and calorie-expenditure ratios, which we will cover in greater details in chapter 13. However, considering hunger and obesity in tandem points to the global food regime and its inherent inequality as the structural underpinnings of these problems, which may have more to do with food supply than demand.

As we learned in chapter 3, our contemporary food regime is controlled by large corporations that determine what we grow and what we eat. The drive for profit dictates that we produce food at an ever-larger scale, in single-crop monoculture, using technology and chemicals to maximize output, paying workers as little as possible, depleting soil, polluting nature, and producing growing amounts of greenhouse gasses (see chapters 7 and 8). It also explains why food is transformed into highly processed calorie-dense but nutrient-poor consumer products sold through a handful of powerful retail companies. Environmental, social, and health considerations are secondary, if not downright irrelevant, to the corporate imperative of earning profits and staying competitive in the global economy. Fair trade and ethical consumerism are the only exceptions that owe their success to their ability to generate profit by expanding markets (see chapter 5). In short, food is not produced to feed people, but to make profit.

Even in the most affluent countries where consumers are faced with seemingly limitless food options, choices are constrained. For example, in the United States, in 2018, 13 percent of the population was food insecure and almost three-fourths was overweight or obese, leading to chronic conditions such as diabetes, heart disease, and certain types of cancer. What is being offered to mainstream consumers is either unhealthy, unaffordable, or both.

In poorer countries of the Global South, where food tends to be scarcer, choices have also been constrained by the transformation of food provisioning away from subsistence farming and local production toward increased dependence on commoditized and imported food with more volatile prices. Studies document significant dietary changes in places like Mexico, Brazil, and India, where commoditized processed food is gradually replacing traditional crops with dramatic effects on health, including an increase in obesity and diabetes (see textbox 3.4). A similar trend has been observed in the Pacific Islands, where obesity rates are the highest in the world (see textbox 6.3).

These trends suggest that the problem is not that we do not produce enough food as Malthusians tell us, but rather that we produce too much of it and that most of it is not good for us. As Holt-Giménez argues, the overproduction of food is particularly troubling because:

> it keeps us from ending hunger, it depletes our resource budgets, and creates dangerous, and unmanageable, pollution loads. Not only is the environment the destination for the food system's effluvium; also, our bodies are recipients for its toxic levels of salts, sugars, fats, additives, and chemicals. Like the oceans, the aquifers, and the atmosphere, we have become yet another toxic sink for the food system's crap (2019, 41–42).

This expands the political ecology of hunger summarized in the previous section to a broader set of food issues, including malnutrition in all its forms. In this context of overproduction, it becomes impossible to talk about malnutrition without talking about inequality. Whether we are considering food insecurity, micronutrient deficiencies, or overconsumption, people earning lower incomes and facing greater discrimination are more likely to suffer. For example, in the United States, people in poverty are eight times more likely to be food insecure than those with income twice as high as the poverty threshold. Similarly, food insecurity rates for Latino, Black and Native people are at least twice those of whites. Statistics about obesity, although less drastic, follow a similar pattern: the adult rate of obesity is higher among low-income women (45.2 percent), Blacks (48.1 percent), and Latinos (42.5 percent), compared

Box 6.3 **Obesity in the Pacific Islands**

Pacific island countries such as Nauru, Cook Island, Palau, Marshal Islands, Tuvalu, Tonga, Vanuatu, Fiji, and Samoa experience the highest rates of obesity in the world–ranging between 46 and 61 percent of their populations. This has been attributed to dietary changes that are caused by the availability of food linked to international trade and aid (Snowdon and Thow 2013). Although traditional diets consisted primarily of locally grown starchy vegetables, coconuts, fruits, fish, and seafood, today's diets are dominated by imported meat, rice, and highly processed snacks and drinks.

Most of these islands consist of atolls—relatively infertile lowlands susceptible to sea-level rise—that have limited capacity to sustain a growing and urbanizing population. As a result, these countries have become highly dependent on international trade and food imports that come primarily from New Zealand, Australia, the United States, France, and increasingly China. Under pressure from these countries and the World Trade Organization, international trade has expanded, reducing the domestic availability of fish and seafood being diverted

to exports. In fact, maritime zones are commonly rented to foreign commercial operations to earn foreign currency. This has raised concerns regarding overfishing and the possible collapse of fisheries of tuna and valuable reef species (see chapter 9). To encourage exports, governments have also promoted the cultivation of cash crops like cocoa and sugar, further limiting subsistence agriculture. Meanwhile, the import of processed food, characterized by high fat and sugar content, has increased, along with overweight and obesity.

Island countries have begun to address these trade-related issues by imposing additional taxes on imported sugar and sugar-sweetened food, banning certain products such as high-fat mutton slap meat and turkey tails, and requiring nutritional labels. However, their dependence on imports and high levels of food insecurity makes it difficult to impose these restrictions without raising the price of basic food. Reclaiming traditional food livelihoods through support for subsistence farming and protection of local fishing zones are promising alternatives.

to whites (34.5). Thus, as economic inequality increases both within and between countries, health disparities between the haves and the have-nots continue to rise. In the United States, the geography of hunger and obesity mirror one another, with very localized pockets of hunger and obesity in isolated rural areas, low-income urban neighborhoods, and communities of color.

6.4. Eradicating Hunger and Enhancing Food Security

Famines and chronic hunger are two different conditions on a continuum of food insecurity that require different types of interventions commensurate to the severity of the crisis. As humanitarian crises, famines are typically met with emergency food aid to limit mortality. In contrast, chronic hunger has been addressed by a broader range of approaches aimed at increasing food production, encouraging economic development, or promoting food sovereignty in the long term.

The public often confuses famine and malnutrition and tends to view foreign aid as the only option. Gloomy and deterministic predictions regarding the inevitability of hunger are bolstered by images of starvation that commonly portray emaciated children in baren landscapes. While these images are often meant to encourage empathy and action, they also convey the idea that famines and food insecurity are unsolvable crises caused by external factors that we cannot control. They also represent children as victims with no agency, who can only be helped by international intervention.

Although such interventions are critical in ameliorating immediate conditions and reducing mortality, there may be more effective ways to address chronic hunger.

6.4.1. Food Aid

Over the past four decades, international humanitarian assistance in the form of food aid has become more organized and helped reduced some of the worse impacts of famine. Multilateral organizations such as the United Nations' World Food Programme (WFP), nongovernmental agencies like the International Committee of the Red Cross, the Red Crescent, Médecins Sans Frontières, Care, Oxfam, and donor nations have worked together or independently to mobilize and distribute food to those in need. In addition to raising funds, important logistical issues must be tackled to ensure that food is properly distributed—a challenge in areas where conflicts or natural disasters hinder transportation. Today, the WFP provides food assistance to about one hundred million people in over eighty countries, coordinating and delivering about two-thirds of all food aid. Yet, food aid today is lower than it was in the 1970s and 1980s.

Food aid has been justified by the most basic understanding of the *right to food*, which requires that food be provided by governments and international agencies when it is not available through regular mechanisms of procurement, for example in areas hit by natural disaster or conflict, in refugee camps, and in detention facilities. Yet, as we will see in the next section, the right to food is not a right to receive free food, but a right to feed oneself with dignity. Charity does not meet this right since it provides people with food, rather than promoting their ability to acquire food through the sort of entitlements considered by Sen.

In addition to ethical reasons for providing food aid, there are geopolitical justifications, such as the preservation of peace and world order by preventing state collapse, civil war, and terrorist activities. Indeed, one of the major criticisms of food aid is that it is often used as a political tool, with the interests of donors taking priority over those of recipients. For example, food aid can be used to dispose of a domestic food surplus, open foreign markets, push for reforms, strengthen alliances, and support political groups. As we learned in chapter 3, the United States' Marshall Plan and Food for Peace programs, which were created after World War II to direct domestic wheat surplus to countries in need, have contributed to global dependence on US wheat and solidified the United States' economic position as the breadbasket of the world. In the 1970s and 1980s, many donor countries began to "untie" aid from their domestic commodity production, instead purchasing food from the recipient or a near-by country and working with the World Food Program to distribute it (Clapp 2012). However, the United States continues to tie its aid to domestic commodities, as do new donors like Brazil, China, and Russia that seem to use food aid as a mechanism to establish or expand foreign markets. As a result of these geopolitical considerations, some countries receive more aid than others and some of the neediest nations, like Yemen, Syria, and Venezuela today, may not receive sufficient assistance.

Another critique comes from the notion that food aid depresses prices and thereby discourages food production, with farmers abandoning their fields and letting food rot. This line of argument views food aid as distorting markets, destabilizing local economies, creating disincentives to work, and worsening food scarcity. Politically, aid has also been blamed for undermining and corrupting local institutions, putting in their place a set of NGOs and international aid agencies that stymies local initiatives and civil society development. This sort of criticism has prompted some observers, both on the donating and receiving end, to argue against aid. More often, it has inspired donors to attach strings to food assistance and demand that it be used to support agricultural development, trade liberalization, and democratization—what

Essex (2012) describes as the *neoliberalization* of food aid. For example, food-for-work programs have become common in places like India, Ethiopia, and Indonesia, where people are required to participate in agricultural development projects to receive assistance. Similarly, work requirements have been advocated in the United States as a way to reform and shrink the Supplemental Nutritional Assistance Program (SNAP), also known as Food Stamps.

A related criticism focuses on the institutionalization, professionalization, and consolidation of nongovernmental organizations (NGOs) that have become powerful actors in dispensing aid and influencing policy. This transformation has taken place in a context where governmental and intergovernmental organizations lack the resources and will power to address crises, leaving room for the nonprofit sector to step in with support from philanthropies and corporations. NGOs, however, tend to be less accountable than public agencies. Their work is more about charity and benevolence than it is about an obligation or mandate to ensure that all people have access to food. The focus is on "feeding people" by efficiently channeling surplus to those in need, rather than on empowering people to feed themselves. In the past three decades, large food banks have emerged to address hunger in many countries, including the United States, where they have allowed the government to reduce its commitment to tackle poverty and inequality (see textbox 6.4).

Box 6.4   Food Banks

In many affluent countries where the food industry generates a large food surplus while contributing to poverty via the exploitation of labor, food banks have become a way to solve both problems by redirecting "left-over" food to "left-behind" people through organized corporate donations and charitable distributions (Riches 2018). The repurposing of waste has become a big business that is increasingly controlled by the charitable arms of corporations, eager for positive publicity. In the United States, food banks emerged in the 1960s and grew dramatically in the 1980s and 1990s at a time when the state was withdrawing from the provision of social services. Today most food banks in the United States are part of the Feeding America network. According to Warshansky (2018), this network prioritizes the needs of food companies seeking to boost their ethical legitimacy (and benefit from tax breaks) and government agencies trying to support farmers and minimize welfare expenditure. The needs of food insecure people and their dignity are often ignored. For instance, there is evidence that a large proportion of food distributed by food banks through food pantries and other programs is unhealthy—high in fat and sugar, low in nutrients, and full of preservatives and other potentially toxic ingredients.

The US food bank model appears to be spreading globally, as illustrated by the European Food Banks Federation and the Global FoodBanking Network, which is active in over forty countries, including Brazil, South Africa, Argentina, India, and Mexico.

Poppendieck (1998) raises concerns that the growth of food banking has depoliticized hunger by giving the impression that it can be solved without government interventions by volunteers and private donations. She argues that, rather than addressing the underlying causes of hunger such as inequality, poverty, and the lack of housing, good jobs, and medical care, the shift of food assistance to the nonprofit sector has given policy makers a license to dismantle the safety net of the welfare state by assuming that kindhearted individuals and communities will take care of the hungry.

In her book *Feeding the Other*, de Souza (2019) shows how asking for food is a stigmatizing experience in which the hungry—mostly poor people, women, and racial minorities—are being marked as irresponsible, immoral, and lazy. This neoliberal narrative that defines who is a worthy citizen is developed in the space of the food pantry through food insecure people's interactions with primarily white, formally educated, and relatively affluent volunteers who are perceived as responsible and caring. This distinction between "us" and "them" allows even the most caring volunteers to ignore their privilege and responsibility in a highly unequal society.

One of the most enduring critiques of emergency food assistance is that it fails to address the structural conditions that contribute to food crises, keeping the global food regime unchanged and allowing the persistence of class, race, and gender inequalities in shaping the ability to control and access food. While recognizing the urgency and necessity of aid in extreme situations, these critics emphasize the need for anti-hunger policies that address the deep roots of malnutrition.

6.4.2. Anti-Hunger Strategies

The long-term work of eliminating hunger is a lot more challenging than providing humanitarian assistance and gets less support from donors, who prefer to see immediate results. The challenge also comes from deep-seated debates between those who believe that technology and economic development will reduce hunger, those who argue for market reforms to improve distribution and reduce externalities, and those who support systemic change in the political and economic relations surrounding food.

Many experts believe that technology can solve the world food crisis by increasing productivity, solving environmental problems, and boosting the nutritional content of food. The Green Revolution was premised on this idea. Today, bio engineering, including the genetic modification of plants to increase their mineral and vitamin content, is being pushed by international organizations and corporations holding patents on these products, as a way to increase yields and address micronutrient deficiencies. But so far, the results of these efforts are mixed at best: while the production of genetically modified (GMO) corn, soy, and wheat has increased, it has also depleted soils, lowered biodiversity (and hence diet diversity), and driven small farmers who cannot afford this technology off the land. Additionally, much of this gain in production is being diverted to biofuels and livestock feed instead of feeding the hungry. Recognizing the environmental impacts of the heavy chemical use associated with the Green Revolution (see chapters 7 and 8), some have pushed the idea of a "doubly green revolution" which would build on the presumed success of the first Green Revolution and make it more environmentally friendly through green technology. Although perhaps less taxing on the environment, this approach prioritizes growth as the primary solution to hunger, without much consideration for what is produced and for whom.

Arguably, markets could be reformed and regulated to alleviate poverty and environmental damage. Numerous reports by international organizations such as the World Bank emphasize gender equity, sustainable development, and waste reduction as steps toward increasing food security. At the same time, these reports also argue for the need to expand markets and encourage free trade, failing to recognize the contradictions between these various goals. Sustainable development is still predicated on the notion of economic growth and faith in markets, as many critics have pointed out. Reducing food waste has received significant attention as a promising way to free up resources, given that about a third of the world's food is left uneaten (see chapter 8). However, efforts focus primarily on the consumption end, encouraging retailers, restaurants, and households to plan more effectively and to recycle surplus into ethical commodities or food aid. Limited attention is given to the reasons why producers flood markets with unnecessary food products that remain out of reach for many. Similarly, numerous gender equity initiatives aim at increasing women's participation in commercial agriculture through education and improved credit markets to access land and inputs. Yet, subsistence agriculture might offer more stability than commercial farming, especially if the latter leads to indebtedness and greater dependence on foreign markets. In short, these well-intended projects often ignore the power imbalances between the participants they aim to lift up and the more powerful actors who continue to control the industrial food system and run business as usual.

If what is at stake is the overproduction rather than the underproduction of food, then promoting growth and increasing productivity are not the most pressing issues. Furthermore, if we agree that the primary causes of hunger are political, then solving hunger requires that we confront the structural forces that underlie food insecurity and injustice by simultaneously fighting poverty, inequality, and environmental destruction. As Holt-Giménez (2019, 90) puts it, "a food system that depends on unlimited growth, exploitation, periodic financial crises, and that is running out of resources, can't be 'fixed': it must be transformed."

This perspective aligns with the most recent understandings of the right to food, defined by former UN Special Rapporteur on the Right to Food Jean Ziegler (2008, 9) as:

> The right to have regular, permanent and unrestricted access, either directly or by means of financial purchases, to quantitatively and qualitatively adequate and sufficient food corresponding to the cultural traditions of the people to which the consumer belongs, and which ensure a physical and mental, individual and collective, fulfilling and dignified life free of fear.

Upholding this right requires that we take steps to create a world free of hunger. In addition to tackling the most urgent crises, we must take measures to remove the social, political, economic, and environmental obstacles that prevent people, including women and Indigenous people, from obtaining adequate food or the means for its procurement. In addition to major legal, economic, and political reforms, we must also reconsider how we grow food.

Frances Lappé and Joseph Collins's book *World Hunger: Ten Myths* (2015) debunks several myths regarding food insecurity, including the assumptions that only industrial agriculture, GMOs, free markets, and free trade can solve hunger and that organic and ecological farming is incapable of feeding the world. Along with former UN Special Rapporteur on the Right to Food, Olivier De Schutter (2010), they advocate for agroecology as a solution to food insecurity, poverty, and environmental destruction.

Agroecology is a scientific approach to sustainable farming that applies ecological principles to agriculture to preserve soils, water, and biodiversity. This includes practices such as crop rotation, manure applications, biological pest control, water-harvesting, and polyculture, which have been performed by communities that hold intimate, historical, and localized knowledge of nature (see chapter 7). As such, this approach privileges community-based knowledge over scientific knowledge that is controlled by governments and corporations.

Agroecology is also a political project that supports farming livelihoods and decentralizes power away from international corporations into the hands of small farmers. It mobilizes farmers to regain control of knowledge, seeds, land, water, biodiversity, and culture. As such, the principles of agroecology align with those of food sovereignty.

Food sovereignty emerged in the 1990s as a tool to frame peasant struggles and motivate a social movement that unites small farmers in dismantling the global, corporate, and industrial food system. The concept was first mentioned in the 1990s by members of *La Via Campesina*—an international peasant rights organization—to criticize food security policies that failed to address the underlying causes of hunger (Patel 2009). In 2007, more than five hundred representatives of landless peasants, pastoralists, fisher-folk, and urban farmers from eighty countries cosigned the Nyéléni Declaration, which defines food sovereignty as follows:

Food sovereignty is the right of peoples to healthy and culturally appropriate food produced through ecologically sound and sustainable methods, and their right to define their own food and agriculture system. It puts those who produce, distribute and consume food at the heart of food systems and policies rather than the demands of markets and corporations. . . . It ensures that the rights to use and manage our lands, territories, waters, seeds, livestock and biodiversity are in the hands of those of us who produce food. Food sovereignty implies new social relations free of oppression and inequality between men and women, peoples, racial groups, social classes and generations. (La Via Campesina 2007)

The enactment of food sovereignty requires addressing power relations such that everyone is able to exercise their rights to shape the food systems and policies impacting them. In short, democracy and participation are pillars of food sovereignty, which help uphold the right to food.

The city of Belo Horizonte in Brazil exemplifies a comprehensive approach to ending hunger that is grounded in democratic participation, agroecology, the right to food, and food sovereignty (see textbox 6.5). Other examples include seed banks, cooperative farmer enterprises, community savings and credit groups, collective urban farms and kitchens, and grain reserves that have been created in places such as Ecuador, Burkina Faso, Honduras, Colombia, Haiti, Cuba, and Nepal. The

---

Box 6.5   Ending Hunger in Belo Horizonte, Brazil

Belo Horizonte is the fourth-largest city in Brazil and a model for successfully engaging residents in the fight against hunger through government leadership. Its turn-around has been documented by Lappé (2011), Chappell (2018), and Rocha and Lessa (2009). In the early 1990s, poverty and food insecurity were widespread in Belo Horizonte: one out of five children were not getting enough food. By increasing democratic participation and fostering social inclusion, the city government in collaboration with farmers, entrepreneurs, and consumers was able to develop a series of food security innovations that creatively met residents' most pressing food needs. These have resulted in a more than 50 percent cut in infant mortality and an increase in the consumption of fruits and vegetables, even among the most impoverished and marginalized populations. Rather than a single policy, Belo Horizonte's approach to end hunger included several related initiatives that were often built on existing informal practices. Importantly, these initiatives were developed by a new city agency—the Municipal Secretariat for Food Policy and Supply—created in 1993 to support the right to food for all citizens. Committed to social justice and inclusion, the agency engaged thousands of citizens in planning for a sustainable city with equitable food access.

The core strategy consisted of programs to support small farmers, with the idea that enhancing farmers' livelihoods would lower the cost of fruits and vegetables for urban consumers, slow down urbanization, generate income for farmers, and promote sustainable practices and biodiversity. Spots in public spaces were set aside for farmers' stands throughout the city, allowing farmers to sell directly to consumers. City land was made available for entrepreneurs willing to open stores selling basic, healthy, and local food items below market price. The city also launched several large "popular restaurants" where affordable meals made with locally grown food are served daily to thousands of residents—rich and poor. Community gardens were started in urban neighborhoods. Free meals were provided in schools and nurseries. The city experimented with a dozen of programs, some more successful than others. In 2010, the overall cost of these initiatives represented less than 2 percent of the city's budget, about $10 million US dollars per year or a penny per day per resident.

The success of Belo Horizonte has inspired cities around the world to create food policy councils and engage citizens in participatory planning around food. The main lessons are the importance of encouraging participation in democratic governance, supporting local livelihoods of farmers and food entrepreneurs, committing to equity, and integrating environmental and social objectives.

goal of food sovereignty initiatives is to strengthen local livelihoods and resilience strategies, rather than undermining them.

We will address solutions to obesity in chapter 13. However, it is important to acknowledge again that hunger and obesity are not the opposite of each other. Instead, as we learned in this chapter, there are caused by the same dynamics that put profits over human health. Addressing the so-called obesity crisis requires that we recognize these connections and consider how to reform or create alternatives to the current food system to produce nutritious and healthy food for all.

6.5. Conclusion

In a world of food abundance, it is troubling that so many people would face food insecurity in its various shapes, including insufficient caloric intake, nutrient deficiencies, overweight and obesity, and starvation. While hunger is most common in poor countries, especially in Africa and Asia, malnutrition is more widespread in wealthier nations, even if it disproportionately affects poor and socially marginalized people.

Several arguments have been put forward to explain the prevalence of hunger: excessive population growth for the earth's limited capacity; external shocks such as natural disasters, economic downturns, and conflicts; and heightened vulnerabilities due to a lack of entitlements. Some scholars, however, argue that the problem does not lie in the lack of food but in its overproduction. More specifically, they posit that the corporate, industrial, capitalist, neoliberal, and global food regime we described in chapter 3 causes food insecurity by destroying food livelihoods and robbing people of their ability to control how they grow food and feed themselves. Within that framework, poverty, environmental destruction, climate change, and conflict are not exogenous to the food system, but inherent to it. Hunger is thus an embodiment of structural inequalities that prevent certain people from accessing life-sustaining food—a process equated to structural and silent violence.

Addressing food insecurity depends on the severity of the situation and what policy makers believe to be causing it. In the case of a famine, immediate short-term humanitarian assistance is needed. This, however, does not target the underlying causes of food insecurity, which must be addressed through long-term strategies. While some experts advocate for population control, technological fixes, and/or market reforms, others support a more radical approach that restores food sovereignty and upholds the right to food. They argue that we produce more than enough food to feed the world, but that deepening economic inequality and social marginalization are preventing people from accessing it with dignity.

ENVIRONMENTAL GEOGRAPHIES
OF FOOD

Food's Ecological Pillars
Soil, Water, and Biodiversity

7

Outline

7.1. Agroecosystems and Nature's Inputs
7.2. The Environmental Impacts of the Contemporary Food System
7.3. Environmental Threats to Food Security and Safety
7.4. Sustainable Agriculture
7.5. Conclusion

Learning Objectives

- Investigate the importance of soil, water, and biodiversity in agroecosystems.
- Identify the main threats posed by the contemporary food system, including the effects of agricultural expansion and intensification.
- Assess the impact of food-related environmental degradation on yields, food security, and food safety.
- Define sustainable agriculture and envision more sustainable ways to produce food, including technological innovations, organic farming, and agroecology.

The ecological pillars of food systems are soil, water, biodiversity, and climate. Wherever these are threatened, so is food security. In this chapter we focus on the first three pillars and analyze their geographic relationships to food production, including crop cultivation and animal husbandry. Given the significance of climate change, the urgency of addressing it, and the importance of food as a contributor and possible solution to the escalating climate crisis, it is addressed separately in the next chapter. Fishing and its relationship to marine resources are also given their own chapter, allowing us to explore aquatic ecosystems independently.

The earth's capacity is finite. We currently have about 13 billion hectares of land area, of which 3.3 billion are used for raising animals and 1.6 billion are used for cultivating crops. Much of the remaining land is not easily accessible or useable. Although 70 percent of the earth's surface is covered in water, only 3 percent of it is fresh and two-thirds of that is not available for immediate use. The number of current species on earth is debated: scientists estimate that between 5 million and 1 trillion different species of plants and animals coexist, with only 1.7 million officially recorded. Of these various species, we only use a very small fraction for food.

Over the past three hundred years, we have exploited land, water, and living species as if they were unlimited "resources" up for grabs. The rate of exploitation has increased in recent decades as new technology convinced us that we could control nature. However, it is now becoming increasingly clear that our disregard for nature and ecosystems is beginning to have serious and perhaps irreversible consequences for planetary health. Whether due to population growth, changing diets, or profit motives, the expansion and intensification of agriculture discussed in the previous chapters have had significant impacts on agroecosystems, threatening their health and ability to sustain food production in the future. These impacts are not geographically uniform, with some agroecosystems more stable and resilient than others.

This chapter begins with an overview of the importance of water, soil, and biodiversity in agroecosystems and an examination of the threats posed by contemporary agriculture and food production. We then consider how food-related environmental degradation compromises food security and safety. We conclude the chapter with an exploration of potential ways to make food systems more sustainable.

7.1. Agroecosystems and Nature's Inputs

The concept of *agroecosystem* introduced in chapter 2 denotes the socio-environmental characteristics of farming systems. Referring back to figure 2.3, we understand agroecosystems as functional units constituted by interrelated physical, social, political, and economic systems that shape our ability to turn nature into crops and livestock for food. In this chapter, we focus specifically on the physical or environmental system without losing track of the other three.

Humans use a variety of *inputs* to produce food in combinations that depend on availability and methods of production. These include the trilogy of labor, capital, and land. The latter is a broad term that has traditionally been used to describe natural resources such as soil, water, and biodiversity. Land also provides energy, including nonrenewable energy such as coal, oil, and natural gas found underground and renewable energy like wind, sun, water, and biomass. As we will see later in this chapter, looking at nature as an input or a resource is problematic. But before we get to this critique, let us first highlight the significance of soil, water, and biodiversity in food production.

7.1.1. Soil

Soil is a mixture of minerals, dead organic matter, living organisms, air, and water essential to feeding plants. Its productivity is determined by its composition and its chemical and biological properties, which form overtime. It is shaped by geographically unique physical factors like climate, landscape relief, and underlying parent material and therefore varies greatly around the world. Soil is dynamic; it is always evolving, forming itself, eroding, or breaking down. While soil is often understood as renewable, the process of soil formation takes thousands of years and might be hampered by several factors, including climate and land use changes related to food production.

Healthy soils allow air and water to circulate to the roots of plants and contains high levels of organic matter, a diverse population of microorganisms like bacteria and fungi, and many nutrients such as nitrogen, phosphorous, and potassium. These nutrients feed plants and animals (including worms, insects, birds, and small mammals), which may eventually become nutrients in our food. When plants shed fruits, leaves, or twigs or when animals die, they decompose and return nutrients to the soil in the form of humus. However, when these are removed as a source of food, that renewing cycle is broken and the soil becomes less healthy. Thus, soil management is essential to soil function and food security. In addition, as we will learn in the next chapter, soil is extremely important in sequestering carbon—one of the major greenhouse gases contributing to climate change. The importance of soil has led some scientists to advocate that, as a living thing, soil should have rights and be protected just like air and water are protected in the United States by legal frameworks like the Clean Water Act and the Clean Air Act.

7.1.2. Water

Water is another indispensable element of agroecosystems. Not enough water (drought), too much water (flood), and poor-quality or polluted water have a negative impact on soil health and plant productivity. Although there is an abundance of saltwater on earth, most plants and animals, including human beings, need fresh water to survive. The 3 percent of all water that is fresh is stored on the surface of the earth (e.g., in glaciers, ice sheets, snow, lakes, ponds, rivers, creeks) and in the ground (i.e., aquifers)—much of it inaccessible at this time. Water used in agriculture comes from rain, surface inflows, or ground water, with the latter two requiring some form of infrastructure to be harvested and utilized. There are important geographic differences in access to freshwater that, in turn, influence the potential for agriculture, food security, economic development, and human health.

Water is intimately related to climate and weather since it originates in the atmosphere in the form of rain, snow, and even mist before hitting the earth's surface and eventually becoming available for use. In other words, the *water cycle*, in which water evaporates, forms clouds, and returns as precipitation, shapes how much water is available. When water withdrawals exceed water recharges, which is more likely to happen in heavily irrigated areas, water systems are compromised. Climate change, especially increases in the severity and frequency of weather events, also makes water availability less certain. Pollution from human activities, including the use of chemicals in farming and the lack of wastewater sanitation, also reduces the health of watersheds and the availability of fresh usable water.

Globally, agriculture accounts for 70 to 80 percent of freshwater withdrawals, even though only 20 percent of cultivated land is irrigated. This figure tends to be higher in the Global South where a larger share of the economy is based on agriculture. Competing demands linked to population growth, urbanization, agricultural intensification, and economic diversification have been putting pressure on water resources—a problem exacerbated by climate change. Today, 40 percent of the world's population lives in areas where water is scarce and competition between uses such as drinking, sanitation, agriculture, and industry is heightened. One out of three people in the world do not have access to safe drinking water. This troubling situation has prompted the United Nations (2020b) to declare 2018–2028 the Water Action Decade in an effort to prevent a looming crisis.

7.1.3. Biodiversity

Biodiversity is essential to the good functioning of ecosystems, including the ability to produce food. According to the FAO (2019c), biodiversity for food and agriculture "includes the diversity of animals, plants and micro-organisms at the genetic, species and ecosystem levels that sustain structures, functions and processes in and around production systems and provide food and non-food agricultural products." Not only does it include the plants and animals (both raised and wild) that we consume as food, but it also encompasses the many living organisms that sustain production systems through pollination, pest control, nutrient cycling, erosion control, water purification, and other important *ecosystem services* (see textbox 7.1).

The health of ecosystems and its ability to provide services and sustain human life depends on this biodiversity and the complex and still poorly understood interactions between plants, animals, soils, water, and climate that occur at a variety of geographic scales. For example, a particular crop might benefit from earthworms that live in and maintain the soil, weeds that keep away certain pests, bees that pollinate the plants, and climate-regulating services that ensure proper temperatures and humidity, each of which interacts with larger environmental systems.

Ecosystem Services

The concept of ecosystem services is widely used by environmental scientists to measure the various contributions of ecosystems to human well-being. Food and water are among the most essential and basic "services" that nature provides. Other services include regulating climate, producing oxygen, and contributing to spiritual and cultural life.

The notion of ecosystem services was popularized by the *Millennium Ecosystem Assessment* called by the United Nations in 2001 to assess the effects of ecosystem changes on human well-being and inform conservation and sustainable use of those systems. Published reports produced with input from hundreds of experts introduce the idea of ecosystem services to show how recent and dramatic changes to ecosystems are having negative impacts on human well-being that are likely to get worse in the future, making it difficult to meet development needs (Millennium Ecosystem Assessment 2005). Specifically, the reports outline a series of ecosystem services divided into four categories: *supporting* (primary production and soil formation), *provisioning* (food, fiber, fresh water, genetic resources, and biomedicals), *regulating* (air quality, water filtration, climate resilience, erosion prevention, pest and disease control, runoff and flood reduction, pollination, waste treatment, and carbon sequestration) and *cultural* (spiritual, aesthetic, recreational, and educational values).

Approximately 60 percent of ecosystem services are being degraded or used unsustainably.

Although food related provisioning services such as crops, livestock, and aquaculture have been quantitatively enhanced in the past fifty years, this increase has caused significant losses in other services including lowering water availability for other uses, degrading water quality, reducing biodiversity, increasing the frequency and magnitude of floods, encouraging soil erosion, and causing climate change. Indeed, food production is a major contributor to the degradation of ecosystem services, particularly in the world's poorest areas. Those losses have important economic, public health, and social consequences that, according to the report, reduce the "freedom of choice and action" of those most dependent on these services.

The concept of ecosystem services is useful to begin accounting for the economic value of nature and devising incentives and penalties to regulate its use. For example, policy makers have been advocating for "payments for ecosystem services" as a mechanism to make consumers pay for the services they receive from nature and encourage conservation. However, it has been criticized for promoting a managerial and neoliberal approach to environmental issues by focusing on maximizing the production of certain ecosystem services through market valuation, ignoring the politics of commodifying and attaching monetary value to nature as a "resource" and the power imbalances between market participants locally and globally (Castree 2008, Kosoy and Corbera 2010).

7.2. The Environmental Impacts of the Contemporary Food System

Not only does agriculture depend on these related ecological pillars, but it also has a profound impact on their health and functioning. In the process of producing crops and livestock for human consumption, the physical characteristics of agroecosystems are modified, leading to alterations in soil and water chemistry and impacting surrounding life, which in turn, affects our ability to produce food and support human life.

As noted in chapter 2, it is important to acknowledge that *agroecosystems* are socially produced ecosystems in which soil, plants, and animals are manipulated via labor- and capital-based technologies, such as crop rotation, irrigation, seed selection, breeding practices, and the use of chemicals, for the purpose of increasing biomass or output. By most accounts, contemporary food production, distribution, and consumption are exerting an immense toll on ecosystems. This tension between humans' use of nature for production and the ecological processes required for nature to renew itself is what Marxist scholars and political ecologists call *metabolic rift* (Foster 1999). From this perspective, ecological deterioration is caused by capitalism. Without referring to the theory of metabolic rift, many observers have attributed environmental degradation to the industrialization of agriculture, the pressure to increase food production to meet population growth, and the globalization of the food system. Both the *expansion* and the *intensification* of agriculture have interrelated effects on water, soil, and biodiversity.

7.2.1. Agricultural Expansion: Deforestation

The expansion of agriculture into new territories drives the clearing of land through deforestation, causing a variety of problems. Data indicate that agriculture has been responsible for 80 percent of global deforestation in the past two decades. Over the past fifty years, croplands and pastures have increased by 320 million hectares (FAO 2020c), which is more than the agricultural land of India and Argentina combined. Indeed, agriculture has a much more significant impact on deforestation than urbanization, logging, and mining combined. The problem is particularly acute in developing countries, such as Brazil and Indonesia, that have adopted economic development strategies promoting agricultural commodity exports. For example, between 2008 and 2018, the soy cultivation area in Brazil has increased by 63 percent, while the palm oil footprint in Indonesia rose by 36 percent (see textbox 7.2) and the land dedicated to cocoa in Côte d'Ivoire grew by 75 percent (FAO 2020d). The most common *drivers* of deforestation are soy, cattle ranching, and palm oil (Boucher et al. 2011). The majority of newly expanded agricultural lands is used to raise animals or produce animal feed such as soy or corn.

Besides the significant climatic impacts that will be presented in the next chapter, there are numerous environmental problems triggered by deforestation. Trees play an important role in keeping water and nutrients in the *soil*. When trees are removed, topsoil is likely to erode, wash away with heavy rains, and leach nutrients, leaving a hard and less productive surface and potentially leading to desertification and collapse of ecosystems. It is estimated that half of the earth's topsoil has been eroded during the past 150 years. This topsoil had taken centuries to form and cannot be replaced in the short term. Each year, seventy-five billion tons of soil are eroded from arable lands worldwide (FAO 2017a), producing an estimated annual financial loss of $8 billion and a reduction in global food production of thirty-four million tons (Sartori et al. 2019). These losses, however, are not only caused by food-related deforestation: other human factors, including urbanization, contribute to this trend, along with droughts, climate change, and other physical processes causing erosion. Yet, by all accounts, food production is the primary driving factor.

Deforestation also increases *water runoff* as water flows over the surface of the earth carrying heavy metals, oil, nutrients, pathogens, and sediments into streams, polluting and clogging life-sustaining waterways. For example, in Madagascar, where the central highlands are being cleared to make room for agriculture, topsoil has been washing away at a very fast rate (as much as four hundred tons per hectare per year). During tropical storms, rivers, estuaries, and the surrounding Indian Ocean turn blood red—a phenomenon astronauts noticed from space and described as the island "bleeding to death." In addition, runoffs create gullies that remove entire sections of fields and destroy valuable infrastructure. Runoff also reduces the amount of water absorbed into the ground to feed plants and replenish aquifers. As more water rises to the surface because of deforestation, it brings with it salt deposits, increasing

In the last forty years, Indonesia has become the largest producer of palm oil, generating 40.5 million tons in 2018—more than half of the global output. Palm oil production has increased fourteenfold since 1980. Today, it represents 40 percent of all vegetable oils and is used primarily in processed food, cosmetics, and biofuels. It is in high demand because it is cheap, does not contain trans fats, and has marketable qualities, including texture, flavor, and a long shelf life. As a result, it can be found in almost half of all packaged food, including chocolate, cream cheese, and baked goods.

Palm oil plantations have been the leading cause of rain forest destruction in Indonesia, where twenty-eight million hectares of forest have been lost since 1990. In a 2007 report, the United Nations Environment Programme (UNEP) warned that 98 percent of the rain forest would be destroyed by 2022 if deforestation continued at the same rate as the decade preceding publication of the report. This destruction is threatening orangutans, gibbons, tigers, elephants, wild pigs, and hundreds of endangered species of plants and animals that cannot survive in the "green deserts" of plantations. For instance, the population of orangutans in Borneo has been halved between 1999 and 2015 (representing the death of about one hundred thousand animals) as the tree canopy in which they live is disappearing (Voigt et al. 2018). In a recent Greenpeace campaign, Mondelez—the company behind chocolate and cookie brands such as Cadbury, Toblerone, Chips Ahoy, and Oreo—was accused of having destroyed seventy thousand hectares of great apes' habitat between 2015 and 2017 alone. Another 750 to 1,250 orangutans are killed each year in conflicts with humans linked to plantation expansion.

In addition, the degradation and burning of Indonesia's peatlands are causing soil erosion and air pollution. The fertilizers and pesticides used on

new plants are polluting rivers and drinking water. Deforestation and peatland draining are also releasing huge quantities of carbon from the soil into the atmosphere, driving climate change. In fact, Indonesia is now in the top tier list of contributors to climate change, behind the United States and China. Indonesian palm oil plantations alone are said to have contributed 4 percent of global anthropogenic greenhouse gas emissions between 2000 and 2010. Oil palm plantations are also imposing significant health, economic, and emotional costs on Indigenous people whose land they occupy in violation of their rights.

Because of the dramatic environmental and social impacts of palm oil, it has received significant attention from organizations like Greenpeace and the World Wildlife Foundation that have pushed for sustainable practices. The Roundtable on Sustainable Palm Oil (RSPO) was created in 2004 to promote the production and consumption of sustainable palm oil. The organization represents multiple stakeholders including palm oil growers, processors, and distributors, environmental and social NGOs, retailers, and investors. While some advocates support bans on palm oil, others argue that it meets demand more efficiently than other types of vegetable oil because it uses less land—0.26 hectares per ton compared to 1.25 for rapeseed, 1.43 for sunflower, and 2 for soybean oil (IUCN 2018). The RSPO proposes to increase sustainability by halting deforestation and protecting existing forests, imposing rules for newly developed areas, creating an ecolabel certification program, offering training and assistance to small holders, enforcing human rights (such as communities' right to free and prior informed consent to new projects), and promoting decent living wages. Today, just about 20 percent of Indonesian palm oil is certified sustainable—not enough according to environmental NGOs.

the salinity of soils and decreasing their productivity. Water cycles are also altered by imbalances between decreasing rainfall and rising evapotranspiration, impacting the severity of weather events such as droughts, contributing to climate change, and threatening crops (see figure 7.1).

Another devastating effect of deforestation is the loss of *biodiversity* caused by the destruction of habitat provided by trees and forest canopies. Indeed, according to the FAO (2019c), deforestation and habitat loss or alteration are the two main causes of food-related biodiversity losses. New roads, fields, and pastures divide existing

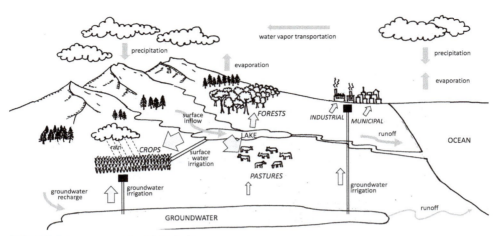

FIGURE 7.1 Water cycle and agriculture.

Source: Author.

habitat and interfere with species reproduction. Species introduced through agriculture may also interfere with existing ecosystems. It is estimated that about 80 percent of the world's land species live in forests, with tropical rain forests being the most biodiverse. Unfortunately, rain forests are endangered everywhere, especially in the Amazon—the largest in the world. In the past fifty years, more than seven hundred thousand square kilometers—or 17 percent—of Brazilian Amazonian rain forest have been destroyed, with almost ten thousand square kilometers cleared in 2019 alone (Butler 2020). Because so many species of animals, plants, and organisms have yet to be documented, it is difficult to estimate the full effect of deforestation on biodiversity loss. Studies have shown negative impacts on numerous individual species of animals such as butterflies, birds, gorillas, chimpanzees, elephants, and many smaller creatures that provide important ecosystem services. Plant ecology is altered by the loss of tree canopy and the spread of alien and invasive species. Wild relatives of crops and livestock are also endangered. In addition to the plants and animals that we are using directly as food, a multitude of smaller living organisms that play an important role in sustaining food production through ecosystem services like pollination, pest and disease prevention, soil retention, nutrient and water cycling are endangered by deforestation. Even fisheries and coral reefs are damaged by deforestation-related water runoff that pollutes waterways and oceans.

Deforestation is also having an indirect impact on *disease*. As the primary hosts of zoonotic pathogens like Ebola lose their habitat and disappear, these dangerous microorganisms can turn to humans who are now in closer contact (Olivero et al. 2017). At the same time, other species like bats and mosquitos strive in recently anthropized environments bordering wilderness, leading to the spread of diseases like the Zika virus, COVID-19 (see textbox 3.3), and malaria through direct contact, domestic animal infection, or contamination by urine or feces (Afelt et al. 2018).

Indigenous communities who live in forests experience the most immediate effects of deforestation and biodiversity loss, which reduces their access to medicinal plants, timber, nuts, fruits, and game on which they depend to sustain their way of life. Conflict between tribes and farmers often erupts as they compete for land access. Eager to promote economic development, governments typically support farmers, ignoring the rights of Indigenous people (see textbox 1.1). For example, in Brazil, President Bolsonaro has opened the Amazon to large-scale farming by removing constitutional protections for Indigenous communities and cutting funding for agencies responsible for the enforcement of the rights of Indigenous people, which he described as "obstacles to agri-business." Some fear that these policies will result in an "ethnocide" in the Brazilian Amazon (Londoño and Casado 2020).

7.2.2. Agricultural Intensification

Given the constraints on expanding cultivated areas, farmers, agrifood corporations, and governments have focused on increasing yields on existing land. This approach has been especially common in places where population growth is high and fertile land is limited. For example, the Green Revolution has played a central role in transforming agriculture in India (see textbox 2.2). Since the 1950s, China also increased its agricultural productivity to compensate for growing population and urbanization. Constrained to feed almost one-fifth of the world population on less than 7 percent of the arable land and about 10 percent of renewable freshwater resources, China relied heavily on fertilizers, pesticides, and irrigation to boost food production, which has grown faster than population.

The *intensification* of agriculture refers to the increasing use of technology to raise land and labor productivity. Such technology includes machinery, agrochemicals, and irrigation. Intensification has been made possible by the seemingly unlimited availability of fossil fuel used to manufacture agrochemicals and power equipment to plow soils, sow seeds, apply fertilizers and pesticides, harvest crops, and irrigate increasingly large fields, lowering the demand for farm labor, except for technicians and machine operators. Instead of relying on biological cycles, rotating crops such as clover and beans, and applying organic matter to fields like manure and fish heads, farmers began relying on agrochemicals to replenish the soil and continue growing the same single crop without any interruption. Agricultural intensification also involves the use of hybrid and genetically engineered (GMO) seeds to generate greater output, more resistant crops, and more frequent harvests. When it comes to livestock, intensification is often associated with concentrated animal feeding operations—known as CAFOs—which feed thousands of animals in a confined space. All these technological innovations lead to what is often described as "factory farming," which requires the separation of animals and crops in order to promote specialization and economies of scale, resulting in the rise of monoculture. Like the territorial expansion of agriculture into forests and grasslands, technological intensification has significant and related environmental consequences.

The intensification of agriculture has had dramatic impacts on rural *landscapes* (see figure 7.2), none of which is more noticeable than the increase in *monoculture—*

  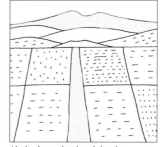

Semi-natural habitat with extensive agriculture

Declining grassland and riparian habitat with agricultural intensification

Limited grassland and riparian habitat with intensive agriculture

FIGURE 7.2 Landscape changes and crop biodiversity losses caused by agricultural intensification.
Source: Author, inspired by ECA (2019, 5).

an approach to farming that poses a major threat to *biodiversity*. Today, commercial farms typically specialize in a single crop or animal to be traded to food processors and distributors. Most farmers in a given region grow the exact same species of crop, often purchasing seeds from the same large company such as Monsanto, Dupont, or Syngenta, displacing native varieties of plants. For example, 89 percent of US corn acreage is planted with herbicide resistant GMO corn. Other varieties are becoming much less common and could potentially disappear if they stopped being cultivated. Even animal feed is grown separately from dairy farms and cattle operations. While cows used to graze on local grasses and continue to do so in smaller and subsistence farms, those raised in concentrated animal feeding operations are being fed corn, soy, and alfalfa grown in distant locations.

Globally, 75 percent of our food comes from just twelve species of plants and five species of animals, with wheat, rice, and maize providing more than half of the world's food energy (FAO 2019c). Soybean, sugar cane, potatoes, sweet potatoes, oil palm fruit, millet, barley, cassava, sorghum, and plantains are other important plant species. For meat and animal products, we primarily rely on pig, chicken, cattle, sheep, and goat. Within these dominant species, we are increasingly relying on a small number of plant varieties (also known as landraces) and animal breeds. Many varieties and breeds used for food have either disappeared or are becoming less common. The genetic diversity of our food supply or *agrobiodiversity* has been shrinking with the industrialization of agriculture. The FAO (2010) estimates that during the past century, particularly since the intensification of agriculture and the Green Revolution, 75 percent of crop diversity has been lost. As we will see in chapter 9, aquatic biodiversity is also severely threatened by a similar process of industrialization in fishing and aquaculture.

The increased use of technology leads to *overexploitation* and overharvesting. The abandonment of crop rotation and intercropping (i.e., growing two or more mutually beneficial crops on the same field) and the shortening of fallow time (when land is allowed to rest) in favor of continuous monoculture, dependent on irrigation, synthetic fertilizers, and pesticides, is a form of overexploitation that quickly depletes *soils* of their nutrients. Similarly, overgrazing by herbivore livestock damages soils and makes them prone to erosion. It can also lead to a loss of grassland biodiversity and, in arid areas, desertification. Overharvesting of wild plants and animals also poses important threats to biodiversity. For instance, the American bison—the dominant animal in North America in 1800—became almost extinct when Euro-Americans turned it into a tradeable commodity, starving Native people into submission and eventually making room for cattle ranching. Another significant form of food-related overexploitation is overfishing—an issue to which chapter 9 is devoted.

A major problem with industrial agriculture relates to *tilling*—the process of preparing the soil for sowing or planting by agitating and loosening the top layer. It softens and aerates the soil, incorporates crop residue into the soil, and kills weeds. This important part of cultivation was historically done by humans (often enslaved people) or by animals pulling wooden plows. More recently, however, tilling has been done by increasingly powerful and heavy tractors that can drag or push large metal plows. The negative effects of deep and frequent mechanical tilling on soil and water processes are similar to those of deforestation discussed in the previous section. It reduces organic matter and microorganisms such as worms and important microbes and encourages erosion, which contributes to nutrient loss and water runoff.

A key concern with agricultural intensification is the use of *chemical fertilizers and pesticides*—a problem that seems to be worsening as losses in biodiversity, soil, and water reduce the healthy functioning of agroecosystems. Indeed, there is a vicious cycle in which the use of chemicals to support monoculture degrades ecosystems, requiring ever greater application of synthetic fertilizers and stronger pesticides and herbicides. The growing use of genetically modified organisms (GMOs)—many of which were designed to be used with specific fertilizers and herbicides—has also caused an increase in the use of agrochemicals.

Scientists have long known that soil nutrients are important for plant productivity. By the middle of the nineteenth century, they had identified nitrogen, phosphorous, potassium, and other minerals as important fertilizers. While phosphate and potash could be mined and added to the soil, atmospheric nitrogen could only be fixed into the soil naturally by leguminous plants and sun energy. Decades later, German scientist Haber discovered a way to create synthetic nitrate (or ammonia) by mixing atmospheric nitrogen and hydrogen at very high temperature. His discovery was first used to make explosives for the German army during World War I and only later became a common fertilizer, after being commercialized by the chemical giant BASF under the leadership of Bosch. In the United States, at the end of World War II, the military surplus of ammonia was repurposed as fertilizer, signaling the beginning of widespread use in agriculture (see figure 7.3) and the switch from a farming system fed by sun energy to one fed by fossil fuel (Pollan **2006**). Haber and Bosch went on to win the Nobel Prize for their discovery, which some credit for having fed over three billion people (Smil **2001**). Others however are much more critical of the impact of fertilizers and other agrochemicals on the environment and the broader food system.

The negative effects of fertilizers and pesticides, including insecticides, fungicides, and herbicides, have been documented in numerous studies. A classic is Rachel Carson's *Silent Spring* which was first published in 1962 and helped shape the environmental movement in the United States and the creation of the Environmental Protection Agency (EPA) by drawing attention to the harmful effects of pesticides on

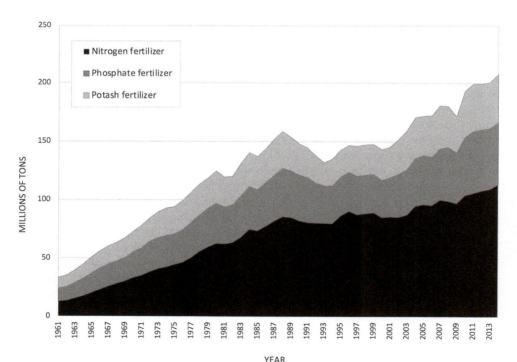

FIGURE 7.3 Global fertilizer production.

Source: Author with data from FAO (2020e).

human and animal health. In general, agrochemicals are criticized for polluting the soil, water, and air and threatening biodiversity and human health.

Less than half of nitrogen fertilizer and phosphorus applied to fields is actually absorbed by crops as nutrient. The other half spreads into the broader environment, including streams, ground water, and the air. Nitrate (a soluble component of synthetic nitrogen) leaches into ground water and pollutes drinking water, posing significant risks to human health, including cancer. Runoff of nitrate and phosphorus also causes *eutrophication* of waterways, leading to excessive growth of plants like algae and animals dying due to an ensuing lack of oxygen. This is contributing to the expansion of *dead zones* in coastal areas, where algal blooms kill sea life by depleting oxygen. For example, the northern part of the Gulf of Mexico is home to a twenty-three-thousand-square-kilometer dead zone primarily attributed to nitrate, phosphorus, and animal waste leaching from Midwestern farms through the Mississippi River. The enclosed Baltic Sea has the world's largest dead zone (more than sixty thousand square kilometers) caused by fertilizer pollution, sewage, and overfishing and exacerbated by climate change.

Pesticides like DEET and Roundup are widely used around the world despite their known risks to human health. Farmers who mix and apply agrochemicals are most likely to suffer from poisoning and exposure, but others are also affected as those chemicals enter the food chain and are ingested by consumers. Pesticides have been linked to cancer, Alzheimer's disease, birth defects, kidney and liver diseases, and attention deficit disorder. They are also harmful to animals other than the pests they target, including birds and beneficial insects such as bees, spiders, and beetles. Like fertilizers, they leach into water and evaporate into air.

Pesticides that remove specific weeds, insects, or fungi alter the ecosystem by threatening the species depending on those and making room for other species to take over. So-called "super bugs" and "super weeds" have grown with pesticide use. Unfortunately, these invasive species are often removed with further pesticide applications, creating a vicious circle that exacerbates the negative consequences.

The use of agrochemicals is regulated by environmental agencies with varying degrees of power. While a few harmful chemicals such as methyl bromide, glyphosate, and atrazine, have been banned in certain countries, these bans are not evenly enforced. Chemical companies lobby aggressively to ensure that their products can be distributed, regularly suing public agencies, funding their own research to challenge inconvenient scientific evidence, and using marketing campaigns to change perceptions. Therefore, numerous known dangerous chemicals remain in use, especially outside of Europe, which tends to have stricter regulations.

Another important element of agricultural intensification is *irrigation*—the deliberate artificial application of ground, surface, or treated water to soils. Gravity irrigation uses gravity to distribute water across a field, using tubes, canals, and/ or ditches. This describes most surface irrigation, including flood irrigation for rice fields. Pressure irrigation, which is becoming more common, uses pressure to deliver water through sprays, sprinklers, or drips emitters. One of the most common pressure irrigation systems in the United States is the center pivot system in which water is distributed overhead via sprinklers attached to a long pipe that rotates in a circular pattern around the central tower. While some irrigation systems require manual labor, more sophisticated versions are entirely automated.

Today, over 70 percent of freshwater withdrawals are for agricultural purposes. Only about 20 percent of cultivated lands are irrigated, with the remaining being rain-fed. The proportion of irrigated land doubled in the last fifty years. Most of the world's irrigated land is in India and China. While we are withdrawing about 9 percent of our renewable water annually at a global scale, there are important variations

Agricultural Sector Water Risk

- 0 to 1
- 1.01 to 2
- 2.01 to 3
- 3.01 to 4
- 4.01 to 5

FIGURE 7.4 Map of agriculture-related water risk.

Note: Water risk is an aggregated measure based on risk indicators of agriculture-weighted water quantity, quality, and regulation.

Source: Aqueduct Global Maps 2.1 (Gassert, Landis, Luck, Reig and Shiao 2014)

by countries and regions. On average, poorer countries are using water for agriculture at a greater rate than wealthier countries—a phenomenon shaped by both climate and the economic importance of agriculture. In Europe, where fresh water is generally more plentiful, only 6 percent of renewable freshwater is being withdrawn, with just 20 percent of that amount used for agriculture. In South Asia, where irrigation is common, 57 percent of renewable water is being withdrawn, primarily for agriculture. And in North Africa, the Middle East, and Central Asia where water is very scarce, almost all renewable water is withdrawn for agriculture. Withdrawal rates above 20 percent start putting pressure on water systems, and rates above 40 percent are considered unsustainable (FAO 2011c). Figure 7.3 depicts the geographic distribution of agriculture-related water risk and highlights areas with high levels of stress.

Irrigation is more reliable than rain and thus is perceived as an important way to increase productivity, especially in regions like South Asia with highly variable rainfall, long dry seasons, and recurrent droughts. Lower agricultural productivity in Africa is often explained by international agencies like the World Bank as a result of limited irrigation. However, while bringing some stability, irrigation is also causing significant environmental disturbances. About 60 percent of withdrawn water flows back to local systems (often polluted and contaminated), but the remaining 40 percent is lost through evapotranspiration. Increased evaporation can cause precipitation. At the same time, the water that goes back to the ground is likely to lead to salinization, drainage problems, and soil saturation known as waterlogging. Stagnant waters, by providing habitat for mosquitos and parasites, may also increase diseases such as malaria, dengue, yellow fever, and bilharzia. Irrigation, which is not always properly applied due to mechanical problems and poor water management, also causes additional runoff that washes away agrochemicals, salt, soil nutrients, and sediment into waterways.

Surface water withdrawals for agriculture, including the creation of reservoirs and canals, in a specific area decreases the flow of water and affects river ecology downstream. Similarly, water pollution from agricultural runoff has consequences downstream. Because many important river basins cross national boundaries, these issues can lead to international tension and conflict. Groundwater depletion is also impacting people who rely on the same aquifer but do not have the same wells, as illustrated by the example of tensions between commercial and subsistence farms in the Ica Valley of Peru (see textbox 7.3).

Different types of food require different amounts of water. Animal products, including meat and dairy, necessitate much more water than most fruits and vegetables. This is because, throughout its life, livestock is fed large quantities of food, including corn, soy, or alfalfa that demand water and are increasingly grown on irrigated land. Indeed, about 40 percent of all grain harvested globally is used as animal feed. Additionally, large amounts of so-called "gray water" are required to dilute the runoff generated by feedlots. The concept of *water footprint* seeks to measure how much water is being used to produce food. Figure 7.5 reveals significant differences in the water footprint of various types of food and highlights the burden imposed by meat and dairy products. Even when considering their nutritional value (liter per kcal), these items consistently have a higher footprint. Thus, it should come as no surprise that vegetarian diets (which exclude animal sourced food) have a water footprint that is 25 percent lower than the average diet and represent the most effective way to reduce the consumption of water, saving about 750 liters per day per person. Less drastic changes in diets can still have a significant impact on water conservation. For instance, the footprint of a reduced animal food diet is 18 percent lower than the average and that of healthy diets (which follow national dietary guidelines) is 6 percent lower (Harris et al. 2020).

When comparing footprints, it is important to consider the type of water being used. *Green water* comes from rain, *blue water* comes from surface or ground water

| Box 7.3 | **Asparagus Farms in Peru's Ica Desert** |

Peru is the world's largest exporter of asparagus, generating about $450 million in revenue annually by exporting 99 percent of what it grows to the United States, Europe, and Asia. Asparagus farming was promoted in the mid-1990s as part of an agricultural diversification program sponsored by the United States Agency for International Development (USAID). Over the years, it received significant technical and financial support from international lending institutions, such as the World Bank, that saw asparagus exports as a promising way to promote economic development, create jobs, and generate foreign currency. Most of the production is taking place in the Ica Valley, along the coast of southern Peru, where over one hundred square kilometers of desert have been reclaimed. None of this would have been possible without foreign investments that enabled the development of irrigation and, along with the privatization of land and water, contributed to consolidation and integration within the industry. A few large companies own most of the land and water rights and have expanded vertically into processing activities (e.g., packaging, canning, and freezing).

Irrigation relies entirely on groundwater, which is being withdrawn at a much faster rate than it can be replenished. As old wells dry, new deeper wells are dug to access water. Large producers, especially those who benefit from international loans and have purchased water rights, can afford to explore and drill farther out and down. Smaller producers, however, must purchase water that is becoming increasingly expensive. As aquifers retreat, soils are becoming drier and water drains faster, requiring more frequent applications. Even drinking water consumption has been curtailed, posing a public health risk (Laurence 2010). To address this situation, the government imposed restrictions on new wells. However, these restrictions are not effectively enforced and three-fourths of wells in the area are estimated to be unlicensed or illegal. The continued withdrawal of water reflects a *collective action* problem where everybody knows that they should reduce their water consumption, but nobody has an incentive to stop pumping water if others continue to do so.

This growing water scarcity in the Ica region is causing conflict between large corporate exporters and Indigenous farmers (James 2015), who blame each other. Many also blame development agencies for ignoring the environmental costs of export-led economic policies. Several NGOs have drawn attention to this issue and are working with large retailers in exporting countries to foster more sustainable practices (see Progressio 2010).

irrigation, and *gray water* refers to the volume of freshwater required to assimilate the load of pollutants released. Inefficient use of irrigation, especially on low-quality soil prone to runoff, raises blue and gray water footprints. Similarly, the heavy use of fertilizer and pesticides also raises gray water footprints. This means that production methods matter and can be adapted to reduce water consumption. For example, animals that are raised on pastures consume less blue and gray water than those raised in industrial farms where they require external feed and generate more pollution. According to Mekonnen and Hoekstra (2012), 1 kilo of grazing beef requires 16,140 liters of green water for rain-fed pastures, 213 liters of blue water, and 0 liters of gray water, compared to 10,992 liters of green water, 933 of blue water, and 1,234 of gray water for industrial beef. While the latter uses less water overall, its heavy use of green and gray water imposes a much larger environmental cost than pasture-raised animals.

The idea of water footprint is related to the notion of *virtual water* that is hidden in our food. Although people drink 2 to 4 liters of water per day, they "eat" 2,000 to 5,000 liters of virtual water indirectly through their food depending on their diet (Hoekstra and Chapagain 2007). Because so much food is traded globally, water is virtually exported around the world. Between 1984 and 2011, international food trade and the associated trade of virtual water almost tripled (d'Odorico et al. 2019), representing approximately a fourth of all production. Often, water is exported from

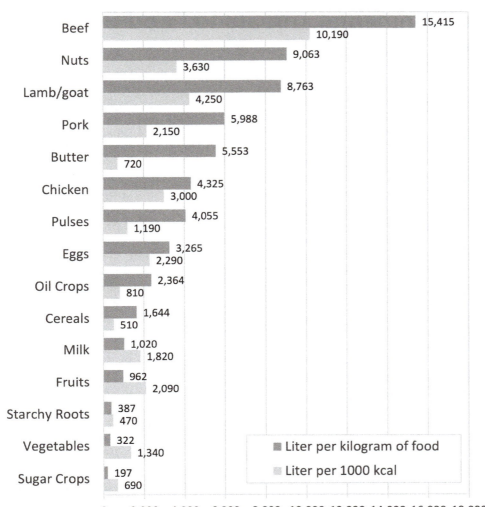

FIGURE 7.5 Water footprint of selected foods.

Source: Author with data from Mekonnen and Hoekstra (2012).

regions that suffer from water scarcity. For example, California has been experiencing droughts for decades. Yet it is the largest food producing state in the United States and a major exporter of food products. This includes alfalfa grass grown in the irrigated desert of the Imperial Valley and exported to places like the United Arab Emirates and China. Peruvian asparagus are also a good example of virtual water exports from a region where water is very scarce (see textbox 7.3).

The concept of water footprint comes from the idea of *ecological footprint* that measures the total global acres, including land and water areas, needed to support the world population. As we will learn in the next chapter, another related concept is that of *carbon footprint*, which measures the amount of greenhouse gases released in the production of specific items. Footprint accounts facilitate comparisons between countries, types of food, and growing methods. However, they have been criticized on methodological grounds, partly because it is very difficult to measure all the resources used from farm to fork and data are limited.

7.3. Environmental Threats to Food Security and Safety

The industrialization and intensification of agriculture is taking a toll on nature and imposing a disproportionate burden on people whose livelihoods depend on farming and who live in areas where the environment has been stretched to its limit. The stress that industrial agriculture is imposing on land and water systems is threatening our ability to produce food in the future. In other words, modern agriculture is *unsustainable*.

On the surface, based on basic economic indicators, the last fifty years look like incredible progress. The world's agricultural output has almost tripled, while cultivated land increased by only 12 percent. Much of this rise in production has been attributed to irrigation, with irrigated land doubling over the period and producing more than 40 percent of total food. Although the amount of cultivated land per person has decreased from 0.45 to less than 0.25 hectares, the amount of food per capita has increased by 33 percent to almost three thousand kilocalories per day, suggesting that agricultural intensification has resulted in significant improvement in productivity. However, a closer look reveals a host of serious environmental problems described in the previous sections, including loss of biodiversity, water pollution, and soil depletion, that put this success story into question and compromise food production and livelihoods.

The FAO (2011c) estimates that almost half of the world's agricultural land is degraded, meaning that *ecosystem services*, including food, are declining because of water scarcity, soil erosion, and biodiversity losses. Globally about two billion people and almost two billion hectares of land are affected by land degradation. There are several methods to estimate land degradation, including expert surveys, audits, models, and remote sensing of the environment. Although estimates of the extent of land degradation differ across methods, all find that losses are concentrated in Asia (especially India, China, and Indonesia), Sub-Saharan Africa, Central America, and South America (see Gibbs and Salmon 2015). These areas are homes to much of the world's population and places where agriculture represents not only a major source of food, but also a source of foreign currency.

As more areas are reaching their limits in terms of production capacity, yields have begun to stagnate or even decline after decades of growth. Indeed, evidence suggests that we are losing ground on agricultural productivity improvements achieved in the 1960s and 1970s and that yields have not improved for several staple crops, including maize, wheat, rice, and soybean, since 2000 (Ray et al. 2012). This is especially true for some of the largest grain producers in the world. For example, China and India are experiencing significant yield stagnation or decline with at least a third of their maize, wheat, rice, and soybeans. The United States has not witnessed any yield increase in 36 percent of its wheat cropland area. Rates are even worse in France and Germany where about 80 percent of wheat croplands are not seeing any yield improvements. Unfortunately, such stagnation or declines often encourage greater reliance on irrigation and agrochemicals, which in turn further weakens soil and water ecosystems, leading to a widening *metabolic rift*. The typical response is the *technology treadmill* (see chapter 4) in which more and more capital and inputs are applied to sustain yields.

The decline in agrobiodiversity associated with monoculture and the heavy use of fertilizers and pesticides is decreasing *resilience* among farmers. Crop failures have a much more drastic effect under monoculture, when there are no other crops to fall back on, than they do under polyculture, when complementary crops provide some level of security. The Irish potato famine that began in 1845 and killed approximately one million people through starvation over the following five years was caused by an infestation of phytophthora that destroyed about three-fourths of the potato crop (see chapter 6). A single varietal of potato had been introduced in Ireland by the British landed class and had become the primary source of food. Once the crop failed, there was practically no other food available—a problem exacerbated by the fact that Irish food commodities continued to be exported to England during the famine.

Invasive alien species, which fill the vacuum left by biodiversity losses caused by agricultural intensification, pose a major threat to food production (Nellemann, et al. 2009). Globally, sixty-seven thousand pest species, including nine thousand insects and mites, fifty thousand pathogens, and eight thousand weeds, attack crops. Up to 70 percent of them are "alien" in the sense that they have been introduced by human activity. Alien invasive weeds and pathogens are estimated to be responsible for yield reductions of 8.5 and 7.5 percent respectively, with some of the worse losses in Africa. This too is often met with more lethal pesticides applied in larger quantities.

Food safety is also compromised by farming methods. The heavy use of agrochemicals leaves residues on the food and pollutes water used for drinking and manufacturing food products. In addition, the rise of large feedlots is adding large quantities of untreated or poorly treated waste and chemicals into waterways, which are then tapped into for irrigation causing the spread of pathogens like E. coli and salmonella to crops. As a result, food safety has become an issue of concern, with much food unsafe for human consumption and recalls of contaminated food increasingly common (see chapter 13). For instance, in 2018, Arizona romaine lettuce contaminated by E. coli sickened over two hundred consumers in thirty-six states, resulting in ninety-six hospitalizations and five deaths. An investigation by the US Food and Drug Administration (2018) revealed that the pathogen had been transmitted through manure-tainted water from an irrigation canal. The water was either directly applied to crops or used indirectly to dilute pesticides or fertilizers. Because the lettuce was processed in large plants and mixed with lettuce from multiple sources, tracing the origin of the pathogen was nearly impossible.

These various threats to food security are not caused by external factors, such as natural disasters or conflict. As discussed in chapter 6, they are caused by a political and economic system that has encouraged agricultural expansion and intensification while ignoring its environmental costs for too long. Stopping and reverting this destructive trend is becoming urgent and various strategies are being debated and implemented by international agencies, governments, nonprofits, and grassroots organizations.

Sustainable Agriculture 7.4.

Our food system needs a major overhaul, otherwise the earth will soon become incapable of supporting its population. *Sustainable agriculture* seeks to meet society's present food needs, without compromising the ability of future generations to meet their needs. This means preserving nature's inputs while being economically and socially viable. Sustainability is generally understood as comprising three interrelated elements: healthy environments, economic viability, and social equity. Although most people would agree that sustainability is a positive thing, the concept is ambiguous and debates regarding how to achieve it abound.

Because of overlapping environmental, economic, and social concerns, sustainability often requires trade-offs and inevitably leads to prioritization of one domain over the other. For example, a more environmentally friendly growing method, might be less profitable and lead to increases in the price of produce, such that it becomes affordable only to the most affluent members of society. Yet sustainability requires that we attend to all three of these aspects and consider how they may reinforce or work against each other.

The complexity and dynamic qualities of ecosystems alone, including the interconnections between water, soil, and biodiversity, means that maintaining the status quo (as the notion of sustainability implies) is impossible. Thus, some scholars are now downplaying sustainability and focusing instead on *adaptation* and *resilience* of ecosystems, studying how they can evolve in mutually beneficial and stable ways.

Although the concept of sustainability seems universal, consensual, and all-inclusive, it is deeply political. As Greenberg (2013) argues, sustainability asks us to define what in our present we want to sustain and what we are willing to discard to achieve a utopian vision of the future. As such, sustainability is a normative and political concept that involves societies envisioning, negotiating, and deciding what needs to be sustained and for whom. Instead of a universal notion of sustainability, Greenberg suggests that there are many *sustainabilities* that reflect the values of individuals, communities, agencies, and industries involved in sustainability projects in particular places and times. She distinguishes between *eco-oriented* sustainabilities that privilege "big nature" and environmental preservation, *vernacular* sustainabilities that focus on how people live and work with nature in the everyday, *justice-oriented* sustainabilities that prioritize environmental equity across race, ethnicity, gender, and class, and *market-oriented* sustainabilities based on green technology, corporate responsibility, and ethical consumerism (see chapter 5). Unless the capitalist, corporate, and neoliberal food regime we described in chapter 3 is altered, most sustainability initiatives will likely reflect market-oriented perspectives at the expense of other visions, particularly those related to social justice.

7.4.1. Sustainable Intensification

Today, several sustainable agriculture scientists argue for *sustainable intensification* which would increase productivity while reducing negative environmental impacts. This is often premised on the notion that new green technologies—the "green green revolution"—can solve inefficiencies and minimize the use of external inputs such as water and agrochemicals. For example, there has been concerted efforts to improve irrigation infrastructure (e.g., drip system) to reduce the use of water through more precise and targeted applications. Bioengineers are also conducting research on new plant varieties, including salt-tolerant crops that could be irrigated with plentiful saltwater instead of scarce freshwater. New presumably less dangerous fertilizers and pesticides have been developed. GPS and sensor technologies are being used to determine the ideal rate of irrigation and agrochemical application. Robotic bees have been tested by the US Department of Agriculture as alternative pollinators to live bees. Scientists are even considering alternatives to photosynthesis as the primary way to transform carbon dioxide into edible biomass.

This approach is sometimes known as *precision* or *low-impact agriculture*, suggesting that science-based management and more efficient technology can increase productivity and reduce negative impacts. Instead of challenging the *productivist* model, it refines and promotes it by attempting to maximize output while minimizing certain negative environmental impacts.

Because this type of technology is expensive, it is unaffordable to most farmers, especially small-scale farmers in low-income countries. Even in more affluent regions where agriculture is already capital-intensive, farmers may be hesitant to invest in technology that will likely raise costs in the short run. Thus, the adoption of green technology often requires policy intervention in the form of assistance, incentives, or regulations. For example, international development agencies like the World Bank have been advocating for sustainable intensification of agriculture in Africa, where "productivity gaps" suggest that current production methods are not maximizing output. The bank has invested funds in modernizing irrigation in Malawi and Tunisia, among other countries, with limited success. Such international development aid typically comes with strings attached, including trade liberalization and the removal of agricultural subsidies. The Alliance for a Green Revolution in Africa (AGRA), a program funded by the Bill and Melinda Gates Foundation, has invested millions of dollars in Ethiopia, Nigeria, Rwanda, and Tanzania to identify areas suitable for intensification, develop appropriate technology, including genetically modified seeds and digitally enabled innovations, train farmers, and monitor performance. This approach is gathering criticism from experts in ecology and food security. A recent multi-agency report entitled "False Promise" shows how it is reducing crop diversity, increasing farmers' debt, and failing to address malnutrition in the name of calorie production (Wise 2020).

7.4.2. Organic Farming

Organic farming seeks to replace chemical fertilizers and pesticides with biological processes that enhance soil nutrients, foster cycling of resources, and preserve biodiversity. While organic farming requires complex crop management, it is often summarized as the elimination of synthetic pesticides and fertilizers and other materials such as hormones and antibiotics. This reductionist approach that defines organic by what it is not rather than by what it is, has been fostered by the commercial popularity of organic as an organized "cure-all" alternative to commercial agriculture. In that context, the organic label became a valuable way to advertise to consumers that food is produced according to certain principles verified through a *certification* process.

For example, in the United States, the most common certified organic label is provided by the US Department of Agriculture. Different criteria are used for crops, meat and animal products, and processed food. In general, however, organic means that prohibited substances or methods are not being used. To be called organic, produce must be grown on soil to which no prohibited substances have been applied for at least three years and meat must come from animals allowed to graze, fed organic feed, and not administered antibiotics or hormones. None of these can be genetically modified. Processed food must contain a certain proportion of organic ingredients and no artificial preservatives, colors, or flavors. Different labels, including "100 percent organic," "organic," and "made with organic" reflect the proportion of organic ingredients in processed food.

Instead of promoting a form of farming that replicates natural ecosystems as often claimed, organic farming ends up being defined by compliance with a set of regulations established by a third party (e.g., government agency, nonprofit). As Guthman (2014b) argues, in places like California, organic is big business. Indeed, many large farming corporations have diversified into organic to meet growing consumer demand, with very minimal change to the process of production that remains akin to factory farming (i.e., heavily mechanized monoculture that relies on off-farm inputs and cheap labor). It is important to acknowledge that, although not certified, many subsistence farmers around the world are effectively growing organic food since they do not use any of the prohibited inputs. Indeed, the process of certification is expensive and time-consuming, creating barriers of entry for small producers with limited budgets and reproducing existing inequalities.

What organic has succeeded in doing is appropriating the image of nature to sell commodities. The global market for organic food and beverages was worth more than $100 billion in 2017, with Europe and the United States being the largest markets. It has enjoyed steady growth in the past two decades. Between 2009 and 2019, organic food sales grew from $21.3 to $50 billion in the United States, where it is now found in three out of four conventional retailers and represents 5 percent of all food purchases. Fresh produce is the top-selling category (43 percent), followed by dairy (15 percent), beverages (11 percent), and grain products (9 percent). Denmark, Switzerland, and Sweden have the highest organic shares of food purchase at about 10 percent. Yet organic remains more expansive than conventional food, even though premiums have fallen recently. For instance, the average price of organic milk—one of the single most popular organic items—is 88 percent higher than the price of conventional milk ($4.76 per gallon compared to $2.53 in 2018). The difference appears to be driven as much by high consumer demand as by production costs.

7.4.3. Integrated Agriculture

A more comprehensive approach to sustainable agriculture seeks to integrate a wide range of methods to manage pests, enhance soils, and conserve water. This typically requires a return to polyculture, where various crops and farm activities support each other. By-product or waste from one activity becomes input for another, enhancing natural processes and reducing the need for artificial inputs. The goal of integrated agriculture is to sustain a system of healthy production in the long run, rather than maximize output in the short run.

Integrated agriculture has several key components. One of its building blocks is *integrated pest management*, which encourages natural pest control mechanisms as opposed to pest eradication via the use of chemical pesticides. In that framework, *crop protection* can be achieved through proper plant selection, removal of diseased or problematic plants, and the use of beneficial fungi and bacteria. Animal husbandry can promote weed management and organic fertilization by having animals eat certain weeds and using manure to fertilize desirable plants. Another key component is *crop management*, which includes careful selection of compatible and locally appropriate crops, as well as their rotation over time. Crop rotation and intercropping of cereals and legumes have been common throughout the world, until commercial monoculture began to replace it. It is still being practiced among subsistence farmers and is being reintroduced to increase sustainability. For example, pigeon pea has been promoted as a complementary crop for rice, millet, or maize in South Asia. In Latin America, maize has traditionally been grown with beans and squash—a combination known as the three sisters (see textbox 7.4). In Europe, various types of vegetables (e.g., cabbage and potatoes) are combined in horticulture. Specific plants help rebuild the soil, create nutrients for other plants, provide pest control, and absorb carbon dioxide. So-called hedgerow crops—a planted row of shrubs and/or trees that borders a field—are also useful in preventing soil erosion, managing pest, conserving water, and providing habitat for beneficial animals. For example, rose bushes traditionally lined vineyards, serving as an early warning system for fungal disease, allowing farmers to respond appropriately. Shifting annual crops to perennials has also been advocated by proponents of integrated agriculture, who consider food forests or *agroforestry* an important and productive element of rural landscapes. Along with no-tilling methods, the mixing of crops and animal husbandry also encompasses *soil* and *water management*.

Proper monitoring and management are critical to the success of integrated agriculture, which must be adapted to local conditions and respond to unique

Box 7.4 **The Three Sisters**

Native Americans have combined maize, beans, and squash to maintain soil, generate natural fertilizers, minimize water use, and enhance production for over five thousand years. Throughout Mesoamerica, these three companion crops have been grown together in *milpas* where tall corn stalks provide support for beans, corn flowers attract beneficial insects, beans fix nitrogen in the soil, and squash provide cover that keeps weeds away and reduces evaporation and erosion, illustrating the many benefits of *intercropping*. In North America, the Iroquois call them the "three sisters" and view them as "sustainers of life." Planting techniques are adapted to the environment through the use of raised beds or well-drained mounds of soil and the potential addition of supportive crops like the "rocky mountain bee plant" added by the Anasazi to attract bees for pollination of the beans and squash.

These three crops form the basis of a balanced diet, including carbohydrates from corn, protein from dried beans, vitamins, and oil from squash and its seeds. Many Indigenous dishes combine these ingredients, highlighting the connection between food cultures, Indigenous knowledge, and sustainability.

Today, milpas are gaining attention as a sustainable farming practice at the same time as they are being threatened. In a recent book entitled *Milpa: From Seed to Salsa,* Dahl-Bredine and others (2015) advocate for the globalization of milpas as a sustainable approach to farming, but warn that climate change, the propagation of GMOs, the consolidation of land, and new laws prohibiting the exchange of seeds, are threatening milpas in places like Oaxaca in southern Mexico.

circumstances. The intimate and traditional knowledge of farmers is an important asset that must be recognized and cultivated, as it is often more valuable than scientific knowledge and technical expertise. Indeed, there is no universal blueprint for integrated agriculture; it must be based on local ecosystems.

7.4.4. Agroecology and Food Sovereignty

Agroecology simultaneously refers to a scientific discipline relying on ecological principles to study, design, and manage agroecosystems, a set of agricultural practices aimed at enhancing the ecological, socioeconomic, and cultural sustainability of farming systems, and a social movement promoting *food sovereignty* (Altieri 2018). Agroecology shares many practices with integrated agriculture, including soil management through animal and green manure, legumes, and cover crops; weed and pest management through inter-cropping, mulching, careful selection of crops, and encouragement of beneficial and predator species; water management through water harvesting and gray water filtering; and landscape management through hedges, woodlands, natural windbreaks, and river buffers. What distinguishes agroecology from other forms of sustainable agriculture is the combination of science, everyday practice, social movement, and political action. Specifically, agroecology is rooted in the experiences of small-scale food producers: it values their knowledges and cultures, fosters participation and local leadership, and confronts the structural inequities that impoverish farmers and inhibit the *right to food*. As such, it rejects capitalist industrial agriculture and dependency on imported and costly artificial inputs, which are viewed as causing the environmental, economic, and cultural dispossessions facing peasants. In its place, it proposes a social process in which agricultural knowledge is continuously coproduced and shared to maximize positive synergies between species and minimize externalities. Thus, agroecology is a people-centered and knowledge-intensive practice and, as such, does not promote generic solutions.

Around the world, especially in Brazil, Mali, Mexico, Cuba, Nicaragua, and Guatemala, small-scale farmers are bypassing the corporate food system and avoiding the use of external inputs. They are rebuilding and diversifying agroecosystems to mimic natural local ecosystems, reintroducing complex biodiversity, enhancing nutrient cycling, and raising crop productivity. Instead of isolating specific components such as soil or water and seeking ways to augment those artificially, agroecology focuses on rebuilding interactions between biological components of agroecosystems to improve their collective functioning. For example, the negative environmental impact of industrial rice production could be curbed and even reversed by agroecology (see textbox 7.5). *Agroforestry* and *silvopastures* (the integration of trees, forage, and animal grazing) are important forms of agroecology in areas where forests are a major source of food.

Box 7.5 **Agroecological Rice Systems**

Billions of people eat rice daily. Globally, rice provides more calories than any other crop. To keep up production, the cultivation of rice has become increasingly industrialized—mostly a result of the Green Revolution. However, industrial rice farming is extremely damaging for the environment. Aside from the significant release of greenhouse gases like methane, which affects climate change, modern rice paddies require irrigation and large quantities of water. For example, in India and the Philippines, one kilo of rice requires three thousand liters of water on average. Significant amounts of water are wasted through seepage and percolation. Continuous flooding and inadequate drainage are causing waterlogging, salinization, and nutrient problems. The expansion of paddies and the growing use of fertilizers and pesticides are reducing biodiversity. Together these changes are threatening the capacity of existing rice paddies to generate enough food and support local livelihoods in the long run.

Agroecology offers solutions by reintroducing biodiversity and enhancing synergies between various components of the rice agroecosystem. Common practices include the addition of ducks and fish to rice paddies, the intercropping of multiple varieties of rice and/or lotus, and crop rotation. Ducks eat weeds, insects, and snails, eliminating the need for pesticides. In addition, duck droppings fertilize the soil and their constant motion prevents sedimentation. Fish can also be introduced in rice paddies, complementing the role of ducks by eating plankton and insects and contributing fertilizer. Fish and ducks also provide food, including protein, for humans. Intercropping of rice varietals reduces vulnerability to pests, and the addition of plants, such as milk vetch and azolla, help restore the soil by adding green manure and fixing nitrogen. Figure 7.6 below illustrates the benefits of polyculture, including nutrient cycling, pest management, food diversity, and increased productivity (see also figure 7.7). These improvements do not require large plots of land or expensive inputs, but demand careful management based on local knowledge.

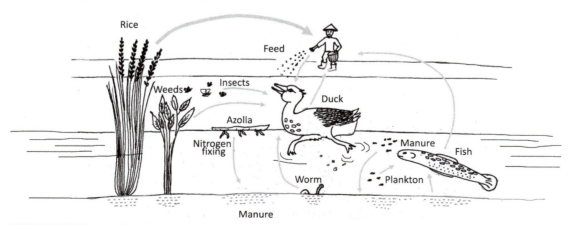

FIGURE 7.6 The benefits of agrobiodiversity in a rice paddy.

Source: Figure created by author, inspired by Third World Network and SOCLA (2015, 11).

Agroecology has become popular in the Global South where it resonates with small-scale farmers, including women and Indigenous people whose contributions to food production have often been dismissed. It has led to the formation of farmer-to-farmer horizontal networks in which farmers share information and strategies locally and globally. For example, La Via Campesina is a worldwide movement advancing farmer autonomy and self-sufficiency through the principles of agroecology and *food sovereignty* (see chapter 6). In Brazil, a peasant movement led to the adoption of a national plan for agroecology and organic production, which supports small farmers and encourages sustainable practices while promoting the consumption of healthy foods in cities (see textbox 6.4). These types of community-oriented approaches, including urban agriculture, offer a path toward economic, social, and environmental sustainability by addressing local needs through stronger networks of trust and reciprocity connecting farmers and consumers. These networks, which facilitate the exchange of knowledge between different generations, sectors, cultures, and traditions, are a crucial building block of agroecology that is intended as a social process resulting in the continuous cocreation of agricultural knowledge and the reconfiguration of power relations. The growing number of agroecology schools set up and run by peasant organizations illustrate how such knowledge can be encouraged and supported.

International organizations are beginning to acknowledge the benefits of agroecology. The 2009 IAASTD (International Assessment of Agricultural Knowledge, Science and Technology for Development) report, prepared by scientists and stakeholders from many countries commissioned by the World Bank and the United Nations, posited agroecology as a sustainable model of agriculture. The United Nations Special Rapporteur on the Right to Food submitted a special report to the General Assembly in 2010 advocating for agroecology to enhance availability, accessibility, sustainability, and nutritional adequacy of food as well as farmer participation. Also, the FAO (2014) held a large forum in Rome to promote agroecology as the basis of food security by turning "vicious cycles" of hunger, poverty, and resource depletion, into "virtuous cycles" of enhanced productivity, biodiversity, resilience, and livelihoods. A second forum was held in 2018 with the goal of "scaling up" agroecology to achieve the sustainable development goals set by the United Nations.

Observers fear that scaling up agroecology might dilute its core principles, especially those related to peasant autonomy and self-determination. Some warn that you cannot simply turn conventional agriculture into agroecology by adopting a few soil or pest management practices. Instead, agroecology requires a new systemic approach that begins with farmers' knowledge and is rooted in local ecosystems. This long-term participatory approach is frustrating to international policy and philanthropic organizations that prefer quick and easily replicated solutions. Thus, despite consensus about the benefits of agroecology, there has been relatively limited financial support for developing, transitioning to, or enhancing such systems.

7.4.5. Which Approach Is Best?

Comparing the sustainability of different approaches is a difficult task that requires agreement about the meaning of sustainability and availability of qualitative and quantitative data to measure its various aspects. Unfortunately, the environmental costs or benefits of food production are not always easy to measure. As a result, conventional actors continue to emphasize productivity indicators to vouch for the superiority of their approach, claiming that organic, integrated, or agroecological farming cannot generate the amount of food needed to feed the world. As we learned, however, this focus on yield is misguided and dangerous. While the adoption of sustainable methods may lead to tradeoffs in the form of reduced productivity in the short-run, evidence suggest that, if properly managed, it can generate enough food for the current population while preserving the ability of future generations to feed themselves. Whenever additional criteria are considered, agroecological methods score higher than conventional or even organic methods.

To visualize the benefits of sustainable farming methods, researchers have used radar diagrams that reveal performance on several relevant indicators. To this end, numerous assessment systems have been developed, including MESMIS (Marco para la Evaluación de Sistemas de Manejo de recursos naturales incorporando Indicadores de Sostenibilidad), SOCLA (Sociedad Científica Latinoamericana de Agroecología), and CAET (Characterization of Agroecological Transition). The latter, developed by the FAO (2018), uses ten agroecology indicators to assess progress toward sustainability. By plotting scores on these indicators, it is possible to compare farming methods and evaluate the impact of various changes on sustainability (see figure 7.7). The size of the area created by connecting the dots represents the aggregated benefit of each approach in terms of sustainability. In these assessments, efforts are typically made to engage farmers in choosing relevant indicators, measuring performance, mapping findings, interpreting results, and designing adaptations.

7.5. Conclusion

Food production, particularly since the rise of industrialized agriculture, has caused significant damage to the earth's natural environment. This metabolic rift is witnessed in the rapidly growing losses of soil, biodiversity, and water reserves. Unless we change how we grow food, our already stretched agroecosystems are likely to fail, causing widespread food insecurity.

The environmental costs of agriculture are not evenly distributed. On the one hand, places are endowed with different quantities and qualities of land, water, and biodiversity. On the other hand, they are utilizing these resources differently based on population needs and economic development strategies. Places that have embraced industrial agriculture and international trade as paths to development, including Brazil and Indonesia, are experiencing severe food-related environmental degradation that threatens future yields. Places with limited resources and/or growing populations are struggling with maintaining production levels without depleting existing resources.

These problems are complicated by climate change, which is intimately connected to the way we produce food, as we will learn in the next chapter. In this context, it is urgently vital that we rethink our agrifood system to address its destructive tendencies and make it more resilient and sustainable. Multiple approaches to sustainable agriculture have been advocated, ranging from green technological innovation addressing specific shortcomings to systemic approaches integrating biological, social, and economic subsystems. It is becoming increasingly clear that sustainability will not be achieved through technological fixes alone. Instead, it requires that we tailor solutions to local circumstances and pay attention to the social, political, economic, and cultural factors that have caused humans to neglect nature while producing food. This includes distributing costs and benefits more equitably and encouraging the use of local knowledge to develop adaptable and flexible solutions that encourage economic self-sufficiency, promotes environmental stewardship, and supports culturally appropriate diets.

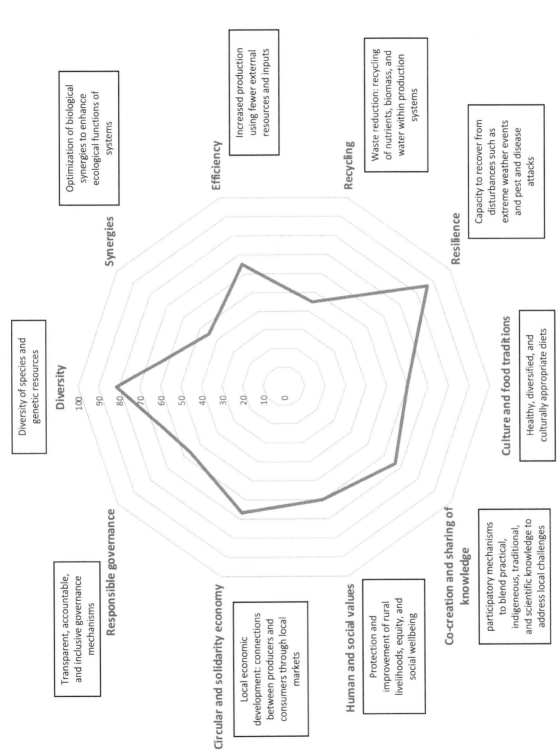

FIGURE 7.7 Radar diagram displaying the benefits of agroecological rice production systems in Thailand.

Source: Author based on FAO (2018).

Food and Climate Change 8

Learning Objectives

- Provide an overview of the nature, causes, and effects of climate change.
- Examine the role of food in creating greenhouse gas emissions, distinguishing between various stages of production, transformation, distribution, and consumption.
- Compare the carbon footprint of different foodways and the effects of dietary change.
- Assess the geographically and socially differentiated effects of climate change on food security and their role in promoting human migration and conflict.
- Appreciate the interdependencies between climate and other pillars of agroecosystems including soil, water, and biodiversity and their relationship to the global industrial food regime.
- Contrast the promises and limitations of various approaches deployed to address the food-related climate crisis.

In the past century, the earth's climate has changed at rates unseen in millennia. Average surface temperatures have risen by about 1 degree Celsius. Oceans, too, are warming, and sea levels have risen by eight inches. Ice sheets and glaciers are melting and extreme weather events such as intense rainfalls and record-high temperatures have become more common. These dramatic changes are the result of human activity, including food production, that generates heat-trapping greenhouse gases. These recent increases in anthropogenic greenhouse gas emissions worry scientists about the future of the planet and humanity. In some parts of the world, climate change is already having severe effects on food security, increasing the risk of conflict, involuntary migration, and further environmental degradation.

Over the past three decades, conversations about climate change have typically focused on transportation and energy. Yet, food accounts for approximately a third of all greenhouse gas emissions. As scientists, producers, and consumers begin to understand the major role that food plays in the climate crisis, they are also looking at food as a solution to this very crisis.

As this book emphasizes, food is a system. To understand the relationship between food and climate, we must look at the whole system, from seed to landfill, going through farms, chemical labs, distribution centers, trains, factories, supermarkets, restaurants, home kitchens, and media companies. We must also consider all actors involved, including farmers, elected officials, regulators, corporate shareholders, researchers, consumers, and activists. We must also acknowledge geographic interdependencies between places and across scales.

Greenhouse gases (GHG) are emitted at each link of the *commodity chain* (see chapter 5). As many observers have noted, the modern industrial diet consists primarily of fossil fuels. By analyzing the *lifecycle* of commodities, it becomes possible to account for their *carbon footprint* and make informed decisions about production and consumption. Nowhere is the climate problem more urgent than in the meat industry, which is receiving growing scrutiny.

This chapter begins with a brief introduction to climate change. Although scientists around the world agree unequivocally that climate is changing and that these changes are primarily caused by human activity, misinformation regarding the causes and effects of climate change abound and infiltrate politics. It is therefore important to provide a general overview before delving deeper into the carbon footprint of food at various stages of the commodity chain. We then explore sources of greenhouse gas emissions through the lifecycle of food and variations between types of food, places, diets, and methods of production. In the following section, we turn our attention to how climate change is impacting food security and migration. We end the chapter by considering various approaches to addressing the climate crisis and discussing their promises and limitations.

As we learn about climate, it is important to remember that it is one of the ecological pillars of food production, along with water, soil, and biodiversity. Indeed, many of the environmental problems we discussed in the previous chapter are connected to climate change, both contributing to it (e.g., deforestation, soil erosion, overexploitation of water) and being exacerbated by it (e.g., droughts, salinization, biodiversity loss, low yields). It is also essential to view all these environmental issues as embedded in the global industrial food regime that shapes human interaction with nature, including how we value nature and choose to preserve or plunder it.

8.1. A Climate Change Primer

In the past fifty years, scientists have been gathering increasingly detailed and sophisticated data on the changing climate of the earth. One of the most respected depositories of such information is the Intergovernmental Panel on Climate Change (IPCC) created by the United Nations to assess the science related to climate change. Hundreds of scientists from around the world collaborate in preparing regular reports summarizing the state of knowledge. The last comprehensive (or synthesis) IPCC assessment report was published in 2014 and the next one is due in 2022, although sections of this sixth report are being released early. The panel also recently issued a special report entitled *Climate Change and Land* (IPCC 2019) that draws attention to the food system. This section is primarily based on evidence gathered in these reports.

Climate change refers to observed changes in climate systems and their impacts on natural and human systems. As noted in the introduction, observed changes include rising surface temperatures, warming and acidifying oceans, rising sea levels, and melting ice and snow. As with all statistical methods, there are differences in measurements. However, despite these differences, scientists can say with confidence that these changes have taken place at an unprecedented rate since the industrial revolution, but more intensely in the past fifty years. All of these are well documented with observational data from various sources in the 2014 IPCC assessment report.

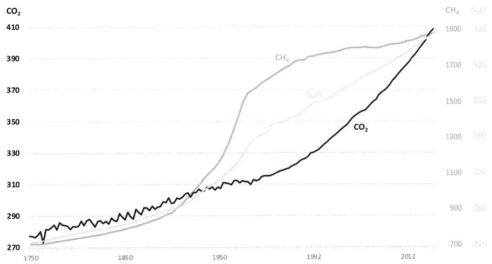

FIGURE 8.1 Global atmospheric concentrations of carbon dioxide (CO_2), methane (CH_4), and nitrous oxide (N_2O), 1750 to 2018. Note: CO_2 is in ppm, CH_4 and N_2O are in ppb.

Source: Author with data from Ritchie, Roser, and Mathieu (2020).

Scientists have been asking what is causing these changes. Since the first report in 1992, evidence that these changes are caused by human activity has continued to grow, such that the human influence on climate is now considered "unequivocal." Specifically, industrialization, including mining, fossil fuel combustion, and deforestation, has caused large emissions of greenhouse gases (GHG) like carbon dioxide (CO_2), methane (CH_4), and nitrous oxide (N_2O). Since 1750, total greenhouse gas emissions (standardized into CO_2 equivalents) have increased steadily, with more than 40 percent of emissions accounted for during the last fifty years (see figure 8.1). Despite attempts to mitigate climate change, the decades between 2000 and 2020 witnessed the largest increases in greenhouse gas emissions, suggesting that we are heading in the wrong direction toward abrupt and irreversible changes in the not-so-distant future.

The reason GHG emissions are so worrisome is their impact on temperatures, through what is known as the enhanced greenhouse effect shown in figure 8.2. The emissions of CO_2, CH_4, N_2O and other lesser-known gases into the atmosphere thickens the layer that traps solar radiation around the earth, preventing heat from escaping back into space and enhancing re-radiation of heat energy back toward the earth's surface, which causes temperatures to rise. Although some of the CO_2 is absorbed by vegetation and soils, deforestation and soil erosion counteract these effects by releasing sequestered CO_2. Oceans also absorb a significant amount of CO_2. However, this leads to acidification and threatens marine ecosystems and fisheries as we will learn in chapter 9.

We must reduce anthropogenic greenhouse gas emissions to slow climate change and keep the increase in global temperature below 2 degrees Celsius—a figure most scientists agree would trigger catastrophic change. To this end, several international agreements, such as the 1997 Kyoto Protocol, the 2009 Copenhagen Accords, and the 2016 Paris Agreement, have set and updated specific targets. Understanding how much CO_2 equivalents various human activities cause is important in identifying industries and practices where the greatest impact could be made. According to the IPCC (2014), greenhouse gas emissions are primarily caused by electricity and heat production (25 percent), agriculture, forestry, and other land use (24 percent), industrial production (21 percent), and transportation (14 percent). This estimate,

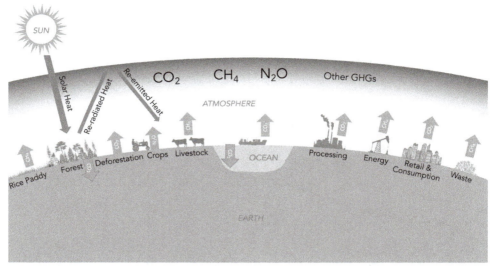

FIGURE 8.2 Agricultural greenhouse effects and CO_2 cycle.
Source: Author.

however, does not tell us much about the role of food outside of agriculture. For example, when soy produced in Brazil is shipped to feed pigs in China, this releases CO_2 through transportation. When the slaughtered meat is turned into packaged sausages, this would be considered industry. When the processed meat is refrigerated, the CO_2 it generates falls under the electricity sector. And eventually, when leftovers and uneaten meat end up in landfills, it is not clear exactly to what economic sector the methane it produces is attributed. As we are about to learn, measuring the *carbon footprint* of food is a complicated task, but scientists have made significant progress during the last few years.

8.2. The Carbon Footprint of Food: Lifecycle Analysis

The notion of *carbon footprint* is used to measure the CO_2 equivalents embedded in specific commodities throughout their lifecycle, from input production to waste disposal. In the past decade, scientists have conducted lifecycle assessments (LCA) of commodities such as smart phones, furniture, fashion, and food to estimate their carbon footprint. Such computations require a series of steps and assumptions. First, scientists must identify the different stages in the lifecycle of commodities and measure the main greenhouse gases released at each stage. Even though food production methods have become increasingly industrialized and standardized, commodities have unique lifecycles depending on where they are produced, how, and for whom. As we learned in chapter 5, global supply chains can be extremely complex and span multiple continents. In that context, obtaining data on every stage of the life of a food commodity is much more difficult than under subsistence farming and localized food systems, where the production and consumption stages are fewer and more proximate. As lifecycle assessments become more sophisticated, we are gaining new knowledge on how geographic differences in the way people produce, distribute, and consume food influence GHG emissions.

Secondly, scientists need to find a common metric to account for different types of gases, with various warming potentials over different periods of time. For instance, methane and nitrous oxide are much more potent than carbon dioxide. To this end, scientists often used carbon dioxide equivalents (CO_2eq), which adjust quantities of specific emission by their global warming potential over a period of one hundred years.

Thirdly, researchers must account for the potential absorption of carbon in the process of production, which would reduce the overall greenhouse effect. For example, farmers around the world are engaged in "carbon farming" that takes carbon out of the air and sinks it into the soil by avoiding tilling, selecting specific plants and trees, composting, using cover crops, and rotating livestock through small paddocks—methods that have been promoted by agroecology (see chapter 7). Shellfish can also have similar positive effects by filtering water and removing carbon from the environment (see chapter 9). Thus, we need to look at *net* greenhouse gas emissions.

Fourthly, if computing aggregate carbon footprints for specific countries for comparison purposes, it is important to distinguish between production and consumption. Because of international trade, some countries produce carbon-intense foods that are exported to other countries. For instance, Indonesian palm oil is exported to countries like India, China, the Netherlands, Spain, Italy, and the United States. In that case, it is complicated to assign responsibility for anthropogenic GHG emissions. Production methods in developing countries are often blamed for causing climate change, ignoring unsustainable consumption habits in more affluent nations. If plantation owners are guilty, so are food manufacturers who profit from low-cost ingredients and consumers who crave cheap and highly processed food.

Because of these challenges, recent estimates of the overall share of anthropogenic greenhouse gas emissions attributed to food range from 19 to 37 percent. An extensive study by Poore and Nemecek (2018), which aggregated data from numerous lifecycle assessment studies on thirty-eight broad food categories from more than thirty-eight thousand farms and sixteen hundred processors around the world, reports that 26 percent of anthropogenic greenhouse gas emissions are linked to food—a total of 13.7 billion metric tons of CO_2eq. This study focuses primarily on farm-to-retail and does not account for consumer behavior, including storage, transportation, preparation, and waste. The IPCC (2019) estimates the food's share of GHG emissions to be between 21 and 37 percent. More recently, Crippa et al. (2021) showed this contribution to be about a third of all emissions. Regardless of the exact percentage, food is clearly a major contributor to climate change, perhaps more important than any other economic sector.

Breaking down the emissions from each of the main stages throughout the lifecycle of food, as illustrated in figure 8.3, highlights the most significant sources and potential areas of intervention. Most of the greenhouse gas emissions (58 percent) originate in the *production* process itself, whether it is crop cultivation or livestock production. This includes emissions linked to ruminant enteric fermentation, fertilizer use, irrigation, rice production, fossil fuel burning to power equipment, manure management, and manure or crop residue left on the fields. In terms of specific greenhouse gases, the largest share of emissions consists of nitrous oxide (mostly from fertilizers and cultivation), followed closely by methane (enteric fermentation, rice production, and manure management). Carbon dioxide net emissions in the production stage are relatively small because releases from on-farm energy use are partly canceled by the sequestration of carbon into the soils through photosynthesis of cultivated plants.

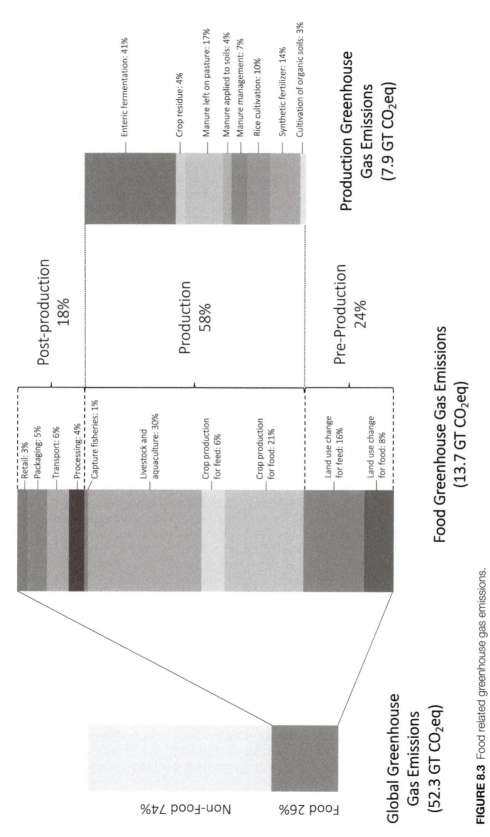

FIGURE 8.3 Food related greenhouse gas emissions.

Source: Author with data from Poore and Nemecek (2018) and FAO (2020f).

Important variations exist between regions. For instance, N_2O emissions linked to fertilizers are especially high in China and India, where agrochemicals have played a key role in agricultural intensification (see chapter 7). CH_4 released in enteric fermentation and from manure left on pastures is a significant component of GHG emissions in Latin America and Africa where cattle feeding on grazing land takes longer to mature. In contrast, in Europe and the United States, where animals are raised in concentrated farming operations and given drugs and feed to grow faster, manure management represents a larger share of total food-related emissions than elsewhere. CO_2 emissions are higher in Latin America, Asia, and Africa where deforestation for agriculture is significant.

The *preproduction* stage is the second most significant contributor. It includes land use change for agriculture, such as deforestation, savannah burning, and forest and peat land degradation. Together these represent 24 percent of food GHG emissions, or more than 6 percent of overall emissions. Many studies confirm that most land clearing is motivated by the expansion of livestock production, especially beef. Palm oil, soy, and other commodities also play an important role (see textbox 7.6). Latin America, Southeast Asia, and Africa are areas that are most severely impacted by deforestation. In addition to the environmental effects on biodiversity, soils, and water cycles discussed in chapter 7, one of the major consequences of deforestation is the release of CO_2 from soil carbon sinks as well as the removal of trees that would otherwise contribute to carbon sequestration into the soil through the roots of plants and the organic decomposition of dead plants.

Burning of overgrown savannah, felled and dried forests, and crop residues, which is a common way to create pastures for animals and prepare fields for crops, generates large amounts of CO_2. As recent events in Brazil indicate, these fires can become unmanageable blazes and lead to extensive forest destruction. In August 2019, seventy-six thousand simultaneous fires were observed in the Brazilian Amazon by satellite imagery—80 percent more than the previous year (Borunda 2019). A total area of almost eight thousand square kilometers had been cleared by the end of the dry season in October. Other similar fires were reported in Peru, Bolivia, and Paraguay that year.

At the *postproduction* level, which is often described as the supply chain stage, most of the GHG emissions consist of CO_2 associated with processing, packaging, and transportation to retailers. This postproduction stage tends to be more important in wealthier countries, where food is more likely to be processed, packaged, and refrigerated. Refrigeration in food stores is particularly inefficient, especially when doorless freezers and refrigerators are used. Although transportation has received significant attention from activists and policy makers, it represents a relatively small share (just 6 percent) of food-related GHG emissions. The focus on reducing "food miles" may be a result of the fact that early research on climate change privileged CO_2 and viewed transportation as the leading cause. A better understanding of GHG emissions, including the role of methane and other potent gases, has revealed other components of the food system, such as deforestation, meat production, fertilizer applications, and waste, that make relatively larger contributions to climate change.

Poore and Nemecek's (2018) estimates shown in figure 8.3 do not include emissions generated after food reaches retailers and very few studies include data on the consumption stage. Garnett (2011) estimates that, in the United Kingdom, energy use related to transporting, refrigerating, and cooking food in homes and catering businesses increase food GHG emissions by 13 percent. These emissions, however, are likely smaller in developing countries where refrigeration, transportation, and use of fossil fuel are less prevalent.

One postproduction aspect that has generated significant attention in recent years is *food losses and waste* (FLW). Food losses occur at many stages of the production process, including harvesting, storage, processing, and transportation, when damaged, contaminated, or excess food is discarded. Food waste takes place in the postproduction stages after it reaches retailers and consumers. Food waste disposed in landfills decomposes anaerobically and generates direct emissions of methane (CH_4) and carbon dioxide (CO_2). Waste disposal generates less than 1 percent of the global carbon footprint of food. The largest impacts, however, are indirect and come from the embedded GHG emissions generated throughout the lifecycle of food products purchased by consumers, restaurants, and caterers that are not eaten. Globally, it is estimated that about 24 percent of food is lost or wasted, representing about 4.4 billion metric tons of CO_2eq or 8 percent of all GHG emissions (Lipinski et al. 2013)—more than emissions from all air traffic or any single country except the United States and China.

Postproduction waste is the largest source of food waste and loss, especially in affluent countries. Yet it could be almost entirely avoided with better planning and management. The further products are in their lifecycle, the more embedded GHG emissions they carry. For instance, unharvested tomatoes left on the vine may represent a loss of potential food, but they also easily decompose into compost. In contrast, a discarded jar of expired tomato sauce represents much more embedded GHG and natural resources. Geographically, production-stage food losses are more common in low-income countries where the lack of equipment, infrastructure, and refrigeration leads to losses in the production, handling, and storage stages. In contrast, consumption-stage food waste is much more extensive in high-income countries, where consumers and retailers discard large amount of uneaten food. In the United States, consumer waste accounts for 61 percent of all food lost and wasted, that is 26 percent of all food produced, worth approximately $1,500 per capita per year. Average food wasted at the consumption end is significantly lower in other parts of the world and decreases with income: 11 percent in Europe and in industrialized Asia, 4 percent in Latin America, 2 percent in South and Southeast Asia, and 1 percent in sub-Saharan Africa (Lipinski et al. 2013). In addition to contributing to climate change, food loss and waste also consumes about one-quarter of agricultural water and lands, adding to the tremendous environmental stress discussed in the previous chapter.

Accounting for these various stages in the lifecycle of food, from pre- to postproduction, draws attention to important geographic differences in the way food is produced, processed, stored, distributed, prepared, and consumed around the world. If we consider food production, Latin America and Asia are major contributors to GHG emissions, especially if we include land use change. However, if we look at food consumption, a different picture emerges in which the GHG emissions of more "efficient" regions like North America, industrialized Asia, Oceania, and Europe are higher because of food imports and consumer waste. This points to the importance of considering diets and food cultures in addition to farming approaches, supporting calls made in chapter 5 for reconciling production and consumption in the study of food.

8.3. Diets and Carbon Footprints

Because different types of food have very different carbon footprints, it follows that dietary patterns and food practices influence GHG emissions. Data presented in figure 8.3 shows that, globally, livestock production causes the majority of greenhouse gas emissions in the production stage, especially once the indirect effects of animal feed production and changes in land use are taken into account. Thus, in recent years, scientists and environmental activists have raised the alarm regarding the growing global consumption of meat.

Indeed, meat—especially red meat—is the most carbon-intense food in the human diet, whether we look at CO_2eq per pound, per calorie, or per unit of protein. Figure 8.4 shows the mean carbon footprint estimates of major food items. Because there are different ways to produce each food item, there can be important variations in their carbon footprint. These variations can be seen in the differences between the tenth and ninetieth percentile values, which illustrate low-impact and high-impact poles. For instance, on average, coffee has a relatively large carbon footprint of 28.5kg CO_2eq per kilo. However, the most climate-efficient 10 percent generates less than 5.2 kg CO_2eq, while the least efficient 10 percent causes the release of more than 84.9 kg CO_2 eq, more than the average carbon footprint of beef.

Although some foods, such as meat and animal products, have a larger average carbon footprint than others, where and how they are produced matter. For example, some cheeses have a larger footprint than sustainably produced lamb or beef. Yet, even the most climate-friendly beef has a larger footprint than any seafood, white meat, or vegetable protein (e.g., tofu) alternative. Among non-animal foods, nuts, rice, and vegetable oils have larger footprints than fruits and vegetables. At the most climate-friendly end, certain fruits, nuts, beans, and peas, even have a small negative carbon footprint because they remove CO_2 from the air and require very little fertilizer, especially if they are planted according to agroecological principles as illustrated by the "three sisters" of corn, beans, and squash (see textbox 7.5).

Equipped with this information, it is possible to assess and compare the footprint of specific diets and explore the effects that dietary changes might have on GHG emissions. In the average current global diet (see figure 8.5), meat and dairy accounts for three-fourths of all food related GHG emissions—that is almost a fifth of total GHG emissions and more than what is released by the entire transportation sector (e.g., cars, ships, trucks, airplanes). Cow products represent more than half of these

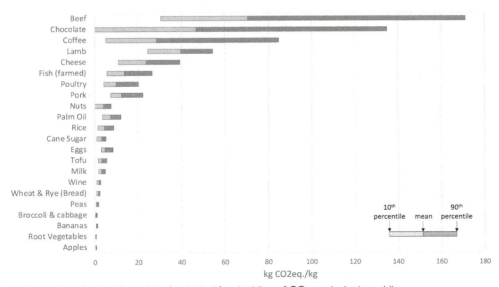

FIGURE 8.4 Carbon footprint of selected foods: kilos of CO_2 equivalent per kilo.

Source: Author with data from Poore and Nemecek (2018).

emissions. Indeed, in terms of GHG emissions, it is argued that one quarter-pound hamburger patty is the equivalent of driving a car for 80 to 100 miles. Unfortunately, beef production and consumption have been rising steadily since the 1960s, especially in countries like China and Brazil, where dietary patterns are increasingly resembling those of the United States and other OECD countries, where most protein consumption comes from animal sources (Ranganathan et al. 2016). Today, given its population size and quickly expanding middle-class, China is the single largest consumer of animal-based food.

These numbers suggest that reducing meat consumption globally might have a very positive effect on the reduction of GHG emissions. This is especially true in the United States where per capita consumption is the highest. Figure 8.5 summarizes findings from a recent World Resources Institute study (Ranganathan et al. 2016) that compares the effects of current US and world diets with a low-calorie diet (designed to eliminate obesity and halve the overweight population), a reduced red meat diet (that cuts red meat consumption by 75 percent), a so-called "Mediterranean" diet (that relies heavily on fruits and vegetables, whole grains, nuts, olive oil, and a variety of animal-based foods), and a vegetarian diet (that excludes animal meat but includes eggs and dairy).

These comparisons reveal several interesting facts. First, if US consumers were to start eating like the rest of the world, GHG emissions from food would be cut in half. This is because most of the world consumes significantly less animal-based food and more rice, root vegetables, and tubers. Global per capita calorie consumption is also lower at 2,433 (compared to the 2,904 average for the United States). Second, reducing the overconsumption of calories to what is deemed a healthy level would not only improve American health but also have a mild positive environmental impact. So would a switch to a more diverse diet like the Mediterranean one. Third, adopting a vegetarian diet would have the greatest positive impact on GHG emissions. Unfortunately, the World Resources Institute study did not include any information on vegan diets that eliminate all animal-based food, including eggs and dairy. Other studies suggest that such a diet would be even more impactful in reducing GHG emissions. Fourth, even small changes, such as reducing beef consumption, would have a significant positive impact.

These charts are based on average figures. However, it is important to remember that GHG emissions vary dramatically for any given product based on how and where it is produced. Thus, while reducing beef consumption is an effective way to cut GHG emissions, a switch from conventional to more sustainably produced animal products could also help move the needle in the right direction. Conversely, if switching to a vegetarian diet means greater consumption of unsustainably grown plant-based foods, then the positive impact may not be as large as expected.

Another issue not reflected in figure 8.5 is the postproduction emissions that result from processing, storing, transporting, and preparing food. Because the data reflect consumption, waste is also ignored. Adding these lifecycle stages could alter the effects of different diets on total per capita GHG emissions.

On the surface, changing diets and food practices seems like a pretty simple way to address the climate crisis. However, many of these habits are deeply engrained in our cultures and identities (see chapter 10). For instance, suggestions by representatives of the US Congress that we ought to collectively reduce our beef consumption as part of the proposed Green New Deal have been met with fierce resistance, including a politically motivated misinformation campaign claiming that "Democrats want to take away [your] hamburgers." Changing diets will take a combination of education, incentives, and regulations as we will discuss below.

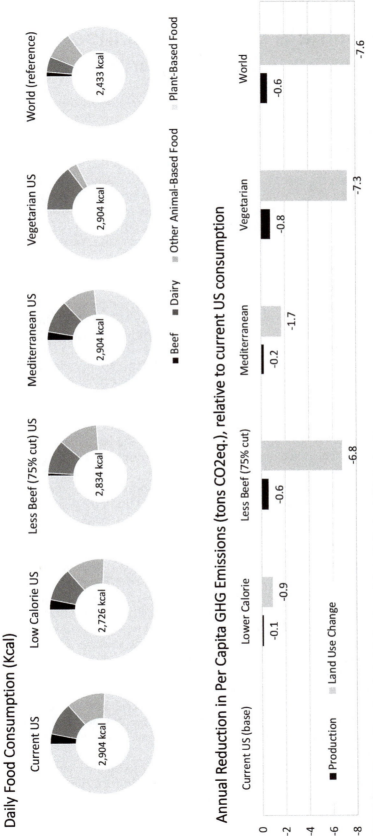

FIGURE 8.5 Per capita carbon footprint of various diets and effects of dietary change.

Source: Author with 2009 data from Ranganathan et al. (2016).

8.4. The Climate Crisis: Food Insecurity, Involuntary Migration, and Conflict

Climate change, whether caused by the food system or other economic sectors, is already having significant impacts on agriculture, exacerbating the degradation of land, water, and biodiversity discussed in the previous chapter. It is useful to distinguish between short-term and long-term effects. The former are caused by sudden-onset events like tropical storms, floods, and pest invasions that usually do not last but are predicted to occur with increased frequency and intensity in the future. The latter result from gradual environmental change such as rising temperature, soil salinization, ocean acidification, desertification, changes in precipitation, and sea level rises—sometimes described as a form of slow violence. Although one is temporary and the other seems more permanent, both disrupt production, weaken food security, and threaten food-based livelihoods.

Globally, researchers have observed reduced yields in crop commodities like corn, soy, and wheat in the past two decades, which they very confidently link to climate change (IPCC 2019). For example, in India, temperature warming has reduced wheat yields by more than 5 percent between 1981 and 2009. In Africa, numerous studies report losses of productivity in staple crops like maize, wheat, and sorghum due to extreme drought conditions, especially in the Sahel region that stretches horizontally across the continent from Senegal on the Atlantic side to Ethiopia on the Red Sea. In Europe, some countries seem to have benefited from rising temperatures and/or increased precipitations (e.g., England, Ireland, Poland, Czech Republic), while others have experienced drying trends (e.g., Italy, Greece). Australia has witnessed drastic losses in productivity since the 1980s due to rising temperatures and declining rainfalls. In contrast, Northeastern China has experienced rising rice and wheat yields, which may be partly attributed to rising temperatures.

Under current climate trends, global crop yields are projected to continue to decline, with important variations between regions and more severe negative impacts expected in South and Southeastern Asia, the Middle East, Africa, Oceania, Central and South America, the Caribbean, Southern Europe, and most of the United States—practically everywhere around the world except for a few northern regions and other outliers. Rough predictions suggest that, in tropical regions, every single degree Celsius rise in temperature will lead to a 10 percent reduction in crop yields. The IPCC found that global warming of 1.5 degrees Celsius above late-twentieth-century levels, which are on the lower range of global warming predictions, would result in negative effects on global yields that are likely to accelerate after 2050. It would also alter land use patterns because some areas would become unsuitable for certain crops while others would become suitable. For instance, wine production in North America and Europe is predicted to shift north, with new vineyards being established in England and Canada, while those in California and southern Europe will continue to struggle with warmer temperatures and less water. Livestock may also suffer from warmer temperatures that decrease forage quality and quantity, increase water needs despite lower availability, intensify heat stress, impair reproduction, and create new diseases, resulting in less milk production, slower growth, lower weight, and even mortality.

The stress and uncertainty brought about by climate change is having a severe impact on food security and livelihoods, which in turn induces migration. Smallholder farmers who depend on crops and animal production to feed and support their households are particularly vulnerable to climate change, especially when there is little diversification in farming activities, few other sources of income, and limited social protections. Throughout the world, farmers are beginning to feel the effects of climate change.

For example, the region of northern Africa and the Middle East is particularly vulnerable to climate change. Although there are geographic variations, researchers are expecting rain precipitations to decrease, extreme events such as prolonged droughts to become more frequent, and temperatures to rise by 4.8 degrees Celsius by the end of the twenty-first century (Lewis, Monem, and Impiglia 2018). These changes have dire consequences in terms of food production and food security. The IPCC (2014) projected a 10 to 20 percent reduction in crop yields in the region by 2050. In places like Yemen, Sudan, Oman, Mauritania, and Libya, up to 80 percent of livestock populations are vulnerable to climate change, threatening the livelihoods of herders. Fish catch in the southeastern Mediterranean is also expected to drop by 50 percent in part because of warming oceans and overexploitation (see chapter 9).

These trends pose major threats to already weak food systems, which have been undermined by poor resource management, limited security in land tenure, low investment, conflict, and social inequality, including gender inequality (see textbox 8.1). These conditions will likely lead to greater food insecurity and depressed food livelihoods. Unfortunately, they will also promote less efficient use of resources, such as more water withdrawals and forest clearing, reduce investment in diversification and climate mitigation, trigger survival strategies such as selling livestock, and increase dependence on imports, which will likely exacerbate the problem. All of this heightens social tensions and the risk of conflict. In that context, migration is often the only option to survive.

These trends worry the United States Department of Defense that views climate change as a source of instability and a "threat multiplier" that "may increase the

Box 8.1   Gender and Climate Change

Gender, along with age, race, and other social factors, influences how climate change is experienced and the ability to respond and adapt to its effects. Women, especially in developing countries, have less power than men in accessing resources, controlling assets, making decisions, participating in politics, and having their work recognized and valued. As a result, they are disproportionately vulnerable to climate change.

For example, in many parts of the world, women are responsible for fetching water. As droughts and rising temperature increase water scarcity, women must walk longer distances to gather water. Yet, because they are also responsible for taking care of children, they may not be allowed or able to travel far. This threatens their subsistence farming activities and might force them to take low-pay wage work. In many households, women work in subsistence agriculture while men cultivate cash crops. When climate pressures increase, tensions arise between these two sectors, often resulting in a reallocation of scarce resources toward cash crops, depriving women of their livelihoods and increasing household and community food insecurity.

When climate disasters hit, men are more likely to migrate than women, who must stay behind to take care of children and the elderly. While this migration might bring remittances, it also increases the burden and vulnerability of women who must take on men's work in addition to their own work under very difficult circumstances that also make women more vulnerable to sexual harassment and violence.

Because of their role as primary keepers of traditional knowledge in their communities, women hold valuable information on how to adapt to climate change, including through agroecological practices. However, their knowledge is not often valued in policy circles where they rarely have a place at the table. Evidence from North America and Europe suggest that women are more concerned by climate change and willing to take actions to combat it. Empowering women to contribute to policy and make decisions in everyday activities would strengthen adaptation capacity.

frequency, scale, and complexity of future missions" (cited in Sova, Flowers, and Man 2019, 4). For instance, the effects of climate change and food insecurity in the Sahel have contributed to social, economic, and political instability, fueling groups like Boko Haram, al-Qaeda, and Al-Shabab. Such unrest is likely to further destabilize livelihoods and encourage migration.

It is estimated that almost thirty million people are displaced each year because of climate disasters—a number that has been rising and is expected to surpass four hundred million by 2050 (FAO 2017b). Sudden-onset disasters like floods and tropical storms trigger mass movement of people over a short period of time, many of whom return later to rebuild. For example, a recent typhoon in the Philippines, a drought in Pakistan, a cyclone in Samoa, a hurricane in the Bahamas, and giant bush fires in Australia—all of which are linked to climate change—have destroyed harvests and infrastructure, severely hindering rural livelihoods and forcing millions of people to migrate. Slow-onset climate disasters, such as rising temperature, declining rainfall, and rising sea levels, have more gradual effects on migration through the destruction of agricultural livelihoods and the rise of food insecurity. These impacts are more complicated to assess than those of sudden-onset climate disasters. Because other interrelated factors such as political instability, ethnic tensions, poverty, and conflict exacerbate *climate migration*, it is not always recognized as such, making it difficult to generate consistent estimates and the political will to assist *climate refugees*.

It has been argued that food is the primary driver of migration—the urge to survive and feed one's family. While food insecurity produces unvoluntary displacement, such displacement increases vulnerability and causes hunger among both those migrating and those left behind—often women and children who have less control over resources. Managing flows of international migration, so that migrants are safe and receive necessary assistance wherever they go, will require international cooperation. The resurgence of anti-immigration sentiments in many high-income countries, particularly in the United States and certain European countries, suggests that this may be a difficult task. Ironically, those most opposed to migration also tend to be skeptical of climate change, refusing to adopt policies that would reduce GHG emissions and potentially prevent future migration.

Migration is becoming an increasingly important adaptation and survival strategy in many regions experiencing negative effects of climate change on their food systems, including the Middle East and North Africa, Small Island Nations, and the Northern Triangle region of Central America which comprises Guatemala, Honduras, and El Salvador (see textbox 8.2). Since January 2020, the United Nations' refugee agency recognizes climate as one of the key drivers of unvoluntary displacement, granting climate refugees protection to the extent that climate change and disasters violate their basic human rights to life. While some climate refugees migrate internally, often temporarily or seasonally, others travel across international borders, hoping to find a better life in cooler places. Being admitted in a country as a refugee is generally a very challenging process that often takes many years. Given the lack of protocols and understanding about the effects of climate change, being granted refugee status on the basis of climate has proven to be even more difficult.

8.5. Reducing Food's Carbon Footprint

Incriminating evidence regarding the contribution of food to GHG emission is reframing climate policy by drawing attention to the ways we produce, distribute, and

| Box 8.2 | **Climate Change and Central American Migration to the United States** |

Central America has been severely affected by climate change, including extreme El Niño events in the so-called "dry corridor"—a region covering most of El Salvador, Guatemala, Honduras, Nicaragua, and parts of Costa Rica and Panama. El Niño is a cyclical climate phenomenon that occurs every three to seven years when surface temperatures over the Pacific Ocean increase and alter atmospheric flows around the globe, with more intense effects in particular regions such as the Pacific Coast of Central America. With global warming, El Niño events have increased in frequency and severity, leading to recurrent droughts, followed by excessive rains and severe flooding that negatively affect agricultural production, especially on already degraded land. Scientists are predicting that yields of many crops grown in the dry corridor will continue to decline in the next fifty years, causing massive food insecurity.

In that context, the increase in northern migration from Central America should not be surprising. About 250 thousand people migrate from Guatemala, El Salvador, and Honduras each year, taking a dangerous journey across Mexico, often traveling on top of freight trains. The number of apprehensions of Central American migrants at the US-Mexico border increased from 50,000 in 2010 to 408,870 in 2016. While pundits often blame violence for the surge in migration, evidence gathered by the United Nations' World Food Programme (WFP 2017) reveal that employment loss and food insecurity linked to crop failures are more significant drivers. Indeed, most migrants are men who worked in agriculture before migrating. Researchers interviewed people who had a household member migrate during the 2014–2016 drought and found that almost half of the respondents were food insecure. The vast majority were engaging in emergency coping strategies such as selling land or other assets and many were becoming heavily indebted, jeopardizing their ability to farm in the future and becoming less resilient. Eighty percent of households earned below-poverty incomes and many relied on remittances from migrant family members to support themselves.

If carbon emissions subsist and international borders remain closed to climate refugees like those escaping Central America's dry corridor, urban areas in these countries will become overpopulated and rife with social tension, agriculture will collapse, and hunger and human suffering will increase (Lustgarten 2020).

consume food. Reducing the carbon footprint of food could go a long way in helping bring climate change under control and thereby improving food security and strengthening rural economies. As is the case with preservation of water, soil, and biodiversity in agroecosystems, there are different visions on how to best address the climate crisis. Scholars and policy makers often distinguish between *mitigation* and *adaptation*. The former refers to efforts to reduce the level of GHG in the atmosphere, while the latter addresses the need to adapt and increase resilience to ongoing global climate change. This section focuses primarily on mitigation.

Some believe that new technologies will allow us to feed the next generations while reducing and possibly absorbing CO_2 emissions. Others are confident that consumers can make a difference by changing their diets, buying locally and sustainably grown products, and reducing waste. However, critics worry that the green economy, where markets and technology come together, does little to alter the political and economic conditions that underlie the quest to grow and earn profit while disregarding environmental consequences viewed as *externalities*. Instead, by giving affluent consumers good conscience, it might actually encourage consumption, especially if the negative impacts of climate change are felt in faraway places. Proponents of this critical perspective often advocate for agroecology and food sovereignty (see chapter 7). Short of systemic change, lifecycle assessments point to several areas, from input production to waste disposal, where GHG emissions could be reduced.

Several *mitigation* measures have been proposed and implemented in various settings, as summarized in numerous publications (see IPCC 2019 and Searchinger et al. 2019). First, to address the effects of land use changes, ongoing initiatives seek to protect and restore natural ecosystems. *Conservation* efforts that protect land from agricultural expansion and prevent deforestation could play an important role in reducing the release of CO_2 from carbon sinks. For example, in 2010, under pressure from the international community, Indonesia imposed a moratorium on new palm oil concessions to preserve primary forests. Regulations against burning savanna could also help curb CO_2 emission. However, given that the world population continues to increase, albeit at a declining rate, and that existing land is becoming less fertile, preventing land conversion to agriculture will likely cause a tradeoff in terms of food security, unless productivity can be increased. Rather than preserve wilderness at the expense of agriculture, some argue that we could use existing forests more sustainably through agroforestry and silvopastures that combine trees, crops, and sometimes livestock to close ecological cycles and preserve resources. In many parts of the world, Indigenous People have done so for centuries. Thus, protecting Indigenous People's territories and practices would help preserve land. *Afforestation* could also rebuild degraded forests while sustaining food production. For example, in Costa Rica, coffee plantations and cattle ranches have been returned to forests through an extensive incentive program to promote agroforestry. A form of "payment for ecosystem services," the program pays farmers a stipend for restoring their land and providing ecosystem services such as carbon sequestration. Money is raised through a fuel tax and the elimination of cattle subsidies, discouraging activities that generate GHG. Today, forest areas have doubled in size, covering about half of the land—up from a quarter in the mid-1980s. Many of these forests include a combination of trees and plants including cocoa, banana, and vanilla (The Food and Land Use Coalition 2019).

Second, the *sustainable intensification* of food production on existing land could also help reduce GHG emissions. Although intensification historically meant greater use of synthetic fertilizers and fossil-fuel dependent machinery, proponents of this approach argue that science and innovations will enable a new and greener Green Revolution in which technology is used to increase productivity and reduce GHG emissions—a sort of "smart agriculture" (Searchinger et al. 2019). Such innovations range from input optimization, including fertilizer microdosing and more efficient irrigation, to methane capture through anaerobic digesters that produce biogas, and bioengineering of animal feed, crops, and even animals. For instance, if transgenic cows could grow faster, reach their optimal weight sooner, and produce more milk, their meat and dairy would have a lower carbon footprint. Presumably, this could be achieved through improved feed, health care, animal management, breeding, and genetic manipulation. Crop productivity could also be enhanced in a similar fashion. These approaches have received significant criticism from people and organizations concerned about ecology, animal well-being, health, and food sovereignty. Yet advocates of smart agriculture dismiss these critiques by claiming inconclusive evidence.

Lab-grown and plant-based meat have received significant attention in the media as technological innovations that could help solve the climate crisis. Booming investment in companies like Memphis Meat, SuperMeat, Eat Just, Impossible Food, and Beyond Meat suggests that such products are likely to become highly profitable as people begin to adapt their diets to include more climate-friendly products. *Lab-grown* meat products, which include animal cells, are slowly being approved for commercialization but remain very costly to produce in large quantities and unaffordable to most consumers. In contrast, *plant-based* meat and dairy are now widely distributed in grocery stores and restaurants, contributing to the growth of veganism in the Global North where one can find meatless burgers and non-dairy cappuccinos at most fast-food restaurants and chain coffee shops. While many are hopeful that such products will help address the climate crisis, others contend that we must look at the bigger picture and consider the industrial farming of components such as soy, the processing required, and the concentrated structure of technology ownership. Some estimates suggest that plant-based meat has a similar carbon footprint as chicken, which is much higher than legumes and vegetables.

Proponents of agroecology and regenerative agriculture contend that better soil and water management could be achieved without having to rely on expensive technologies that promote monoculture, decrease biodiversity, and increase dependency on foreign investments. As discussed in the previous chapter, synthetic fertilizer and pesticide use could be significantly reduced through agroecological practices such as intercropping of plants and trees, crop rotation, and composting, that increase soil organic matter.

Agroecological practices can also promote *carbon removal*. The notions of "carbon farming" and "zero emission farming" suggest that it is possible to sequester atmospheric carbon into the soil and roots of crops to offset most agricultural emissions. This can be achieved by selecting crops known for absorbing carbon; planting perennials, trees, and hedgerows; leaving crop residue on the field; and eliminating or minimizing tillage. Carbon farming has gained recognition in recent years and is being promoted by many organizations, including the World Bank, as one of the most promising ways to reduce the impact of food on climate change. While some support agroecological methods to achieve this goal, others are hopeful that specific crops can be engineered and industrially grown to maximize carbon removal without having to alter production methods.

Cap-and-trade programs have been implemented in various places around the world to encourage farmers to adopt carbon-sequestration practices, by allowing them to sell carbon credits to heavily polluting corporations who need to "offset" their carbon footprints (see textbox 8.3). There are concerns that the expansion of large-scale carbon sequestration projects could reduce the food supply and take land away from small farmers, repeating the ills of concentrated factory farming.

Third, there is much room for improvement in the *postproduction* stages, including processing, packaging, transportation, and refrigeration. Among those, transportation has received the most attention, with many environmentalists focusing their energy on reducing *food miles* by promoting local food systems. The focus on food miles may be misguided, however, if the rest of the lifecycle is ignored. For instance, in the United Kingdom, locally grown greenhouse tomatoes have a carbon footprint three times larger than that of tomatoes imported from Spain, even after accounting for transportation.

The consolidation of retailing into a handful of large supermarket chains (see chapters 3 and 5) and the global expansion of distribution networks and associated infrastructure reinforce our dependence on cheap food imported from distant places and impede the switch to local food. For instance, consumers have grown accustomed to eating strawberries any time of the year. Maintaining year-round supplies requires sourcing from many different countries, especially in the winter, and long-distance refrigerated transportation.

Fourth, the demand for food could be reduced without causing food insecurity. A major criticism of productivist approaches, such as smart farming, is that it takes for granted current consumption trends and the need to increase production to meet future demands. Furthermore, it assumes that climate-friendly production methods will lead to productivity tradeoffs, which in turn will threaten food security. However, as we learned in chapter 6, hunger is not due to a lack of food but caused by systemic forces that make it difficult, if not impossible, for people to secure food. In

| Box 8.3 | **Cap-and-Trade Programs** |

Carbon credit markets, also known as *cap-and-trade* programs, are an increasingly popular approach to reduce carbon emissions. In these policy schemes, actors who reduce greenhouse gas emissions and sequester carbon can sell credits to those who release large amounts of CO_2, offsetting carbon footprints and capping total emissions.

Several national, state, and regional governments around the world have adopted cap-and-trade program in an effort to reduce their total emissions. This provides farmers with a financial incentive to adopt climate-friendly practices to reduce their emissions (e.g., cutting their use of synthetic fertilizers, altering manure management, reducing fossil fuel consumption, and rotating grazing) and/or sequester carbon in the land (e.g., afforestation, no till, crop rotation, and new crops). For example, many farmers in Australia have benefited from the nationally mandated cap-and-trade program. In Kenya, the World Bank created a program to encourage farmers to stop savanna burning practices, providing a credit to those who adopt less damaging practices of land clearing.

The state of California has one of the most extensive cap-and-trade programs in the United States, where no national program exists. It requires emitters across a wide range of industries, from energy to manufacturing and transportation, to offset their emissions by purchasing credits from those who reduce their emissions or sequester carbon. Unfortunately, only a few farming practices are considered for offset credits, including reduction in methane emissions from livestock through dairy digesters. A new protocol for rice cultivation has just been added, encouraging farmers to sow seeds in dry land, drain their fields early, and alternate wetting and drying to reduce methane production. Small-scale farmers who adopt agroecology and other carbon-neutral farming practices are ineligible for credits since their contributions are not recognized as offsets. This points to one of the major criticisms of carbon credit markets: they require the monetary valuation of carbon offsetting activities. Such credits must be measurable, verifiable, permanent, enforceable, and additional. The latter refers to the fact that they must reward activities that go above and beyond "business as usual" to be viewed as worthy of receiving a financial incentive. Implementing these credits often becomes a political exercise that favors certain industries and practices, particularly when data about impacts are limited. Both livestock and rice production generate well-documented methane emissions, which explains why they have been selected for cap-and-trade. However, some worry that the system rewards large-scale producers who invest in new technology and ignores small-scale farmers who together could change the way we grow food.

fact, authors like Holt-Gimenez (2019) and Patel (2012) argue that we produce too much food. Hence, it might be possible to reduce demand without depriving people of calories and nutrients. This could be achieved by reducing food loss and waste, promoting more sustainable (and usually healthier) diets, reducing competition from biofuels that currently sap 27 percent of agricultural production, and limiting population growth. For example, in the previous section, we learned about the potential reduction in GHG emissions that would result from switching from conventional to vegetarian or vegan diets. This could be promoted through a variety of incentives, taxation, regulations, and education campaigns.

Reduction in waste, which mostly take place at the consumer end of the food chain, is another area receiving significant attention. Many initiatives focus on recycling, reusing, and reducing food consumption—the three Rs. Emphasizing how much money consumers waste on uneaten food annually (about $1,500 per person in the United States) might convince them to change their behavior. Working with restaurants and food stores to improve management and help repurpose waste, for example through organized donations to food banks, is another promising avenue. Educating people about expiration dates and the fact that most food remains edible beyond the suggested consumption date is another way to reduce waste, which could be enhanced by changing food labeling practices. Composting could also turn biodegradable food waste into useful organic matter. Several municipalities around the world are requiring consumers to compost household food waste and have organized pick-up to diverting it from landfills, where it would contribute to methane production.

Consumers, especially in countries with a large or expanding middle-class, have an opportunity to make a difference. As Garnett (2011) outlines, we could eat less meat and dairy; reduce our portion sizes while maintaining a healthy diet; eat seasonal, organic, and local foods; select robust foods that require less refrigeration, packaging, and air freight than fragile foods; prepare food for multiple people and for several days; accept less than perfect-looking food; and plan better to limit shopping trips and avoid waste. In addition to reducing GHG emissions, researchers have shown that many of these practices have beneficial health impacts. Yet they remain mostly the prerogative of consumers who have the time, income, and education to adopt them.

In summary, tackling climate change requires a multipronged approach that works at multiple stages of the food system. While there is disagreement regarding how best to achieve reductions in greenhouse gases released in the atmosphere, there is consensus that business as usual is not an option and that even small changes are useful.

Conclusion 8.6.

Food is a major contributor to climate change. Emissions of greenhouse gases occur throughout the life cycle of food, from the input stage to waste disposal. Most emissions are generated in the production stage by fertilizers and ruminant enteric fermentation. In the preproduction stage, deforestation leads to important releases of carbon. In postproduction, once crops and animals leave the farm, significant energy is used to process, package, store, refrigerate, transport, and cook food. Food is lost and wasted at almost every stage, but especially at the consumer end.

Fortunately, food is also part of the solution to the climate crisis. There are many areas where things can be done differently to reduce GHG emissions and even remove carbon from the atmosphere. However, this requires political and social will. Beyond incentives and regulations, it might also require a systemic change that prioritizes planetary health over economic growth. Unless we address these environmental and social concerns, the nexus between climate, food insecurity, and migration is likely to intensify in the coming decades.

Seafood 9

Outline

9.1. Fishing and Aquaculture: Food and Livelihoods
9.2. Overfishing: A Collective Action Problem?
9.3. Aquaculture: A Solution?
9.4. Pollution and Climate Change in Aquatic Ecosystems
9.5. Sustainable Seafood
9.6. Conclusion

Learning Objectives

* Appreciate the importance of seafood in sustaining livelihoods, supplying protein, and shaping costal cultures around the world.
* Analyze causes and consequences of overfishing, highlighting political and economic factors contributing to overexploitation of oceans and fisheries collapse.
* Assess the effects of human activities, including pollution and climate change, on aquatic ecosystems such as coral reefs.
* Explore solutions to protect marine resources and produce seafood more sustainably.

So far, we have paid scant attention to *seafood*—the wide variety of freshwater and marine animals caught or farmed for human consumption, such as finfish (e.g., tuna, salmon, trout), crustaceans (e.g., lobster, shrimp, crayfish), mollusks (e.g., oysters, mussels, clams), cephalopods (e.g., octopus, squid), sea turtles, frogs, and urchins. This neglect reflects a general bias in food studies that privileges land-based agriculture, but also a conscious choice on my part to give undivided consideration to seafood in this chapter, including its relationship to the environment, its cultural and social significance, and its importance for food security. Here I will use the terms *fish* and *seafood* interchangeably to describe any aquatic animals consumed by humans, excluding mammals, crocodiles, alligators, and caimans. While fishing is part of the food system and faces similar structural problems as agriculture, it is also unique in many ways, including its physical, political, and social geographies.

Fisheries and aquaculture are critical sources of food and protein globally, with average per capita fish consumption close to twenty kilograms per year. They also provide income and livelihoods for hundreds of millions of people around the world and are the foundation of countless cultures. Yet, many of the issues we learned about in previous chapters, including social, economic, and environmental concerns associated with the production of food, apply uniquely to seafood. Specifically, there are signs that the world's fisheries are suffering from overexploitation, habitat destruction, and the effects of climate change on oceans, rivers, and lakes. Despite increases in production, particularly in aquaculture, it is estimated that the marine population has been halved since 1970, impacting food security for people whose diet depends on seafood. The decline of many fisheries and the significance of seafood must be understood dynamically within the context of the global food system and the specific local conditions and circumstances in which they are embedded.

This chapter begins by describing the importance of seafood in diets, cultures, and economies. We then turn our attention to the dual problems of overfishing and climate change and highlight how human activities and political economic factors have contributed to them. We conclude with examples of sustainable seafood.

9.1. Fishing and Aquaculture: Food and Livelihoods

For thousands of years, humans have relied on fishing for food. Fish is an important source of protein and fatty acids critical for brain development. In 2018, seafood contributed 17 percent of animal protein consumption globally. Approximately 3.3 billion people obtain at least 20 percent of their animal proteins from seafood (FAO 2020f), mostly in coastal areas, island nations, and African countries with large lakes (see figure 9.1). About two-thirds of seafood comes from marine waters, with the remaining third coming from inland waters, including rivers, lakes, and estuaries, such as the Mekong, Nile, Ayeyarwady, Chang Jiang (Yangtze), Brahmaputra, Amazon, and Ganges river basins and the African Great Lakes, including Victoria, Tanganyika, and Chad. Today almost half of the seafood consumed in the world is farmed, up from less than 5 percent in the 1970s when most seafood came from capture fisheries (WWF 2016). According to the FAO (2020f), in 2018, eighty-four million tons of seafood came from marine captures (47 percent of total), twelve from freshwater captures (7 percent), fifty-one from inland aquaculture (29 percent), and thirty-one from marine aquaculture (17 percent). The most common seafood captures are anchovies, pollock, tuna, and herring. Although crustaceans and mollusks are less important in volume, shrimp and lobster are very valuable in terms of sales. The most harvested aquaculture species are carps of various types, tilapia, Atlantic salmon, shrimp, small-neck clams, and oysters.

Globally, annual per capita consumption of seafood rose from six to twenty kilograms between 1950 and 2018. Seafood consumption is higher and rising at a faster pace than the consumption of beef, pork, or even poultry. However, it varies significantly around the world (see figure 9.1), impacting local economies, cultures, and food security. Nordic countries like Iceland and Norway, with their access to the Atlantic Ocean and the North Sea, have some of the highest consumption. People on small-island nations such as the Maldives, Kiribati, Samoa, Antigua, and Barbados are also heavily dependent on products form the sea. East Asian countries, including Japan and South Korea, and Southeast Asian countries, like Malaysia, Myanmar, Indonesia, and Vietnam, consume more fish than average. Seafood is also an important part of the diet in Mediterranean countries such as Portugal and Spain. In the United States, seafood consumption is well below the global average, having stabilized at around seven kilograms per capita for the past thirty years, with 25 percent consisting of canned tuna. Not surprisingly, low-income landlocked countries like Afghanistan, Mongolia, Bolivia, and Niger have a very low fish consumption.

Within countries, people living in coastal regions tend to consume more seafood. For instance, in the United States, many coastal states are known for their seafood dishes, such as lobster rolls in Maine, crab cakes in Maryland, and crawfish étouffée in Louisiana (see figure 1.5). These specialties reflect fish or shellfish availability and continue to shape local diets, cultures, and economies. Historically,

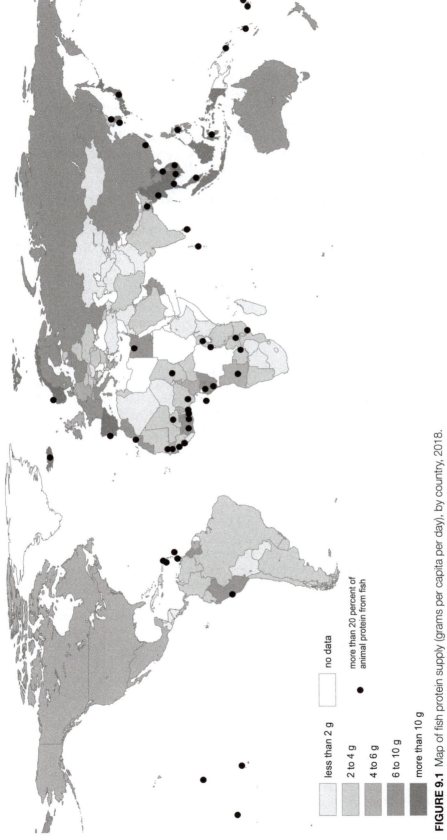

FIGURE 9.1 Map of fish protein supply (grams per capita per day), by country, 2018.

Source: Author with data from FAOSTAT (2020g).

less than 2 g

2 to 4 g

4 to 6 g

6 to 10 g

more than 10 g

no data

● more than 20 percent of animal protein from fish

Indigenous coastal communities have relied on products from their natural environment to feed themselves. A recent study estimated that coastal Indigenous peoples with cultural ties to the ocean eat approximately fifteen times more seafood per capita than non-Indigenous peoples in their countries (Cisneros-Montemayor et al. 2016). For example, in Alaska and along the northwestern coast of Canada, salmon is an important part of Indigenous diets, culture, and spirituality, even as it became a globally exported commodity.

Dependence on seafood is higher in low-income coastal regions where poverty and food insecurity are common and there are few other sources of nutrition and income than fishing. For example, in Senegal, which suffers from high overall food insecurity, fish provides an important source of sustenance (23 kg per capita in 2018) and protein (49 percent of animal protein). It also enhances food security indirectly through employment and income. Other countries where dependence on fish is high include Ghana, Sri Lanka, Bangladesh, Thailand, Indonesia, and many small island states, among others. A few landlocked African countries like Malawi, Rwanda, and Uganda rely heavily on lake fish for protein. When fisheries in these dependent regions deteriorate, the impacts on food security are devastating. In contrast, some of the world's biggest seafood consumers are not "dependent" on seafood to the extent that they have alternatives sources of nutrition and employment.

As fishing and refrigeration technologies improved over time, making it possible to freeze and keep large quantities of fish on ships for days, seafood consumption became geographically disconnected from consumption. In many coastal regions, especially in the Global South, a significant portion of catches and harvests is exported. For example, Peru, Chile, and Ecuador are major exporters of seafood, shipping more than three-fourths of their production abroad, mostly along the Pacific Rim. Vietnam and Thailand, despite their high dependence on seafood for food security, are following a similar path. In the Northern Hemisphere, Norway, Denmark, and Iceland export more than 85 percent of their production. Although exports represent a small share of its total production, China is the leading exporter in the world in terms of volume, trading over nine billion tons in 2017. Most of these seafood exports go from the Global South to the Global North, including the United States, Japan, and the European Union. In the United States, between 70 and 85 percent of the seafood consumed is imported, primarily from China, Thailand, Canada, Indonesia, Vietnam, and Ecuador (NOAA 2020). In Europe, more than half of the seafood consumed is also imported. In Japan, imports make up about half of the annual per capita seafood consumption of twenty-four kilograms. Salmon from Chile, Mackerel from Norway, Shrimp from Indonesia, octopus from Mauritania, and lobster from Mexico are found in Japanese markets and restaurants (see textbox 9.1).

Low-income countries have been pressured to export their most valuable seafood commodities to gain foreign currency. As a result, their food security has been compromised and their foodways altered. For example, as illustrated in textbox 6.3 in chapter 6, fish consumption has dropped and obesity has risen in Pacific islands such as Tonga, Samoa, and Fiji, where growing exports of seafood have forced people to eat more imported processed food. Coastal Indigenous people around the world have lost access to subsistence seafood due to colonization, regulation, and competition with large and often international fishing fleets or aquaculture projects that capture or harvest fish for exports.

In addition to direct sustenance, seafood also supports *livelihoods*. Like in agriculture, industrialization is gradually replacing subsistence labor with wage labor in fishing and aquaculture. In 2018, there were sixty million jobs globally in this primary industry: thirty-nine million people in capture fishing and twenty-one million in aquaculture. Asia has by far the most people engaged in fisheries and aquaculture,

Box 9.1   Tokyo's Fish Market

The Tsukiji market opened in 1935 in central Tokyo and became the largest wholesale seafood market in the world. It was built by the government to replace a smaller market that had been destroyed in the 1923 earthquake and was part of a public program to address concerns about food shortages and price gouging by wholesalers that had led to riots a few years earlier. In 2018, it was closed because of decaying infrastructure and moved to the Toyosu market in a nearby district.

The market included the "inner market" where about fifteen hundred tons of seafood were shipped in daily from all over the world and auctioned each morning to licensed buyers (see figure 9.2) and the "outer market" which consisted of restaurants and stalls selling kitchenware, dried goods, and souvenirs and remains partly in operation for tourism. There were about nine hundred licensed dealers working in the inner market and more than sixty thousand employees, including accountants, fish cutters, forklift drivers, and cleaners. Some of the most notorious types of fish traded at the Tsukiji and Toyosu markets include large tunas and whales, which have become controversial and less common. Each year, new price records for the endangered species are broken, going from $1.8 million for a six-hundred-pound bluefin tuna in 2013 to over $3 million for a similar weight animal at the new Toyosu market in 2019.

The new market is cleaner, larger, more modern, and just a few miles away from the old market. However, many of the long-time vendors were reluctant to move, seeing their relocation as the end of an era. In the same spirit, in the months and weeks preceding the relocation, many Tokyo residents purposely visited the Tsukiji market for one last meal. The tuna auction at the Tsukiji market was depicted in the film *Jiro Dreams of Sushi* as a significant place in the operation of a small sushi restaurant that has earned accolades both in Japan and internationally. Tsukiji will be remembered for its controlled chaos, noise, smells, and overall grittiness. Many are questioning the sustainability of the market and of Japan's seafood consumption in general, especially as the proportion of imported fish has risen and local catches have declined over the past thirty years.

FIGURE 9.2 Tsukiji fresh tuna auction, Japan.

Source: https://commons.wikimedia.org/wiki/File:Tsukiji_Fresh_Tuna_Auction.JPG.

accounting for 85 percent of all fishing and fish farming jobs. Nearly half of global seafood employment is in small-scale subsistence or artisanal operations, but the number of people working in larger industrial operations has been growing steadily. Artisanal fishing is more common in the Global South, including Africa, the Middle East, Central America, and the Indian peninsula. Most fishers and fish farmers are men, with women representing only 12 percent of the workforce in fisheries and 19 percent in aquaculture.

Beyond these direct jobs, the seafood industry generates many more indirect employment opportunities, including manufacturing, selling, and maintaining vessels and fishing gear as well as handling, processing, storing, packaging, and marketing fish post-harvest. It is estimated that nearly 820 million people worldwide obtain livelihoods directly or indirectly from fishing and aquaculture (FAO 2019d). Women are more likely to be employed in post-harvest jobs than in fishing or harvesting, which are often perceived as too physically demanding and dangerous for them. Tasks such as cleaning, peeling, smoking, salting, or drying fish or shellfish, selling at local markets, and repairing nets are often performed by women and children, with or without pay. In many coastal areas, seafood structures the social, cultural, and spiritual life of communities.

Numerous reports have raised concerns regarding labor conditions in the seafood industry. Working on fishing boats is notoriously dangerous, especially in harsh weather conditions. However, a bigger risk appears to come from forced labor, slavery, and human trafficking (ILO 2013). The remote location of increasingly large vessels that venture outside of national waters hides labor exploitation and abuse. Distance from the coast also challenges monitoring and the enforcement of existing laws, particularly in low-income countries where the institutional capacity and political will to regulate may be limited. Vulnerable fishers around the world are trapped into what may be considered modern slavery, forced to work for someone else without any or adequate pay and unable to leave because of confinement, death threats, and debt bondage. For example, there is troublesome evidence about the Thai fishing industry, which has become increasingly industrialized and export-oriented and relies heavily on Burmese and Cambodian migrant workers kept in physical captivity and economic bondage (EJF 2019, Urbina 2015, Walk Free Foundation 2019). Although documenting such conditions is difficult, researchers are showing that other countries, including China, Taiwan, Spain, Sierra Leone, the United Arab Emirates, and the United States (in Hawaii) are engaging in similar practices, taking advantage of vulnerable workers and weak governance. The vulnerability of these workers is exacerbated by the destruction of subsistence and artisanal fishing and the overall falling productivity that larger fishing fleets cause via competition and environmental degradation. In other words, the industry maintains its profitability in the face of declining productivity at the expense of its workers and the environment (Tickler et al. 2018).

9.2. Overfishing: A Collective Action Problem?

The dramatic increase in fish production and consumption noted in the previous section has encouraged and been enabled by *overfishing* of wild stocks and the expansion of aquaculture. Overfishing occurs when fish are caught faster than they can reproduce. Since the 1970s, the number of fish in the ocean has been halved and populations of large fish like tuna have declined by more than 75 percent (WWF 2015). Scientists warn that, if trends continue, none of the fish we frequently consume today will be left by 2050.

After decades of growth in productivity, capture fisheries became relatively stagnant in the mid-1980s when signs of degradation and collapse arose, especially in the north Atlantic and the Mediterranean. According to the FAO (2020f), the proportion of fish stock that is fished within biologically sustainable levels declined from 90 percent in 1974 to 65.8 percent in 2017. Of these, the vast majority are at the *maximum sustainable yield* (MSY) level—meaning that catches are at the highest possible level that can be sustained in the long run. Only 6 percent of global fish stocks are underfished; the rest is either fully fished or overfished. If catches go above the MSY, as is the case for 34.2 percent of fish stocks, fishing becomes unsustainable and the fish population begins to decline, leading to potential extinction. This was the case for the Northwest Atlantic cod fishery that collapsed in the early 1990s (see textbox 9.2). In 2020, wild Atlantic halibut, bluefin tuna, winter skate, goliath grouper, red snapper, European eel, Maltese ray, and over 616 other species are considered critically endangered, up from 379 species ten years earlier (IUCN 2020).

Box 9.2	The Collapse of the Northern Atlantic Cod Fisheries

In 1992, the Northern Atlantic cod fishery of Labrador and northeastern Newfoundland collapsed, after having lost 98.9 percent of its biomass in thirty years (Mason 2002). This prompted the Canadian minister of fisheries and oceans to declare a moratorium to halt fishing and allow the fish to recover. Almost forty thousand workers, including fishers and processors, lost their jobs in Newfoundland, representing the single largest employment loss in Canadian history, devastating an already poor and isolated region. A year later, five more Canadian cod fisheries were closed for similar reasons. In response, the federal government established economic adjustment programs to provide training to unemployed workers and help diversify the economy, with little effect on the anger and frustration of local fishers who felt that their livelihoods, knowledge, way of life, and capacity to manage the crisis locally were devalued, leading to frequent protests against the federal government (Kurlansky 1998).

Having sustained communities for centuries, including Indigenous people and early European explorers who described seemingly unlimited supplies of cod teeming around their boats, the Northern Atlantic cod fishery collapsed primarily because of technological innovations such as trawler boats equipped with radars, sonars, mechanized nets, and refrigeration that dramatically increased catch levels beginning in the 1960s. By 1970, landings were above maximum sustainable yields. Yet few or improper restrictions were imposed by the federal government partly because of inaccurate information on actual fish stocks leading to unrealistic quotas. Despite having banned foreign vessels within a two-hundred-mile zone from its coast in the mid-1980s, fish stocks continued to decline due to the overcapacity of Canadian and American fishing fleets. The mismanagement of the Northern Atlantic cod fisheries is often blamed for the collapse, which is used to exemplify how weak governance and limited or suppressed knowledge can have dramatic environmental, industrial, economic, and social consequences.

Decisions to maintain the moratorium or reopen the cod fisheries became extremely politicized, with crucial information withdrawn from the public and promises entangled in election campaigns. Today, there are signs that cod fisheries have partially recovered but scientists remain divided about the extent of the recovery and the resilience of current stocks, particularly in the face of new challenges associated with climate change and warming ocean temperatures. Confusion is often caused by differences between cod fisheries, with those of northeastern Canada still struggling while those of Iceland and the Barents Sea doing much better.

Meanwhile, dropping cod population made space for other fish. For instance, snow crabs and northern shrimp have become more common and are now providing opportunities for new generations of fishers, many of whom seem to have learned lessons from the recent past and are engaged in sustainable fishing.

Overfishing, which typically targets larger fish with higher market value, has reduced biodiversity and altered the mix of marine organisms living in oceans and underlying complex ecosystems. While stocks of large predator fish like tuna, cod, and swordfish have declined, populations of smaller plankton-eating or forage fish like herrings and sardines have become more important along with zooplankton (e.g., fish larvae) and phytoplankton (e.g., single cell algae). Consequently, the mean *trophic level* of world fisheries, which measures average position of catches in the food chain and correlates with their nutritional value, has dropped steadily since the 1970s. For instance, jelly fish populations have increased in many places as other competitors for zooplankton such as anchovies and sardines have been overfished. Unfortunately, jellyfish are not edible and interfere with productive fisheries—a phenomenon that has been observed in Japan, Chile, and the Mediterranean. These changes in biodiversity mean that we are increasingly "fishing down the marine food web" as Pauly and his colleagues first pointed out in 1998.

What causes overfishing? Poor fisheries management and lack of political will, along with insufficient information, are often blamed. Many describe overfishing as a classic case of *collective action failure*, what Hardin (1968) described as a "tragedy of the commons." In that perspective, fisheries—like other finite common-pool natural resources such as forest and water—are prone to overexploitation because there is a lack of individual incentives to curb exploitation. Although it would be beneficial for the community to limit fishing to sustainable levels, this does not hold for individuals who have much to gain from overfishing in the short run. This is because individuals do not experience the negative consequences directly, individually, or in the present. Instead, the entire community faces the burden of declining fish stock in the future. This problem could be resolved through *collective action*—cooperation among concerned parties who bring together their knowledge, resources, and efforts to manage access, use, and conservation of natural resources. Collective action, however, is difficult to achieve because of conflicting interests and powerful incentives to defect, especially when monitoring is difficult—for example, in areas beyond national jurisdictions, two hundred miles off the coast.

Small fishing communities with strong social cohesion and shared culture have been relatively successful at managing their fisheries through participatory governance and cooperative agreements that people understand, value, and enforce themselves. However, to the extent that fishing operations are increasingly transnational, with large industrial foreign vessels competing against artisanal and often unmotorized local fleets, the concerned collectivity is hardly a fishing village. In that context, the interests of local fishing communities are overshadowed and suppressed by more powerful actors who have less at stake in the degradation or collapse of local fisheries. Collective action would need to scale-up to the international level to tackle the disproportionate impact that large modern fishing fleets have on common resources. Although some countries have begun taking an active role in managing their fisheries and international organizations like the United Nations have adopted several treaties to protect fish stocks and oceans, cooperation, monitoring, and enforcement remain limited and sporadic, especially at high seas.

To understand overfishing, it is important to consider why large fleets have become so powerful. Between 1950 and 2015, the number of fishing vessels increased from 1.7 to 3.7 million (Rousseau et al. 2019). Another million boats are expected to join the global fleet by 2050, exacerbating the current *overcapacity*. Compared to the 1950s, today's boats, whether used for artisanal or industrial fishing, are much more likely to be motorized (68 percent vs. 20 percent) and the average engine power has risen from 25 to 145 kilowatts (Rousseau et al. 2019). These trends have been most noticeable in Asia, including China, Indonesia, Vietnam,

India, and the Philippines. In fact, motorization of vessels in the Global South was often brought about by international development projects to increase productivity. Although declining in absolute numbers, fleets in Europe and North America are more likely to consist of large, motorized vessels. In fact, in 2009, average per fisher-person fish production in Europe was 25 tons, compared to 1.5 tons in Asia, where many boats remain unmotorized (WOR 2013).

The introduction of steam trawlers in England in the late 1800s changed the fishing industry by allowing fishers to reach areas further off the coast and drag wide nets through deeper waters. This and subsequent technological innovations, such as the diesel engine and mechanized hauling, increased productivity throughout most of the twentieth century. More recently, electronic technologies such as radars, acoustic fish finders, and global positioning systems have improved the ability to track fish populations and predict weather. Better refrigeration and storage also allow boats to stay at sea longer and flash freeze their catch.

Bottom trawling, which consists of dragging very large nets over the ocean floor, has become the most common fishing method among motorized boats. This technology is extremely damaging to the environment. First, it catches everything in its path, included unwanted fish called *bycatch*. Bycatch includes fish with no market value, protected species, and occasionally fish beyond set quota. For example, tuna fishers often catch dolphins and sea turtles, which are thrown back in the ocean injured and likely to die. Second, bottom trawlers destroy the sea floor, including century-old coral reef and other marine habitats. Indeed, new technology along with declining fish stocks are driving fishing into deeper water, with two-thousand-meter depths increasingly common (WOR 2013). Thirdly, bottom trawlers create pollution and debris noticeable in the large plumes of troubled water following them. Together, these effects threaten marine ecosystems, further reducing the ability of fish to reproduce.

The ongoing damage to fisheries and the overcapacity of existing fleets are witnessed in declining *catch per unit of effort* (CPUE)—a standard measure of productivity and abundance. Despite increasing efforts (e.g., more boats, bigger engines, improved technology, more time at sea), global catch levels have been stagnating, reflecting declining fish stocks. In fact, fishing boats today catch about 20 percent of what boats would catch in 1950 for the same effort (Stokstad 2019). Declining CPUE is especially problematic in southeast Asia, where both industrial and artisanal fleets have been expanding and motorizing rapidly.

The continued increase in fishing capacity in the face of declining stock is partly caused by economic pressures to produce for exports. Overfishing must be understood in the context of globalization and capitalist expansion. As noted above, seafood is a highly traded commodity. Just like Brazil and Argentina turned to GMO corn and soy crops to fuel export-led economic growth, many Global South countries view seafood as a source of foreign currency and employment. Subsidies, including oil subsidies that lower the cost of fueling industrial vessels, have played a key role in encouraging the development of export-oriented fishing. National government's short-term interest in increasing fishing exports outweighs the long-term imperative of curbing production to manage resources.

In addition to unreported and unregulated fishing, *illegal fishing*—in contravention of applicable laws and regulations—accounts for a significant share of overfishing. The impact of these activities is difficult to estimate, but experts believe that it represents up to a third of the total legal catch for an annual value of up to $23.5 billion (WOR 2013). This sort of piracy can be very lucrative since it avoids taxes and regulations, including catch quotas and species restrictions, reporting requirements, and time-space constraints. Vessels engaged in illegal fishing are often the same as those involved in labor abuse and slavery. Indeed, this phenomenon has become increasingly organized at the transnational scale, connecting multiple parts of the seafood supply chain with criminal activities such as drug and human trafficking (Belhabib and Le Billon 2020). The prevalence of illegal fishing points again to governance weakness; national agencies in poor countries lack the capacity to monitor their own seas and international organizations are not better equipped to monitor high seas. Thus, standards and quotas are poorly enforced and piracy thrives, as on the West Coast of Africa (see textbox 9.3).

To stop overfishing and its negative effects on the environment and food security, we must begin by acknowledging the economic factors underlying the urge to fish beyond biologically sustainable levels and the politics associated with poor governance.

9.3. Aquaculture: A Solution?

As wild fish stocks became increasingly unsustainable due to overfishing, aquaculture expanded around the globe. Today, approximately half of the fish consumed globally is farmed in a variety of enclosures made of nets, pens, cages, or ponds in marine, brackish, or freshwater settings. China is the global leader, producing more than half of all farmed seafood, followed by India and Southeast Asian countries, where subsistence aquaculture has traditionally taken place in rice paddies and is now expanding in other inland and coastal waters.

Aquaculture is sometimes called the "blue revolution" in comparison to the "green revolution" that transformed agriculture (see chapter 2). To some, aquaculture is a solution to food insecurity, unsustainable fishing practices, and eroding livelihoods in capture fisheries. Others, however, question the environmental impacts and health risks of aquaculture, which they view as putting productivity and profits ahead of food security and environmental sustainability. Ultimately, the impact of aquaculture depends on the species farmed, the location, the operational size, and the technology employed.

A major concern with aquaculture is nutrient and effluent build-up. As in concentrated animal feeding operations where large numbers of cows, pigs, or chickens are being "factory farmed" (see chapter 8), large aquaculture operations generate a lot of waste from fecal matter and too much feed, which builds-up in surrounding waters, spurs eutrophication, reduces oxygen levels, stimulates algae blooms, and ultimately creates dead zones where fish cannot survive.

There are also related concerns with effluents from drugs such as antibiotics that are fed to the fish and may contaminate the water and be absorbed by other animals and plants. Scientists have also warned that farmed fish may escape ponds, nets, or cages and compete with wild fish for food. This is especially problematic if the farmed fish are non-native or genetically modified species. Indeed, there are now several fish species that have been bioengineered for aquaculture to grow faster and resist disease. This includes a patented "super salmon" approved by the US Food and Drug Administration in 2015, which anti-GMO activists labelled "frankenfish." Along with salmon, transgenic species of tilapia, trout, catfish, striped bass, and flounder, among others, are now being farmed around the world and could potentially threaten wild fish populations. Many of the health and environmental impacts of these new fish remain unknown currently.

Coastal and riparian ecosystems have been altered by the presence of large fish farms. For example, mangroves have been destroyed in Vietnam, Sri Lanka, Mexico, Bangladesh, India, the Philippines, Indonesia, Brazil, Belize, and other countries to

Box 9.3	**Illegal Fishing off the West Coast of Africa**

In West Africa, seafood provides coastal residents with more than half of animal proteins and over 7 million jobs (see figure 9.3). In a context where legally sanctioned overfishing is already pushing fisheries to their biological limits, illegal fishing is threatening both food security and livelihoods. The region has one of the worst records: total catches are estimated to be 40 percent higher than reported catches, costing governments an estimated loss of $2.3 billion in 2015, and reducing the number of jobs in artisanal fishing by three hundred thousand (Doumbouya et al. 2017).

Several recent reports on Mauritania, The Gambia, Senegal, Guinea, Guinea-Bissau, and Sierra Leone provide evidence of illegal activities, including using banned equipment such as nets with narrow mesh size that result in higher bycatch, fishing without licenses, illegal shark finning, using multiple flags and boat names, forging registrations, and relying on smaller boats to offload and launder unreported catch and avoid going back to port (Doumbouya et al. 2017, Greenpeace 2017). Chinese, Italian, Korean, Russian, Spanish, Comoros, and Senegalese vessels have been caught engaging in these illegal activities. The crew members on these vessels are often working in unsafe conditions and likely to suffer from labor abuse.

The lack of transparency, combined with weak governance, corruption, and internal conflicts, is hindering efforts to curb illegal fishing and emboldens unscrupulous operators who come from as far as China and Russia. Currently, fines help recover about half a percent of the lost value. Since fish stocks do not know national boundaries, cooperation between West African nations is critical to solve the crisis and protect resources for future generations. UN regional organizations like the Sub-regional Fisheries Commission and the Fishery Committee for the Eastern Central Atlantic can play an advisory role but lack regulatory and enforcement power.

Artisanal fishers are finding it increasingly difficult to compete with authorized and illegal fishing operations. A full day at sea on unmotorized pirogues no longer yields enough fish to feed fishers' families, let alone to earn a decent income. As a result, undocumented immigration from West Africa to Europe has increased in the past ten years, with some fishing operations being converted into dangerous smuggling businesses. The Gulf of Guinea has also become a hot spot of piracy (Denton and Harris 2019).

FIGURE 9.3 Senegalese beach crowded with people waiting for fishing pirogues. Catches have been shrinking due to illegal and excessive fishing.

Source: https://pxhere.com/en/photo/1169732.

build ponds for aquaculture (UNEP 2014). In the Philippines, where milkfish, carp, and shrimp farms have grown exponentially, less than a fifth of mangroves that existed at the beginning of the twentieth century remain today. After decades of poorly managed growth, many ponds have become so polluted by nutrient build-up that fish and shrimp cannot survive, leading to a phenomenon called "fish kill" which raises concerns about animal welfare. Under these circumstances, coastal inhabitants are left with a degraded environment that no longer provides them with food and income. The gravity of the situation and the economic significance of fishing has prompted the Filipino government and the Asia-Pacific Fishery Commission to develop sustainability plans and has drawn the attention of several nongovernmental organizations.

Invasive species of farmed fish that escape their enclosures can also cause a problem by compromising other species, spreading disease, or taking over the natural habitat. For example, Pacific oysters were introduced in the 1960s in the Netherlands to be farmed as a replacement to overfished native European oysters—a venture that was quickly abandoned. Since then, however, Pacific oysters spread along the Dutch coast in the 1980s, invading the Wadden Sea and forming reefs. The Wadden Sea is a shallow coastal sea that borders the North Sea and stretches between the Netherlands and Denmark. It is considered one of Europe's most important wetlands for migratory birds. Yet, because Pacific oysters have taken over blue mussel and cockle habitat, birds have lost an important source of food and their population is declining (Smaal, van Stralen and Craeymeersch 2005).

Despite these well-documented negative impacts, there is evidence that certain types of aquaculture can be sustainable and have a positive environmental impact. For example, bi-valve mollusks such as clams, mussels, and oysters filter water, removing nutrients including nitrogen stored in algal cells in a process known as *bio-extraction*. Seaweed, which is not included in the seafood category used here, has a similar restorative effect on ocean waters and is an effective carbon sequester. Expanding these types of aquaculture could help mitigate climate change.

Many also point to the fact that, generally, farmed seafood requires less feed than terrestrial animals like cows, pigs, and chickens, because they spend less energy. For example, one pound of farmed fish or shrimp requires on average between 1 and 2.4 pounds of feed compared to 6 to 10 pounds for cows, 2.7 to 6 pounds for pigs, and 1.7 to 2 pounds for chickens (Fry et al. 2018). However, farmed fish are mostly fed fish meal and oil (extracted from smaller fish like anchovies), while other animals consume primarily plant-based feed such as grains, cereal, and oil cakes that have relatively lower carbon footprints. We might be better off eating wild fish than feeding it to farmed fish in order to eat the latter. For example, 1 pound of farmed salmon requires about 3 pounds of fish meal. After accounting for the type of feed and the amount of consumable protein generated, aquaculture is estimated to have a carbon footprint similar to meat and higher than chicken (Fry et al. 2018).

As with agriculture, many stakeholders are hopeful that technology and new knowledge will help address the problems of aquaculture and enhance its benefits. One proposed solution is polyculture or integrated aquaculture, where waste from one species becomes feed for another. This is the case in some rice paddies where fish waste feeds the rice plants (see textbox 7.5) and in closed circuit aquaponic systems where tilapia waste provides nutrients for lettuce, herbs, or other plants. Scientists are also working on projects to improve infrastructure to limit nutrient build-up, identify new sources of feed such as protein from plants and terrestrial animals, and develop transgenic species adapted to these new conditions. Yet this new technology is typically controlled by corporations headquartered in the Global North and does little to curb the economic incentives to produce more and to overfish.

Pollution and Climate Change 9.4. in Aquatic Ecosystems

Threats to the health of oceans, estuaries, lakes, and rivers endanger the fish that live in them and heighten the risk of human food insecurity. Human activities generating pollution and contributing to climate change are causing much stress and damage to aquatic ecosystems. Waterways polluted by industrial, residential, and agricultural activities drain into oceans and alter water quality in a series of interrelated ecological processes, impacting fish directly and indirectly via habitat degradation.

First, the release of nutrients such as nitrogen and phosphorous from human activities causes *eutrophication* of lakes, estuaries, and oceans by promoting algae blooms that block sunlight, kill other plants, reduce oxygen levels in the water, and hurt fish populations. The most dramatic consequences of eutrophication are *dead zones* and *fish kills*, like those observed in the Gulf of Mexico, the Baltic Sea, and the East China Sea.

Second, the decomposition of aquatic plants caused by pollution produces carbon dioxide that lowers water pH levels and contributes to ocean *acidification*. This damages shellfish, which could counteract some of the effects of pollution by filtering water, and coral reefs that provide habitat for many species of fish (see textbox 9.4).

Thirdly, greenhouse gas emissions and climate change are worsening these effects. Oceans absorb a large share of carbon emissions from agricultural, industrial, and residential activities, which increases acidification. Rising temperatures also exacerbate eutrophication and acidification by promoting nutrient leaching and intensifying bacterial activity in the water. Severe weather events and prolonged abnormally warm waters can also have dramatic impact, such as coral bleaching (see textbox 9.4).

Box 9.4	Coral Reefs

Coral reefs are like the rain forest. They are our oceans' most biodiverse ecosystems. However, they are in great danger. Over the past forty years, coral reefs around the world have suffered from ocean acidification and stress due to events like El Niños which have been intensified by climate change and lead to abnormally high temperatures in the Pacific Ocean. These spikes in water temperature cause corals to expel from their own tissues the colorful symbiotic algae that nourish them, leaving just a white "bleached" skeleton. Although some corals can recover from *bleaching*, it takes at least ten to fifteen years. Many end up dying, especially if they are hit by another heat spell before they can recover—an increasingly probable outcome given shortening time lapses between heat events. Because coral reef is a natural habitat for numerous fish and plant species, bleaching has a significant impact on the sea life that depends on them.

In 2015, the South Pacific island nation of Kiribati experienced a ten-month-long heat spell that destroyed 70 to 90 percent of the 144-kilometer-long coral reef fringing Kiritimati, one of its many small islands, also known as Christmas island (Greshko 2018). Local people depend on fishing for their livelihood, with few land-based economic alternatives. The government also depends on fees paid by foreign fishing vessels for about half of its revenue and as a source of foreign currency. With daily consumption of seafood averaging 656 grams per day per person, fish represents the primary source of nutrient and protein for atoll residents (Lovell et al. 2000). For Kiritimati residents, the destruction of the coral reef is devastating, particularly in the unstable context in which sea levels have risen by eight to ten inches in the past century and a half. Migration is practically the only option for survival, as is the case for thirty million refugees now leaving their home each year because of climate change (see chapter 8). Today, scientists are studying the 10 to 30 percent of corals that survived to try to understand what made them resilient to warmer temperature and learn how to restore reefs (Dance 2019).

While the production of seafood is impacted by climate change, it also contributes directly to greenhouse gas emissions underlying these changes. As noted above, fishing vessels have become increasingly motorized and equipped with technologies requiring large amounts of fuel. As larger ships venture further away from the coast and often travel internationally in search of fish, oil consumption and carbon emissions increase. Aquaculture, which typically takes place in coastal or inland waters could offer a solution, but it comes with its own set of problems including eutrophication and acidification.

Another major cause of concern relates to pollution from plastics (Beaumont et al. 2019). During the past decade, over eight million metric tons of plastic have been dumped into oceans each year—a figure that is continuing to rise with the consumption of single-use plastics such as water bottles, packaging, and grocery bags. While most of this pollution originates in the Global North, commoditization of food and consumer goods in the Global South is causing a rapid increase in the use of plastic and their discarding into oceans. Such plastics are predicted to stay in marine ecosystems in some forms for centuries and have well-documented negative effects on birds, sea turtles, fish, and mammals that ingest or become entangled in them. Millions of animals die annually from strangulation by plastic objects such as six-pack rings. The ingestion of plastic has negative effects on fish health, including their fertility, and could potentially impact the health of humans who ingest contaminated fish or depend on them for food security. The problem is particularly acute in the South Pacific where ocean currents concentrate plastic trash from all over the world.

9.5. Sustainable Seafood

Well-managed capture fisheries and aquaculture can play a key role in supporting food security and the livelihoods of fishing communities. They might also be instrumental in curbing climate change to the extent that seafood can have a lower carbon footprint than other types of protein and help restore the environment. However, to realize this potential and sustain it in the future, it is important to take steps to protect existing resources and curb overfishing.

Since the 1980s, governments have begun acknowledging the importance of managing fisheries and aquaculture resources. However, governance is very complex, with some jurisdictions overlapping and fish oblivious to any political boundaries. At the global level, the 1982 United Nations Convention of the Law of the Sea provides a legal framework that gives authority to states to govern their own coastal waters and requires cooperation with other nations through regional organizations like the General Fisheries Commission for the Mediterranean, the North Atlantic Fisheries Commission, and the South Indian Ocean Fisheries Agreement. Additional agreements have been signed to govern high seas, manage migratory fisheries like tuna, and fight illegal, unreported, and unregulated fishing. In 1995, members of the United Nations unanimously adopted the Code of Conduct for Responsible Fisheries that defines a set of voluntary principles and standards to promote sustainable seafood production.

The UN Sustainable Development Goals adopted in 2015 provides a framework for reducing poverty and inequality while preserving the environment and confronting climate change. Sustainable Development Goal 14 focuses specifically on "life below water" and aims to "conserve and sustainably use the oceans, seas, and marine resources for sustainable development." It sets goals and targets to reduce marine acidity, eutrophication, plastic debris, overfishing, subsidies, and illegal, unreported, and unregulated fishing, and to encourage conservation, ecosystem-based management approaches, and small-scale artisanal fishing.

The challenge with these multilateral agreements and codes lies in their enforcement, which is complicated by power-laden political and economic dynamics that underlie how the ocean is being used. Except for high seas, the management of aquatic resources falls on nation states, with guidance from the Code of Conduct for Responsible Fisheries and in collaboration with regional organizations. Most coastal states have jurisdiction over exclusive economic zones (EEZ) that extend two hundred nautical miles from their coasts and together make up 42 percent of the ocean. Some states, like New Zealand, Australia, and the United States, have been active in establishing coastal management plans, regulating seafood production, monitoring activities, and enforcing regulations. However, other states, have fewer regulations and limited resources to implement them. According to the FAO (2020f), more than two-thirds of UN member countries have yet to put in place a comprehensive policy and legal framework. These are the states where illegal, unreported, and unregulated fishing activity, mostly from foreign vessels, is more likely to occur and threaten the livelihood of local people, forcing them to engage in unsustainable practices as well.

Within EEZs, many coastal communities have come up with their own plan to protect aquatic resources. Different types of regulations have been enacted to reduce overfishing. Most approaches focus on limiting catch and/or restricting fishing capacity. The former reduces output by setting quotas or maximum catch targets known as TAC (total allowable catch), establishing no-take zones, charging landing fees, establishing minimum fish size requirements, and prohibiting bycatch of specific species. The latter limits fishing capacity by requiring fishing licenses and capping their number, restricting vessel size or engine horsepower, ending fuel subsidies, limiting the number of fishing days, and banning the use of certain technology or fishing methods.

Today, scientists recognize the importance of *integrated management* approaches that focus on ecosystems, rather than single species. This is because restrictions on single species often lead to dramatic ecosystem changes that affect other aquatic organisms and have unintended negative consequences. Spatial approaches such as *marine protected areas* (MPAs) have emerged over the past twenty years as a tool to achieve long-term integrated goals of protecting biodiversity while sustaining fish production. In the United States, MPAs are defined in the Federal Register as "any area of the marine environment that has been reserved by federal, state, tribal, territorial, or local laws or regulations to provide lasting protection for part or all of the natural and cultural resources therein." As of 2020, nearly one thousand MPAs had been established in the United States to protect oceans, estuaries, coastal waters, and Great Lakes (National Marine Protected Areas Center 2020), joining more than fifteen thousand other MPAs around the world (UN 2020) and covering 7.5 percent of waters. Because many MPAs are relatively small, their success often rests on the inclusion of coastal populations in decision-making, which ensures better understandings of existing marine resources and their multiple uses (see textbox 9.5).

A major obstacle in limiting fishing is the impact on livelihoods. Fishers often look at regulation unfavorably, seeing them as a threat to their way of life, as cod-fishers did in the North Atlantic (see textbox 9.2). To the extent that these regulations are only enforced for registered vessels operating in proximate and coastal areas but evaded by unreported vessels fishing in international waters or flying under flags of convenience, they unfairly burden smaller local fishing operations and may encourage illegal activities. The unemployment caused by regulation is devastating for coastal communities where most residents are involved in fishing. However, effective management increases the viability of fisheries and strengthens their capacity

| Box 9.5 | **Marine Protected Areas and Indigenous People's Rights** |

Marine Protected Areas (MPAs) present an opportunity to recognize and fulfill Indigenous people's rights to self-development in accordance with their own needs and aspirations, free of oppression, marginalization, and discrimination. Indeed, Indigenous people have historically cared for and conserved aquatic resources through place-based cultural practices, which have been dismantled by colonization, dispossession, and expropriation. Designating specific Indigenous or tribal MPAs or including Indigenous people in the management of existing MPAs could potentially meet ecological goals while attending to cultural, social, and economic aspirations and self-determination rights. Specifically, these MPAs could bring Indigenous knowledge and cultural practices into marine management.

However, to date, there are very few explicitly Indigenous MPAs, mostly in Canada, Australia, and Polynesia. In most cases, governance is shared between tribal and state agencies and Indigenous people tend to play an advisory role rather than a decision-making one. A well-documented example comes from Palau—a Micronesian nation made of 586 small islands located between the Philippines and Papua New Guinea. Fishing in Palau consists of near-shore domestic subsistence fishing and off-shore commercial fishing by foreign fleets from Japan, Taiwan, and China that pay fees to catch pelagic fish like big eye, yellow, and skipjack tunas. In 2003, the government established a series of MPAs through the Palau Protected Areas Network (PAN), with the goal of preserving marine resources, local culture, and livelihoods. The protected areas include mangroves, seagrass beds, and coral reefs, some of which were completely closed off to any activities while others remained open to subsistence fishing and diving—a major source of tourism revenue. This conservation approach builds on the ancestral practice of "bul" in which kin groups would temporarily close fishing "taboo areas" based on their intimate knowledge of fish abundance and punish violators with shaming and shunning. As a modern version of bul, MPAs have garnered support from Indigenous communities that are involved in their management (Gruby and Basurto 2014). Scientists have documented the positive effect of MPAs on Palau, including much larger biomass of both reef and pelagic fish compared to nearby unprotected areas, providing important long-term benefits for tourism, cultural preservation, and subsistence fisheries (Friedlander et al. 2017).

While larger MPAs that cover entire ecosystems appear to be more successful on ecological grounds, there are arguments in favor of smaller MPAs to the extent that they enhance institutional capacity, social inclusion, and community participation. When larger areas are created, they often lack legitimacy among Indigenous people who view them as "ocean grabbing" and reject them as another form of colonialism.

In 2020, the Palau National Marine Sanctuary (PNMS) became effective, expanding the protected area offshore to 80 percent of the Executive Economic Zone (EEZ), banning foreign vessels that used to fish tuna, and maintaining 20 percent of the nearshore EEZ for local subsistence fishing. Although it has been hailed by the United Nations and many NGOs as one of the world's most ambitious conservation initiatives, there are some unintended consequences due to the low capacity of local fleets that cannot make up the share of foreign tuna catches that was sold locally in supermarkets, restaurants, and hotels. As a result, local fishers have intensified their activities in allowed nearshore areas, increasing their catch of smaller reef fish on which tunas and other pelagic fish depend. The future of the sanctuary has become a political issue and is being debated by presidential candidates.

to sustain seafood livelihoods in the long run. A growing number of operations are successfully engaged in sustainable fishing and aquaculture. Thus, to ensure success and support for conservation efforts, it is important to consider the perspective of the fishing community and engage fishers in decision-making to create agreements that are vetted and equitable.

Without centering conservation efforts on equity, it is likely that subsistence fishers and family-owned operations will be unable to adapt or compete. For example, in Iceland, individual quotas were established for cod fishing. Because quotas are transferable, they are traded between fishers, creating an incentive for owners of small operations to leave the industry and sell their rights to larger businesses. As a result, fishing rights are now concentrated in the hands of a few large enterprises (WOR 2013). While this may be more efficient from a strictly economic standpoint, it leads to the destruction of livelihoods and the gradual erosion of coastal cultures and societies.

Because artisanal and Indigenous fishing operations have usually been more sustainable than industrial enterprises, learning from them could help achieve sustainability goals. For example, harpoons, spears, lines, and small nets have been used effectively by Indigenous people throughout the world to catch fish in rivers, lakes, lagoons, and shallow seas. These methods do not generate bycatch or require any fuel. Rather than going out to sea for extended periods of time and returning with large amounts of fish that need processing and refrigeration, Indigenous farmers farm daily and tend to catch or harvest smaller quantities, which they consume the same day or smoke, dry, or pickle for later use. Yet national and regional regulatory frameworks often ignore the ancestral practices of Indigenous communities, which perhaps ought to be treated differently than large commercial operations. Recently, the United Nations declared 2022 the International Year of Artisanal Fisheries and Aquaculture, suggesting that their role in sustaining livelihoods and healthy oceans is slowly being recognized.

Consumers, too, have a role to play in promoting sustainable fishing by making ocean-friendly choices based on information about the environmental and social impacts of seafood production. Given that such a large share of seafood is internationally traded, it is often difficult to know where it comes from and whether it is sustainably produced. In the United States, for instance, about three-fourths of the seafood consumed is imported, mostly from China, Thailand, and Indonesia where environmental and labor regulations are limited and poorly enforced. To help consumers make informed choices, Seafood Watch (Monterey Bay Aquarium 2020) and other similar programs have created science-based consumer guides that categorize types of seafood according to the way they are managed, caught, or farmed and their provenance. For example, the 2020 Seafood Watch national guide ranks salmon from New Zealand as a "best choice," wild salmon from California or farmed Atlantic salmon from Maine as "good alternatives," and wild salmon from Norway and Scotland as fish to "avoid." Yet learning about the provenance and production method of salmon—or any other type of seafood—takes effort on the consumers' part. Sustainable Seafood certification and labeling programs like the one from the Marine Stewardship Council help convey information to consumers. However, like the organic and fair trade labels discussed in previous chapters, the value of such labels depends on the standards they reflect and their implementation. The generally higher cost of certified sustainable seafood makes it a luxury and challenges consumers for whom it has traditionally been a necessity.

9.6. Conclusion

Oceans, lakes, and rivers are a major source of nutritious food and livelihoods globally. The diversity of aquatic ecosystems underlies differences in cultural practices and points to the relationships between environment, food, and social relations. Coastal people have historically adapted to their environment, devising strategies to safeguard fisheries and aquatic habitat. However, globalization and technological innovation have weakened local control by opening oceans to multiple users with expanding fishing capacity and competing interests. Today, seafood is one of the most internationally traded food commodities, with flows going primarily from the Global South, especially Asia, to the Global North. As world prices continue to increase, seafood has become a luxury, threatening the well-being of those who depend on it for subsistence.

Sustaining seafood is critical to maintaining food security, livelihoods, cultural heritage, and biodiversity. In recent years, international organizations, nonprofits, and local governments have begun paying more attention to oceans and establishing legal and policy frameworks at multiple scales to address overfishing, pollution, and climate change. As long as the incentive to produce more and regulatory holes remain, however, aquatic ecosystems will continue to be under threat, hurting the most vulnerable nations and communities.

SOCIAL AND CULTURAL GEOGRAPHIES OF FOOD

Food, Identity, and Difference 10

Outline

Learning Objectives

- Understand how food relates to identity and difference.
- Consider how national cuisines are socially produced notions, political projects, and hybrids that reflect evolving local and global connections.
- Recognize food consumption and taste as central to the making and marking of social class.
- Examine the meanings of so-called ethnic food as a tool of self-identification, immigrant belonging and homemaking, and economic integration through entrepreneurship.
- Realize how food is used as a means of othering—producing and reinforcing differences along the lines of national identity, ethnicity, race, class, religion, gender, and so on.
- Critique the concepts of authenticity and multiculturalism as they relate to food and place.

In one of the most quoted phrases in food studies, Brillat-Savarin (1826) tells us that "we are what we eat." For many, this means that our body and its health are shaped by what we ingest. For cultural scholars, it points to the deep connection between food and identities. Food makes us who we are, but also reveals our selves to others.

Food becomes dishes, meals, or feasts through culture. Anthropologist Claude Lévi-Strauss contrasts raw food, which he describes as "natural/unelaborated," with cooked food that is "cultural/elaborated." Cooking gives natural ingredients cultural meaning, with cooking techniques such as roasting on an open fire versus boiling in a receptacle reflecting the level of mediation between human beings and nature. As such, food is often conceptualized as a *system of communication*, with our peculiar food habits revealing who we are. Food scholars often distinguish between *cuisine*, *gastronomy*, and *foodways*. The first refers to a particular style of cooking characterized by unique ingredients, techniques, and dishes that are often associated with specific cultures or places. In common parlance, gastronomy refers to fine food as prepared by professionals and described in cookbooks. The term is also used to describe the study of the relationship between food and culture. In recent years, anthropologists, geographers, and sociologists have begun paying more attention to everyday food practices to understand culture, rather than focusing on elaborate dishes reserved for special occasions and cultural elites. The notion of foodways encompasses more than cuisine and gastronomy and includes behaviors, practices, and beliefs surrounding food such as timing of meals, serving styles, shopping, preparation, and manners.

Food and foodways can be tools of self-expression and social inclusion, as well as mechanisms of exclusion. Most people experience connections to specific dishes and cuisines associated with their childhood and place of origin. Preparing, eating, or sharing these dishes usually generates pleasure, comfort, memory, and pride. It also allows people to experience, express, and strengthen their belonging to a given community. At the same time, food is often used to denigrate people, as witnessed in the endless list of food-based xenophobic, racist, classist, and sexist slurs. For many people, hunger, food insecurity, and other food anxieties also constrain how food is being used socially and culturally. Thus, the connection between food and identity may not be as simple as "being what we eat;" questions arise regarding how the collective "we" is produced and specific foods come to represent and shape it.

In this chapter, we seek to answer these questions by examining how food, as a form of self-expression and a means of distinction, relates to identity and difference. Identity emerged as an important topic in food studies in the 1990s, following the "cultural turn" that reshaped the social sciences. Until then, scholars had mostly focused on the production of food, with topics such as hunger, agrarian development, and trade dominating the field. However, food is more than feed and acknowledging its cultural meaning in people's lives and its relationship to identity helped shed a new light on old topics while raising new ones.

After outlining a broad framework to think about identity, we focus on three topics that highlight how food relates to identity, place-making, and belonging. First, we investigate national identity and compare cuisines around the world, paying attention to the geographic and political factors that influence national cuisines and regional foodways. We then turn to class and examine how taste is socially produced and used to distinguish and exclude. We end the chapter by focusing on race, ethnicity, and immigration and investigating ethnic food from a variety of perspectives.

10.1. Identity and Difference

The collective "we" in "we are what we eat" is about *identity*, which may be defined as a person's sense of who they are based on group membership. Social class, race, ethnicity, national origin, gender, religion, and sexuality are common aspects of identity that suggest that where and how one lives is important in their perception of who they are. In an increasingly fluid and mobile society, people's identities have become less rigid and more closely tied to lifestyles. Today, people appear to be less bound to traditional forms of identity such as class, race, and religion, which were historically viewed as passed on from generation to generation. Instead, they have a seemingly wider array of identities to "choose" from and a growing number of signifiers to express them, including clothing, food, and other aspects of lifestyles. Yet, the ability to "choose" an identity and transgress expectations and stereotypes is constrained by many factors including individual resources, cultural environment, and prevailing social norms.

What is often missing from discussions about identity is the fact that they are diverse, layered, situated, mutable, and often contradictory. Furthermore, identities are *relational*; they are not intrinsic and fixed attributes determined at birth, but they are acquired and produced through everyday *performance* and social *interactions* with others that reinforce or challenge our awareness of difference and shape our identities. Most of us belong to multiple groups, which vary over time and space. We see ourselves differently depending on whether we are at home, church, work, school, or at a gym, bar, restaurant, or political rally. For example, if you are Black, you may not consider your Blackness as a major part of your identity when spending time at home with family or close friends, but it might become a lot more important in a school or work environment where other people are primarily white or in a neighborhood where your life might be in danger because you look different. Furthermore, your Blackness likely *intersects* with other identities like gender, class, sexuality, national origin, immigration status, etcetera that become more or less important depending on the circumstances. In other words, our identities are dynamic and emerge from the relationships between the way others view us and our own affinities and experiences. They are never fully "discovered" or "attained" as they are always in the making.

In addition to *intersecting* with each other to form unique identities, notions of race, class, gender, and so on keep evolving. This is because they are *socially produced*. For instance, many critical scholars argue that race has no biological foundation, but is constructed spatially, historically, culturally, and politically to distinguish between "us" and "them." This certainly does not mean that race has no real material and symbolic effects, however it suggests that these consequences are not the result of innate or biological group characteristics, but instead are outcomes of racist practices, ideologies, and policies. Similar arguments could be made regarding class, gender, sexuality, and other identities whose meanings have evolved with social norms and structures. For example, notions of femininity have changed dramatically over the past century, such that being a woman means different things in different times and places.

Thinking about identity relationally highlights the tensions between *structure* and *agency*. Our identities are shaped by both: we have some agency in defining who we are and how we express ourselves, but this agency is constrained by social structures like class, race, and gender. The notion of *difference* emphasizes the latter and highlights social processes of "othering" that categorize people based on visible and external characteristics. In other words, identities are adopted by individuals, while difference is ascribed to them by others.

Through mundane and mostly unconscious everyday activities, food symbolizes and materializes differences and identities. What, how, where, and with whom we grow, purchase, cook, and eat food shape and allow us to express who we are while simultaneously making us vulnerable to stigmatization, disapproval, shaming, and discrimination by others.

National Cuisine 10.2.

The idea of national cuisine epitomizes the relationship between food and place in a modern context where nation-states are one of the most important political geographies. Countries can often be identified by specific dishes that are influenced by food availability, climate, cooking traditions, and cultural practices. For example, Italy is known for pasta and pizza, Mexico for tacos and tamales, Peru for *lomo saltado* and ceviche, Jamaica for jerk chicken, India for curry, Japan for sushi and sashimi, Belgium for potato fries and mussels, Thailand for pad thai, Austria for Wiener schnitzels, Ethiopia for *injera* bread, Argentina for empanadas and asados, England for fish and chips or bangers and mash, Senegal for thieboudienne, the Philippines for lumpias and chicken adobo, and so on. While dishes like these are typically presented as unique and "traditional," they are the product of transnational interactions, political nation-building efforts, and marketing strategies. In other words, cuisines, nations, and identities are human fabrications that are often coproduced.

Cookbooks provide a wealth of information beyond recipes. Photographs of dishes, tables, kitchens, and broader settings, personal stories, descriptions of techniques and ingredients, and assumptions about readers are discursive elements that contribute to a narrative about a particular cuisine and culture.

This is particularly true of "ethnic" cookbooks that focus on lesser-known cuisine from places that readers may never visit. In recent years, "ethnographic cookbooks" that use food to draw attention to social, political, cultural, and geographical differences have become quite popular. Often presented as an individual journey to the roots of an ethnic cuisine or a reflexive autobiography about the writer's origins, these books rely heavily on food aesthetics to weave narratives about place and stimulate geographic imagination. In a context where authenticity has become highly valued, the popularity of recently published books such as *Mamushka*, *Falastin*, *Bibi's Kitchen*, *Kiin*, *Bitter Honey*, and *Fire Island* that respectively showcase the food and culture of Ukraine, Palestine, West African countries, Thailand, Sardinia, and Indonesia, is perhaps not surprising.

These books share important similarities: beautiful photography, personal stories of resilience, strong sense of place, pervasive *nostalgia*, and emphasis on home cooks, with a particular affinity for mothers and grandmothers, street vendors, and artisans over recognized experts. Food is typically presented as the constant positive in a life full of hardship, suggesting that food itself might be the solution to a more peaceful and prosperous life. In many ways, poverty is romanticized as being the source and inspiration of amazing food. As such, these books present a partial ethnography, glossing over the political and economic factors that cause food insecurity and poverty and threaten traditional and Indigenous foodways. This may be partially explained by the fact that ethnographic cookbooks are typically written for educated white readers and by authors who have become successful chefs or food writers after emigrating to the United States or England, indicating a privileged perspective.

Cookbooks and recipes codify and standardize food practices, giving them some rigidity and permanency. In the case of ethnographic cookbooks, this is exacerbated by the emphasis on tradition and claims of *authenticity*. For instance, in the words of its authors (Hassan and Turshen 2020, 2), "*In Bibi's Kitchen* is not about the new or the next. It's about sustaining a cultural legacy and seeing how food and recipes keep cultures intact, whether these cultures stay in the same place or are displaced." Ethnographic cookbooks make explicit what other cookbooks leave up the reader.

National cuisines are embedded in physical geographies that influence the availability of ingredients and partly determine the primary food staples. Early research attempted to classify national or regional cuisines based on lists of typical staples (e.g., rice, potato, corn, cassava) and flavors (e.g., spices, herbs, chilies, fermented sauces, sugar). For instance, Portuguese cuisine is based on seafood, including cod and sardines, olive oil, vegetables, and legumes that can be cultivated or harvested in the region. However, scholars now acknowledge that food cultures evolve over time, with trade and migration, and are shaped by social, political, and economic factors. Instead of categorizing or cataloging foods, researchers try to better understand how national cuisines are created and evolve over time and what that means for people associated with them. Often, they use cookbooks (see textbox 10.1), magazines, and films as a rich source of material from which to study the evolution of cuisine, gastronomy, and popular food culture. *Ethnographic* approaches, which aim at understanding people from their own perspective through extensive qualitative fieldwork, are also used to study people's experiences with food, from the ceremonial to the everyday, and their relationships to identity.

10.2.1. Hybridity

As discussed in chapter 3, trade has moved food commodities around the globe for centuries and shaped regional and national cuisines by adding new ingredients to native foods. For example, grains from the Middle East made their way across the Mediterranean to western Europe where they became an important component of regional cuisines thousands of years ago. The spice trade brought new flavors from Asia into European kitchens. Colonization of the Americas, which many historians view as the start of modern globalization, amped up the movement of food across continents and regions. The so-called Colombian exchange transformed cuisines around the world, bringing among other things tomatoes to Italy, chili peppers to India, potatoes to northern Europe, and wheat, pigs, sugar cane, and bananas to the American continent. French cuisine, which is known for its sophistication, integrity, and rootedness in local traditions, owes its richness and variety to influences from the Mediterranean, the Middle East, Asia, and the Americas (see textbox 3.1).

Today, food cultures are increasingly transnational as commodities, people, and knowledge have become more mobile. For example, curry is considered a national dish in England, döner kebabs have been domesticated in Germany, burgers are now trendy in France, and pizza is ubiquitous in the United States. Indeed, the American experience raises interesting questions regarding the meaning of national cuisine, with many experts pondering whether there is such a thing as *American food*. To the extent that quintessential American dishes like hamburgers, hot dogs, pizzas, tacos, and bagels originated elsewhere, can we really speak of an American cuisine? Indigenous foods such as acorn mush, succotash, corn bread, dried fish, smoked venison, wild turkey, and various berry dishes do not feature prominently in the national culinary register, except perhaps in the "traditional" Thanksgiving dinner (see textbox 10.2). Indeed, they were denigrated as barbaric by European settlers beginning in the early days of the colonial era and gradually eliminated by genocide and dispossession of Native people. Today, there is so much variation in regional cuisines (e.g., Maine lobster roll, New Mexico chili verde, Louisiana gumbo, Kansas barbecue) that no single dish or type of food effectively represents a unified national culture (see figure 1.5). For some, it is this very diversity that is uniquely American, as symbolized in the "melting pot" metaphor so often used to describe the nation. As Gabaccia (2000) argues, what distinguishes American eaters is not what they eat but their willingness to mix culinary traditions, which presumably reflects their cosmopolitan attitudes and multicultural values. *Hybridity* is a term often used to describe cuisines that, in multicultural and postcolonial contexts, borrow from multiple influences.

Race and ethnicity scholars in the United States have criticized the notion of national cuisine for concealing difference and hiding social hierarchies in which the food of Europeans continues to be normalized as superior and other types of food are seen as ethnic and inferior. They also regard claims of cultural hybridity and cosmopolitanism with suspicion, arguing that they assume commensurability between cultures and embrace an assimilationist model in which any cuisine can become Americanized, while ignoring the power relations that marginalize certain groups and devalue their foods. Despite these criticisms, it is clear that cuisines around the world are increasingly hybrid, borrowing ingredients, flavors, and techniques from other places and constantly evolving over time with trade, immigration, and other geographic connections. Recognizing this hybridity and fluidity does not negate the importance of social hierarchies.

10.2.2. Food and Nationalism

In many countries, developing a national cuisine is part of the *political project* of nation building. This is particularly true in a postcolonial context, where newly independent nations have created or promoted national dishes to foster a sense of belonging to political entities that were artificially produced by colonialism. For instance, in

Box 10.2   Thanksgiving Holiday Dinner

The story surrounding the first Thanksgiving holiday is one of the biggest myths of American history. It has received mounting criticism in the past decade, and a growing number of people refuse to celebrate it because of its association with the genocide of Indigenous people (Anderson 2020). Yet, it remains a symbol of American cuisine and tradition.

History books tell us that in 1621, a year after disembarking the *Mayflower* in Plymouth, Massachusetts, the Pilgrims held a harvest celebration attended by members of the Wampanoag tribe, as recreated in numerous artistic renditions, such as figure 10.1. The festive meal presumably included turkey, succotash, corn bread, pumpkin, and cranberries, some of which were brought by native guests for whom there were common foods. The notion that the meal solidified a friendship between Native Americans and Europeans is highly contested, particularly in light of subsequent massacres and land grabs.

It was not until the 1830s that New Englanders begun celebrating what they called "the first thanksgiving," which Lincoln eventually turned into an official holiday in 1861. Since then, "traditional"

dishes have been added to the menu, including pies, sweet potatoes, wild rice, stuffing, green bean casseroles, and other preparations that were not part of the 1621 feast. Some historians argue that even turkey was not on the table, but that venison was served instead.

The popularization of Thanksgiving in the nineteenth century owes much to magazines such as *Ladies' Home Journal* that turned it into a commoditized event that celebrated family values, rural domestic life, and national unity at a time when they appeared threatened by the presence of immigrants, emancipated Blacks, non-Protestants, and a growing urban working class (Wills 2003).

Today, food magazines such as *Bon Appetit* and *Food & Wine* continue to devote their best-selling November issue to Thanksgiving, providing home cooks—mostly women—tips and suggestions on how to prepare a classic New England feast. Indeed, the idea that women can play a leadership role in maintaining traditions and moral values through domesticity was central to nineteenth-century popular writing on Thanksgiving and remains so to this day.

FIGURE 10.1 *The First Thanksgiving, 1621.* Painting by Jean Leon Jerome Ferris, circa 1912.

Source: https://commons.wikimedia.org/wiki/File:The_First_Thanksgiving_cph.3g04961.jpg.

Thailand, pad thai—a dish of rice noodles stir-fried with eggs, tofu or chicken, bean sprouts, tamarind, fish sauce, and other ingredients—was created in the late 1930s at the request of Prime Minister Phibunsongkhram as a symbol of national unity amid political efforts to "modernize" the country. Under Phibunsongkhram's autocratic rule, Thailand adopted a nationalist agenda with a strong anti-Chinese bias. Pad thai was promoted as a symbol of this new era, along with the Western calendar, a new flag, and a new national anthem. Ironically, noodles, which are the main ingredient of pad thai, were allegedly brought to Thailand by Chinese immigrants two centuries earlier and were not considered "traditional" in the past. Today, pad thai is one of the most well-known Thai dishes outside of the country, even though cultural purists like to remind eaters that it is neither "authentic" nor "traditional."

In Global North countries, immigration and cultural change sometimes prompt reactionary efforts to (re)define national cuisine, using food to determine who belongs and who does not. According to Wills (2003), Thanksgiving became a tool to define Americanness in a context where freed slaves and incoming Catholic and Jewish immigrants were seemingly threatening the mainstream white Protestant culture (see textbox 10.1). Because of their high symbolism, holidays and celebrations are particularly important in naturalizing what it means to be American, reinforcing a certain way of cooking and eating. Today, in a context where markets are seemingly flooded with products "from nowhere" and highly processed food, some consumers are willing to pay a premium for products with known and presumably superior origins. Claims regarding the origin of food are thus used in campaigns to support domestic food industries, especially when foreign products are viewed as threatening culture, economy, or safety. For example, at the height of the mad cow disease (or bovine spongiform encephalopathy) epidemic in the United Kingdom in the 1980s, many neighboring countries began labeling their meat with flags of origin to reassure consumers that their products were safe.

The notion of national cuisine is as arbitrary as nation states, which are modern creations that are rarely culturally homogenous. Most nations are made of multiple regions with their own cultural and physical geographies and unique cuisines. Under the homogenizing pressures of nationalist ideology and economic globalization, regional cuisines often become enrolled in strategies of resistance to defend heterogeneity and local cultures. For example, according to Díaz (2012), Yucatecan cooks and chefs resisted Mexico's post-independence drive to create a unified national identity, homogenize regional cultures, and erase indigeneity, as described by Pilcher (1998). Physically isolated from the rest of Mexico, the Yucatán peninsula developed its own cuisine influenced by the Caribbean, Europe, and Northern Africa. One of the most famous Yucatecan dishes is *cochinita pibil*—pork marinated in achiote and sour oranges, wrapped in banana leaves, and cooked in an underground pit. The cooking style has pre-Hispanic Mayan origins, but pigs and oranges were brought to the Yucatán by the Spanish, Habañero peppers which give the dish its heat came from Havana through trade, and bananas were transferred from southeast Asia via Africa. Other famous dishes include *queso relleno*—cheese stuffed with ground pork, raisins, almonds, olives, and spices—and *marquesitas*—crêpes filled with cheese. The main ingredient of both dishes, Edam cheese, was brought to the Yucatán by Dutch traders who came to the region in the nineteenth century for henequen—a valuable fiber extracted from agave to make rope. Enchiladas, moles, refried pinto beans, and other typical Mexican dishes are less common in the Yucatán than in central Mexico. Instead, the cosmopolitan nature of the cuisine is emphasized, symbolizing Yucatecan sophistication, worldliness, and distinction from the rest of the country (Díaz 2012). Despite the existence of rich regional culinary traditions in most countries, a handful of dishes often come to represent the nation, repressing difference and hiding diversity.

As a political project, national cuisine is also used for what is often called *soft diplomacy*, which relies on meals and gifts of food to promote rapprochement. For instance, the menus of official dinners hosted by the president of the United States at the White House are often elaborate exercises in promoting American culture while signaling connections. This seems to be especially important when guests come from countries with renown culinary traditions, such as France, Japan, or Italy. For example, when President Donald Trump and First Lady Melania Trump hosted French President Emmanuel Macron and his wife, the main course consisted of Colorado rack of lamb prepared with a classic French sauce called soubise and served over Cajun jambalaya made with Carolina gold rice. It was paired with wine produced in Oregon and aged in French oak barrels that, in the words of its producer, reflected "French soul and Oregon soil."

10.2.3. National Food in Branding and Marketing

Inventing a national cuisine is an important way to promote tourism and market food products, which may generate significant revenue. In a world characterized by increased mobility, at least for middle- and upper-income travelers of the Global North, being able to offer a unique and authentic experience has become key to attracting tourists. In the past two decades, food has become a primary mechanism for branding and marketing places. National tourism authorities run advertising campaigns showcasing the rich food culture of their countries and travel agencies sell culinary cruises, wine tours, and cooking school vacations. Television shows focused entirely on food and traveling, such as Anthony Bourdain's well-known series *No Reservations* and *Parts Unknown*, have become very popular in the past two decades, producing new *geographic imaginaries* linked to food and changing how elite consumers travel.

Such geographic imaginaries are also helpful to brand and market food, both at home and abroad. Foods that carry a strong national identity and can be labeled according to their place of origin are often more attractive to consumers. For example, many people in the United States associate olive oil with Italy. As a result, Italian olive oil tends to be more expensive than Greek, Spanish, or Turkish olive oil, which are perceived as less authentic. Public agencies, business associations, tourism boards, and corporations in many countries allocate significant resources to branding, labeling, and protecting products for economic gains. Programs like the European Union's Protected Designation of Origin or Protected Geographical Indication serve that purpose by restricting the ability to name products such as wine, cheese, and certain meats according to where they are produced. For example, blue cheese can only be called Roquefort if it is produced in the French village of its namesake. Ouzo, the anise-flavored liquor, can only be produced in Greece or Cyprus. If produced in other places, it must carry a different name. Given the potential economic returns these products generate, appellation schemes often lead to important legal battles.

Another related program is the controversial UNESCO (United Nation Education, Scientific, and Cultural Organization) Intangible Cultural Heritage program which, as of 2020, lists nineteen foods, along hundreds of festivals, crafts, and other traditions, that member nations have requested to be recognized. Among others, foods include the Mediterranean diet, Washoku or Japanese food in general, Turkish Coffee, Belgian Beer, traditional Mexican cuisine, Neapolitan pizza, and the French gastronomic meal. Such claims of originality can be controversial and debated. For example, while kimchi (a pickled or fermented preparation typically made with napa cabbage, chili powder, and fish sauce) has been recognized as North Korean heritage, *kimjang* (the traditional process of making kimchi after the harvest and storing it underground in clay jars) was attributed to South Korea. Similarly, Uzbekistan and Tajikistan—two neighboring central Asian countries—received recognition for two similar rice dishes: *palov* and *oshi palav* respectively. Dolmas—stuffed grape leaves popular throughout the eastern Mediterranean—were attributed to Azerbaijan. Surprisingly, despite a national campaign and growing international recognition, Peruvian cuisine has failed so far to be recognized on the UNESCO list of Intangible Cultural Heritage (see textbox 10.3).

The notion of national cuisine highlights the fact that the relationship between food and identity is produced and used for political, economic, and cultural projects. As such, it is often contested, renegotiated, and reframed, reflecting the relational geographies that underlie nation-states.

| Box 10.3 | **Peruvian Food Heritage and Gastrodiplomacy** |

Since the turn of the millennium, Peruvian cuisine has become known as one of the world's best cuisine, with famous chefs and restaurants receiving prestigious awards and growing attention from international travel and lifestyle magazines. Dishes such as *cebiche*, *tiradito*, *ají de gallina*, *papas a la huancaína*, and *pachamanca* have acquired significant culinary capital globally. Lima, which the Peruvian Commission for the Promotion of Exports and Tourism describes as the "gastronomic capital of the Americas," has become a major culinary destination. To solidify this image and capitalize on it, public and private organizations have submitted several proposals to the UNESCO to recognize Peruvian cuisine as "intangible cultural heritage." Their nomination, however, has been rejected partly because of concerns that it does not represent a "shared" cultural heritage in need of protection.

According to Matta (2016), the dominant contemporary vision of Peruvian food was created over the past two decades by various stakeholders, including ministries, business organizations, farmer associations, chefs, and food entrepreneurs to promote an image of Peru as unified after a long period of economic crisis and political violence. In their view, food could bring people together including Indigenous people and those with African, Asian, and European ancestry; join coastal, urban, and rural areas; merge *comida* (food) and *alta cocina* (haute cuisine); and connect the traditional with the modern. Most importantly, however, food could help brand the nation and promote international relations and economic development—a strategy described as *gastrodiplomacy* (Wilson 2013).

To the extent that this national cuisine—often described by its advocates as "*cocina peruana para el mundo*" (Peruvian cuisine for the world)—has been shaped by elites who have much to gain from its recognition, however, it does not embody the notion of food heritage as "encompassing all food knowledge and skills considered by groups as shared legacies or common goods" (Matta 2016, 338). Specifically, the voices of peasant and Indigenous people have been excluded from debates regarding what constitute Peruvian food heritage while their foods have been reappropriated and reinterpreted as cosmopolitan *alta cocina* by European trained chefs in fine-dining restaurants. What was presented in Peru's candidature to the UNESCO as food heritage was more about gastronomy marketed to elite—and typically foreign—consumers than about everyday foodways, including skills, knowledge, and practices, in need of safeguarding. The UNESCO rejected Peru's application on those grounds. A revised application is currently being considered and is expected to be approved.

10.3. **Food and Class**

Consumption, particularly in postindustrial societies, has become an important element of class formation and marker of social status. During the industrial era, the process of class identification was grounded in the realm of production: the type of work that people did determined their social class, which at the most basic level meant that they were either laborers or capitalists. Class-related wealth and income disparities underlie consumption patterns, with the economically powerful able to indulge in luxury goods such as gourmet foods and drinks that became *symbols* of upper-class status (e.g., caviar, lobster, champagne). Historically, high-income households have typically consumed more meat, cheese, and a greater variety of fresh produce, compared to low-income households, who ate more carbohydrates (e.g., rice, potatoes, cassava, bread) and less meat. Globally, wherever rising income leads to an expansion of the upper and middle class, foodways evolve accordingly. This partly explains the recent rise in the consumption of meat and other animal products in places like China, Mexico, and Vietnam, where the upper and middle classes have grown.

In the postindustrial era, social class appears to have lost some of its connection to the world of production and become more directly tied to consumption that seemingly drives the economy. Faced with "unlimited" choices, individuals have turned to commodities to express themselves. Tastes for food, music, clothing, furnishings, and leisure have become symbols of social class, which is understood culturally and aesthetically as *lifestyle*. These tastes, however, are still shaped by the capitalist system that defines what consumer goods are available and who can afford them.

According to French sociologist Pierre Bourdieu (1984), taste is not just a reflection of social class, it is an integral part of *class formation*. Through socialization, people acquire tastes for particular things, internalizing a social hierarchy ranging from lowbrow to highbrow taste. For example, knowing how to enjoy fine cuisine in a high-end restaurant is not innate but acquired. Eaters must have learned how to read menus and wine lists, recognize rare ingredients, appreciate unique combinations of flavors and cooking techniques, use silverware and stemware correctly, adopt "proper" table manners, and interact confidently with waiters and sommeliers. This intangible knowledge is what Bourdieu calls *social capital*. It confers class distinction by allowing the upper-class to distance itself from other social groups. In rigid societies, those who lack that sort of social capital but still partake in luxurious consumption are easily recognizable as "nouveau riches" (i.e., new money) who are shunned by the upper class and relegated to similar social ranks as those unable to participate because of insufficient income. For the working class, affordability, caloric value, and familiarity of food are presumably more important, underlying the assumption that working-class food lacks "taste" or, as Bourdieu puts it, tastes of necessity. Thus, gustatory preferences and aesthetic notions of "good food" are not objective or universal, but instead influenced by income and cultivated through continuously evolving social norms.

Since the 1980s, when Bourdieu's *Taste and Distinction* was first published, social hierarchies of taste have become less rigid. As more products are made available to consumers, upper-class foods have become democratized. For example, throughout most of the nineteenth and twentieth century, lobster has been a status symbol in most of the Western world. However, fishing, refrigeration, and transportation technologies have made it more readily available and affordable than in the past. Like many luxury foods, such as truffles, caviar, smoked salmon, asparagus, foie gras, oysters, and prized cuts of meat, lobster is now found in many mainstream supermarkets and served at casual restaurants. Exotic products, known only to chefs and connoisseurs, such as harissa sauce, balsamic vinegar, wagyu beef, and fine European cheeses, have also become more common. For instance, matcha—a highly refined green tea powder once reserved for Japan's nobility—is now widely available in Japan and abroad. In the United States, it has become very popular in recent years, with its bright-green color catching attention on social media. This democratization of food has been promoted by food media that normalize the food habits of the rich, turning once extravagant ingredients into familiar ones through images and recipes.

While the lower classes increasingly eat like the upper classes, the reverse also occurred. Popular culture and high culture began blending and the lines between the two became progressively blurred, destabilizing social hierarchies of taste. The resulting culture has been described as *postmodern*—no longer structured by the universalizing tendencies of modernity but instead shaped by the multiplicity of experiences and expressions. Postmodernism is linked to consumerism and the ways individuals make themselves through consumption and lifestyle choices, which act as symbolic codes for many different and often contradictory values.

In that context, *foodies* have emerged as people who construct and distinguish themselves socially through their food practices, even if they rarely use that label to describe themselves. Because food is such an important aspect of foodies' identity, it influences many aspects of their lifestyles, ranging from what and where they eat daily to leisure activities such as entertainment and traveling. As Johnston and Bauman (2010) argue, foodies present themselves as *omnivorous*. Unlike elite consumers of the past, foodies eat "everything" from highbrow to lowbrow. For instance, foodies seem to be particularly attracted to the "simple" food of peasants and the working class, as witnessed in the popularity of country bread, rustic stew, and craft beer (see textbox 10.4). Yet, despite their seemingly democratic and cosmopolitan attitudes, foodies continue to distinguish themselves by their consumption choices, which are presumably more self-conscious, informed, and ethical than those of lower socioeconomic classes, with aesthetic value playing a critical role in defining good

---

Box 10.4   Craft Beer

During the past few centuries in most of the Western world, beer has been the drink of the lower class while wine was preferred by the upper class. Beer is perceived as cheap, bitter, and lacking in diversity and sophistication, in contrast to wine, which offers subtle nuances to trained palates and noses. Furthermore, beer is often associated with drunkenness and a lack of moral values, making beer halls, pubs, and taverns insalubrious places (see textbox 11.1).

In recent years, however, perceptions about beer have changed dramatically. High-end restaurants now list beers alongside wines and offer beer tastings and food pairings. This transformation is primarily explained by the expansion of so-called "craft beers." Typically produced in smaller batches by expert brewers, these beers display greater variety in ingredients, flavors, techniques, and alcohol content. These artisanal and locally brewed concoctions provide consumers

with an opportunity to distinguish themselves by their unique knowledge, taste, and penchant for non-commercial and local products, without the pretention and snobbishness typically associated with wine. Indeed, many craft beer companies embrace the democratic and working-class image of beer with names such as *Working Class Brewery* and *Simple Times Lager*. In the United States, nostalgia has also inspired the return of the beer can and the revival of old-time brands such as *Pabst Blue Ribbon* and *Rolling Rock*.

Despite the appearance of democratization, it is worth noting that craft beers are significantly more expensive than their mass-produced counterparts, making them unaffordable to the working class. During the past decade, craft beers have captured a growing share of the market. In response, large corporations have launched their own versions and purchased many of the smaller local breweries (see textbox 3.5).

food. In that sense, foodies are not truly omnivorous. Although most claim to enjoy hamburgers, they will not be caught with a Big Mac in hand, preferring instead "artisanal" burgers made with special cuts of grass-fed organic beef, brioche buns, and homemade ketchup, preferably served in a vintage setting that nostalgically recalls pre-fast-food times. As we will discuss in the next chapter, it is not just particular dishes that convey status, but the *places* where they are experienced.

The endless search for social distinction through consumption is continuously reframing what is trendy; as soon as a product becomes mainstream, it loses its appeal and must be replaced by a new (re)discovery that will sustain foodies' social status. Social media, where foodies and social influencers post pictures of dishes and reviews of restaurants, have sped-up this trend, such that only "real" foodies with adventurous palates, superior knowledge, keen senses of style, ease of mobility, and fatter wallets can keep up.

The postmodern notion that identity is less a function of rigid class structures than the result of individual projects of self-actualization has been criticized for ignoring the social and economic barriers that many people continue to face in expressing themselves through consumption, let alone feed themselves. For some, rather than signaling a break from capitalism and the emergence of new social formations, it highlights the expansion of capitalism into *hyper-consumerism*—a "late capitalist" stage where goods and services are no longer purchased for their functional value but for their symbolic value, with advertising playing a powerful role in creating such value and thus shaping our identities.

10.4. Ethnic Food

Like class and nationality, ethnicity is both symbolized and produced through food. *Ethnicity* is a fluid and relational category related to immigration, race, and class. It is typically understood as a form of self and collective identity based on language, religion, ancestral origins, and cultural traditions, including food. However, ethnicity is also a way to classify racialized *ethnic others* whose language, skin color, and/or food habits differ from the naturalized and unspoken norms of *whiteness*. As such, ethnicity is closely related to contemporary understandings of *race* as socially constructed. Ethnic food can be considered from both perspectives: as a form of self-identification and a means of othering. On the one hand, ethnic food is an important aspect of immigrant life; it sustains communities and traditions, helps create new homes away from home, provides economic opportunities through entrepreneurship, and forms the basis of cultural encounters with others. On the other hand, ethnic food is often synonymous with cheap and unsophisticated food. It refers to the food of "others"— mostly immigrants and people of color who are viewed as inferior or at best different.

10.4.1. Immigration, Homemaking, and Memory

There is a large literature, both academic and popular, underscoring the importance of food in the immigrant experience. In such writing, *nostalgia* often dominates; ethnic food is romanticized as an escape from the harsh reality of migration and a coping mechanism for settling in a new place. It is the material that preserves traditions, creates communities, builds new homes "away from home," and keeps memories alive. As such, it relates to *resilience*—the ability to withstand and recover from adversity—and *resistance*—the ability to remain unchanged in the face of pressure to do so. Although both concepts come from biological and physical sciences, they are increasingly used by social scientists to understand how people behave and respond to various circumstances, including the traumas of migration, social isolation, poverty, and discrimination. Food might seem mundane and unimportant, but its emotional significance and affective capacity make it a powerful agent of resilience and resistance.

The significance of food to immigrant and ethnic identities differs across groups and evolves over time and space as they negotiate their presence in a new place. There is often tension and contradiction between the desire to cling to familiar foods to maintain a fading connection to distant places; the necessity to adapt to a new context where ingredients and food practices are different; and the appeal of new beginnings and cultural freedoms.

In *Hungering for America*, Diner (2009) shows how Italian, Jewish, and Irish immigrants, who came to the United States in the late 1800s and early 1900s, juggled the memories of hunger back home and the abundance of food they found in America to construct new ethnic identities and settle into new lives. Of these three groups, Italian Americans used food most effectively to strengthen relationships both within their communities and with the rest of society, as illustrated in so many films from *The Godfather* to *Big Night*. Rather than seeking to preserve memories by re-creating the foodways they left behind, Italian immigrants were moved by memories of hunger. The meats, cheeses, produce, and white bread reserved for the upper-class in Italy became widely available, enabling them to create a pan-ethnic cuisine that symbolized a new American Italian self and embodied the "revenge of the poor." Although most Italian immigrants worked in low-wage jobs and earned poverty wages, food in America was more abundant than back home and became the highlight of social life. They ate spaghetti with meatballs, veal scallopini, osso buco, eggplant parmesan, deep-dish pizza, and other dishes that bore little resemblance to what they had eaten in Southern Italy and became classics on restaurant menus. They blended the old and the new and their food evolved along with their identity. Over time, Italian food became mainstream and lost its ethnic association, as witnessed by the popularity of Italian restaurants from fast food to haute cuisine, the large proportion of supermarket freezers taken over by pizzas of all sorts, and the ubiquity of pizza and pasta dishes in school cafeterias and food courts across the country.

Not all immigrant experiences are alike. Circumstances of migration and group characteristics differ, as do the cultural, economic, and political contexts in which new migrants arrive. Since the 1970s, the total number of people migrating voluntarily or involuntarily across national borders has increased, with the majority leaving the Global South for safety and opportunities in the Global North—a reflection of economic globalization and postcolonialism. Yet, as this shift in migration flows occurred and countries like the United States, England, Germany, and Australia became more ethnically diverse and less white, attitudes toward immigrants turned increasingly restrictive and discriminatory, as indicated by the recent rise of nationalist and anti-immigration policies across Europe and the United-States.

Like previous generations, today's immigrants often turn to food for comfort, connections, homemaking, and resistance to marginalization and discrimination. For some, there is an almost reactionary desire to hold on to the past and resist change by structuring and regulating foodways to reflect some sense of ethnic purity. For the majority, however, the notion of ethnic food is fluid and hybrid, like the identities associated with it. *Colonialism* and *postcolonialism* have shaped immigrants' food practices, both in their home countries where imported foods and crops have transformed traditional diets and in migration destinations where demographic and cultural landscapes are increasingly diverse.

For example, in Mexico, the consumption of highly processed and/or imported food, such as industrial tortillas, sodas, and industrial baked goods, has increased dramatically in the past decade (see textbox 3.4). Mexican immigrants to the United States are therefore already familiar with much of the food they encounter upon arrival. In addition, they are usually able to find traditional ingredients relatively easily given the popularity of Mexican food in the United States and the size of the Mexican American population. Thus Mexican food in the United States is a sort of hybrid.

Similarly, as Tuomainen (2009) shows, the food practices of Ghanaian immigrants in London are shaped by their prior exposure to British food in their home country—a result of decades of British colonial rule, which brought them canned and frozen food, Sunday roasts, gin, and Quaker Oats. In the 1950s and 1960s, most Ghanaian immigrants felt comfortable eating British food, which they often viewed as their own and a symbol of social status. However, in the 1980s, as the Ghanaian diaspora became larger and more socioeconomically diverse and a powerful Black cultural movement inspired by growing numbers of Caribbean, African, and Asian immigrants from former British colonies emerged, Ghanaian food became an important marker of identity and a symbol of newly found ethnic pride and confidence. Such food, including dishes like *fufu* and *banku*, is an amalgam of ingredients and techniques that blur tribal and regional differences and occasionally borrow from other immigrant communities, showing that ethnic food is not about the purity of ingredients but about the capacity to affect connections and symbolize shared identities.

The more dire the circumstances of migration and unwelcoming the destination context are, the more challenging—and yet important—it is for migrants to rely on food for comfort and self-affirmation. This is the case for refugees who may be spending years in camps before returning home or settling into a new place. The inability to prepare, share, and consume foods from the homeland is an embodied experience and constant reminder of displacement and dispossession. Dunn (2011), who studied the food practices of ethnic Georgian refugees pushed out of their homes by Russian forces, describes the macaroni distributed by international aid agencies as "anti-food," in that it carried no cultural meaning, came from "nowhere," reminded refugees of what was lost, and prevented them from exercising social roles of providers, cooks, mothers, hosts, and friends. In contrast, smuggled honey became a symbol of home, ethnic identity, resistance, and agency in the face of adversity.

In the literature on the role of food in immigrant communities, it is typically assumed that, with time, as immigrants and their descendants assimilate culturally and economically, ethnic food eventually loses its significance. While preparing and eating ethnic food is an important part of the everyday life of first-generation immigrants, it becomes more symbolic and self-conscious for subsequent generations, for whom ethnic food is reserved for festive occasions. Nevertheless, food is often one of the last aspects of ethnicity to change, long after language, religion, and clothing have been abandoned. What remains are a few iconic dishes that bear little similarity with the original foods but are no less meaningful.

10.4.2. Ethnic Food Entrepreneurship

The establishment of grocery stores, restaurants, street food vending operations, and catering businesses has been a common source of employment and income for immigrants in various times and places. As we will learn in chapter 11, ethnic businesses are an important component of urban foodscapes, improving food access for low-income communities, but also providing exotic destinations for foodies and adventurous eaters. Scholars of ethnic entrepreneurship debate whether the ethnic food economy reflects a proclivity for self-employment among certain groups or an economic necessity when barriers of entry block access to other forms of employment (Joassart-Marcelli 2021). Those who support the former hypothesis often argue that immigrants have "ethnic resources" at their disposal, including business and credit associations that are built on strong social ties. For example, the large number of Chinese and Korean restaurants and grocery stores in many cities, not just in the United States, is often explained by the existence of rotating credit associations that support new businesses, which once established will in turn assist other similar operations, underlying the development of an ethnic occupational niche. Some also posit that certain ethnic groups have an "entrepreneurial culture" that encourages them to pursue self-employment. These ethnic resources are often tied to the existence of ethnic enclaves, such as Little Italy, Koreatown, Little Havana, and Chinatown, where social networks are spatially constructed and maintained.

Other scholars, however, contend that ethnic entrepreneurship is less a function of culture than the result of disadvantageous labor markets and "opportunity structures" that make starting a food business a common means of livelihood in specific places. Anti-immigrant and xenophobic attitudes, restrictive immigration policy, economic downturns, and racial discrimination might limit employment opportunities and force immigrants and minoritized people into entrepreneurship. For example, Turkish immigrants have opened döner kebab shops in England, Germany, and other northwestern European countries. Over time, the dish has become one of the most popular fast foods. According to Wahlbeck (2008), this type of small food businesses allows new immigrants to achieve a respectable social status in a context where other means of social mobility are not available to them.

Given prevalent anti-immigrant attitudes, heavy restrictions, and limited access to credit, many immigrants find it difficult to start and operate formal businesses and often turn to the *informal economy*, especially street vending, to make a living. Informal labor, as defined in chapter 4, includes work arrangements in which labor is unprotected in a context where regulations and safety nets have been put in place to ensure safety and minimize abuse. While informal labor represents a very large share of employment in low-income countries, it is also significant in higher-income countries. For example, in the city of Los Angeles, there are at least fifty thousand street vendors, ten thousand of whom sell food (Liu, Burns, and Flaming 2015). Many are undocumented immigrants from Mexico and refugees from Central America and about 80 percent are women of color. Street vending is heavily regulated and policed, with many vendors being regularly harassed by the police, receiving hefty fines, and having their equipment confiscated for not having the proper permits (Muñoz 2016). Thanks to persistent organizing, the City decriminalized street vending in 2017 and streamlined the permitting process, while keeping regulations and fines in place.

Early immigrants to a destination typically open businesses for their own communities first, playing a central role in supporting ethnic identification, social connections, and the enactment of shared culture. Over time, however, as these groups assimilate and their cuisine becomes recognized outside of their community, they expand their customer base beyond the ethnic economy. As this occurs, they often adapt grocery store selections, restaurant menus, dishes, and flavors to please nonethnic eaters. For example, Indian food is effectively considered "national food" in Britain. The most emblematic dish—chicken tikka masala—was reportedly invented in England by adding a spicy creamy sauce to chicken tikka which is traditionally grilled in a tandoor oven. It figures on the growing list of ethnic dishes that have been adapted and branded to please white consumers, including tacos in the United States, couscous in France, and döner kebabs in Germany.

10.4.3. Race, Multiculturalism, and Authenticity

In contemporary society, people often cross ethnic lines when eating. Foods that have caused derision and disgust in the past are now drawing the attention of foodies and gastronomes. For some, the rising popularity of ethnic food reflects a multicultural and cosmopolitan society where tastes, social norms, and racial hierarchies are less rigid than in the past. Food, they argue, enables cultural encounters, promotes tolerance, and dispels racism.

However, the selective use of the term *ethnic*, as it applies to food, suggests that there remains a powerful differentiation between "us" and "them." In theory, all foods are ethnic, since they come from somewhere and are typically prepared and consumed by people who share an ancestry, language, religion, or other characteristic. The fact that only some foods are described as such reveals the existence of a social process of "othering" by which ethnic food is distinguished from normalized—typically white—mainstream food. For instance, in the United States, French, Japanese, and Italian food are no longer considered ethnic, in contrast to Mexican, Thai, Indian, or Chinese food, to name a few. Whether a cuisine is considered ethnic or not reflects the racialization of its people and their integration in mainstream society. It is therefore relational and situational.

Descriptions of ethnic foods in restaurant reviews, cookbooks, food magazines, and social media are often replete with racist assumptions. In general, ethnic food is viewed as cheap and simple—a form of craft prepared by cooks, including women whose skills and recipes were passed on from previous generations, as opposed to the art of haute cuisine performed by trained chefs, mostly white men. Restaurant décor, music, and service are often ridiculed, but tolerated as part of a cheap meal. Details about the origins, experiences, and struggles of the people who prepare ethnic food are typically left out, unless they contribute to the consumption experience. For instance, national and regional differences tend to be erased under umbrella terms such as Indian food encompassing Pakistani and Bangladeshi dishes and Mediterranean food merging Lebanese, Iraqi, Turkish, and Syrian specialties. Such labels are easier to comprehend and less threatening to consumers, particularly in an Islamophobic environment. As some have pointed out, the appreciation of ethnic food has not translated into an appreciation of ethnic people. For the most part, it is a superficial engagement that celebrates ethnic difference without confronting the racial prejudice and structural factors underlying racial inequality.

The attitude of white people toward ethnic food is complicated by class. Shortly after a particular ethnic food becomes more popular and common, it begins losing its appeal to experts and foodies who deplore its bastardization and lack of *authenticity*. Because the ability to discern authenticity is only accessible to those with the means to travel, acquire knowledge, and experience food in the "right" settings, it reflects an outsider's perspective of what is allegedly original and has become the crucible of taste and distinction in cosmopolitan society. The people who cook ancestral food from their homelands presumably lack the distance to consider questions of authenticity, but instead cook and eat "ethnic" foods in unselfconscious, nonreflective, and habitual ways. Authenticity is a symbolic and performative notion created to please upper-class consumers. As such, it becomes one of the primary paths to recognition by food experts. By constraining what ethnic food ought to be, ideas of authenticity cement otherness and deny immigrants self-identification and creativity. In that cultural context, it may not be surprising that many restaurants praised for their authenticity are owned and operated by white chefs who have had the privilege to discover the food cultures of others and appropriate them for their own benefits. Their version of authentic food elevates affluent diners who distinguish themselves by their impeccable tastes and cosmopolitan attitudes, while simultaneously devaluing, excluding, and erasing immigrants and people of color.

A similar process occurs in the context of culinary tourism in which exotic food is something to be discovered, conquered, and possessed for one's own enjoyment. It appears to be less motivated by a desire to know or experience another culture, than it is about performing open-mindedness, acquiring cultural capital, and reinforcing cultural and social hierarchies.

Conclusion 10.5.

Food is packed with meaning about who and where we are. Rather than reflecting a fixed identity, however, food consumption contributes to making us who we are. To better understand this process, it is useful to think about everyday practices of eating as relational; food connects people to each other and to the world they inhabit. What we eat is shaped by our position in society, geographic origin, cultural background, past experiences, and aspirations for the future. It simultaneously reveals who we are to others, allows us to express our identities, and leads to comparisons and judgment by others.

This chapter described food as more than a symbol, emphasizing its material effects on bringing people of different nationalities, races, ethnicities, and social classes together but also distinguishing and dividing them along these axes. The contradictory inclusionary and exclusionary qualities of food underlie its significance and power.

Food in the City 11

Learning Objectives

- Examine the central role of food in the spatial and social organization of cities and urban life.
- Use the concept of foodscape to investigate how urbanites experience food and place differently.
- Identify how foodscapes are materially and discursively produced by various actors who have different claims to space.
- Explore the relationship between access to food and food insecurity, paying attention to political and economic factors underlying disparities within cities.
- Explore efforts to create alternative and local networks of food provisioning such as urban agriculture and assess their potential to promote food justice.

Food has a deep connection to place. While agriculture and farming are associated with rurality, dining and gastronomy are often tied to urban life. Historians generally agree that cities would not have emerged without agriculture. The domestication of plants and animals presumably freed some people from working the land, allowing them to devote their energy to other pursuits, such as architecture, science, industry, and art, including entertainment and gastronomy, which are closely related. Today, most urbanites purchase food grown in rural areas, whether outside of their city or across the globe. There is a complex food economy in place to feed urban residents, consisting of a vast network of retailers and restaurants dependent on processors, wholesalers, specialty trades, distributors, and advertisers.

Food occupies a central place in the spatial organization of everyday urban life; urbanites spend a significant part of their day navigating the city to shop for food, cook, and eat, whether at home or out. These food provisioning activities differ drastically in space and time and according to socioeconomic status. In developing countries where processed food is less readily available and affordable, those responsible for feeding their households—usually women—must work each day to grow, purchase, and prepare food. These tasks require skillful navigation of the city: planning meals, getting to know suppliers, adapting to market availability, negotiating prices, and checking for quality. Street markets, where vendors sell produce, meat, fish, bread, and all types of prepared foods and drinks, play an important role in meeting

FIGURE 11.1 Lau Pa Sat Market in Singapore. Also known as the Taylok Ayer Market, it is Singapore's largest central-city public market. Under this large Victorian structure, Chinese and South Asian vendors sell a wide range of food, including noodles and satay. Their activities spill into adjacent streets, which are blocked to traffic and crowded with food stalls, eaters, chairs, and tables. Some of these vendors were among the first "street establishments" to be recognized in the notoriously snobbish Michelin guide, shocking many and bringing international attention to the area. Much more expensive and formal French, Chinese, American, and Japanese restaurants can be found just a few blocks from the market.

Source: https://commons.wikimedia.org/wiki/File:Satay_stalls_along_Boon_Tat_Street_next_to_Telok_Ayer_Market ,_Singapore_-_20120629-02.jpg.

people's needs, supporting formal and informal jobs, and creating a public space where people interact, socialize, and engage in politics. With economic development, corporate supermarkets, chain restaurants, and delivery services come to dominate food retailing, seemingly privileging convenience over sociability. Nevertheless, in many cities, food-oriented public spaces continue to be a central node of urban life (see figure 11.1). Shaped by the broader food system, urban foodscapes structure the way we interact with food and with each other through food.

Cities, especially large world cities like Singapore, Shanghai, New York, London, Mexico City, Mumbai, and Los Angeles, distinguish themselves by the diversity of their food economy, with high-end restaurants operating alongside street vendors and other food entrepreneurs who imprint their cultures into the local food landscape. Such diversity is attractive to middle- and upper-class consumers, including tourists, who revel in the many dining options and sensorial stimulations cities offer. Indeed, in a service-oriented post-Fordist economy like the United States' and Western Europe's, food has become an important engine of urban growth. Government officials and developers increasingly use food to brand their city and attract investors, high-income residents, and tourists. Cities strive to distinguish themselves by their food culture, keeping old food traditions alive and creating new ones.

At the same time, access to food within urban environments is highly uneven, particularly since urban dwellers depend primarily on retailers to obtain food. While some neighborhoods are characterized by an overabundance of food choices, others have very limited options for healthy, affordable, and culturally appropriate food. Limited access to food causes food insecurity, which may lead to poor health.

Access to food for urban dwellers has always been a critical issue tied to power. Even in the first cities, social hierarchies—typically justified by religion—determined who controlled the food supply and how it was allocated. Today, power differentials continue to shape food distribution and influence how people experience food environments, including their ability to partake in consumption and capitalize on their food culture.

This chapter builds on the concept of *foodscape* to investigate spatial relationships between food and the city. After defining foodscapes and their key elements, we examine how they are experienced differently by urban residents, some of whom enjoy the pleasures of dining out as integral to the urban lifestyle, while others struggle to meet basic nutritional needs. In the process, we show how foodscapes are constructed by various actors, including governments, corporate retailers, developers, and urban residents of all stripes who work, shop, and eat in the city. We also explore community efforts to transform urban foodscapes and create alternative urban networks of food provision.

Foodscapes 11.1.

The notion of foodscape emerged in recent years to describe the cultural, economic, political, social, and physical landscape in which food is materially and discursively produced and consumed. Attaching the suffix *-scape* to food emphasizes its situated and perspectival aspects and shifts scholarly attention from food itself to the larger geographic context in which it is prepared, advertised, imagined, sold, and ingested.

Foodscape is theoretically grounded in the idea of *cultural landscape*—a key concept in geography. In contrast to physical or natural landscapes which consist of landforms such as mountains, rivers, plains, as well as fauna and flora, cultural landscapes have been modified by human activities to include built structures, domesticated animals, and socialized people. For human geographers, the notion of cultural landscape emphasizes the dynamic interaction between people and their environment. The imprint of human activities is particularly noticeable in urban landscapes, which are constantly changing as people, money, and commodities flow in and out of cities or neighborhoods.

There are several theoretical understandings of cultural landscapes within geography. For many cultural geographers, landscapes are *representations* that reflect unique ways of seeing—a view of spatial arrangements or place from a specific perspective. As in a painting or photograph, the subjects, shades, lights, colors, and angles describe a particular scenery that can be interpreted to reveal what is important and what is hidden. In other words, landscapes are narratives that tell partial stories.

Other scholars, however, are less interested in what landscapes "reveal" than in what they actually "do"—how they become part of everyday life. For them, landscapes are *lived places* that are imbued with power and shape everyday experiences in concrete and material ways. They are simultaneously shaped by social processes and structuring these processes. Those who see *power* in the landscape are often influenced by Marxist theory and argue that capitalism produces highly differentiated landscapes that maintain class inequality and foster capital accumulation. Cities, for example, are structured into different neighborhoods that stimulate consumption, ensure a large supply of cheap labor, isolate industrial and polluting activities, and bring financial returns to investors. For these scholars, landscapes play a central role in reproducing class relations as well as race and gender difference by organizing social life and determining what and who belongs where.

Taking a less deterministic approach, many cultural geographers emphasize how landscapes are experienced, transformed, and negotiated by workers, consumers, and inhabitants who live in them. Focusing on situated everyday experiences, they draw attention to the agency of urban dwellers in navigating, shaping, and giving meaning to landscapes. While attending to the material aspects of landscapes, they also highlight their discursive, symbolic, and affective attributes.

Setting asides scholarly debates, these various perspectives point to the *relational* nature of cultural landscape that emerge and evolve out of interactions between humans, objects, environments, and other related landscapes. For example, a busy commercial street in London with a diversity of ethnic businesses is the lifeworld of immigrants who have come from India, Pakistan, Jamaica, Ghana, or other former British colonies. The streetscape was produced by their efforts to make a living in a new place and helps sustain communities and create a sense of belonging. It is connected to the past upon which it is physically layered and to other places in the city, country, and abroad. In that sense, it is the coming together of multiple relationships, meanings, and experiences that are in constant flux—what geographer Doreen Massey (2005) aptly calls "throwntogetherness."

Foodscapes must be understood as food *landscapes*. They are constructed out of built elements (e.g., kitchens, restaurants, gardens, stores), people (e.g., cooks, chefs, workers, eaters, growers), objects (e.g., ingredients, recipes, pots, spices), ideas (e.g., sustainability, health, ethnicity, domesticity, distinction), senses (e.g., smells, belonging, memories, hunger), and practices (e.g., cooking, shopping, eating). At once physical and discursive, foodscapes incorporate the social, cultural, political, and economic relations that influence the way people interact with food and with each other through food in particular places.

Unfortunately, the theoretical underpinnings of foodscape have been ignored or forgotten as the term became more widely used. For public health scholars and urban planners, it usually refers to the *built food environment* whose elements can be enumerated, categorized, and potentially improved with policy intervention. In that perspective, foodscapes are simply physical containers of supermarkets, fast food restaurants, convenience stores, and other sources of food, offering a very narrow perspective of the relationship between food and place. In contrast, many sociologists and anthropologists focus on the *discursive* aspect of foodscapes—the representation of food and place in magazines, cookbooks, blogs, and celebrity chef television shows—and what they reveal about society and culture.

This chapter engages with the idea of foodscapes as a set of lived and imagined places in which inhabitants relate to each other through food in material and sensory ways. Focusing on urban settings highlights the idea that foodscapes are produced both *physically*, through the labor of those toiling in the many sectors of the food industry and the capital directed at restaurants, food stores, and other food-related spaces and events, as well as *culturally*, through the various ways we use these spaces in our everyday activities and imagine, describe, and package them for consumption. Doing so bridges the economic/material and the cultural/symbolic without privileging one over the other.

11.2. Eating in Public

One of the key characteristics of urban foodscapes is their blending of public and private spaces. While many meals are consumed at home, people also eat in the street, at roadside stands, in taverns, restaurants, and coffee shops, in their cars, and at work. Food shopping also takes place in public requiring some interaction with others and making our purchases visible to those around us, although online shopping and delivery services are beginning to change this. In fact, the line between public and private is a blurry one. We bring home take-out food from restaurants or prepared food from supermarkets, home kitchens increasingly resemble professional kitchens, and the food media has brought chefs and their sophisticated creations into our homes. Meanwhile, some restaurants are located in private homes or decorated to look like them, dining rooms in fancy establishments can be very exclusive, and reservations at popular eateries can be virtually impossible to secure.

In recent decades, eating has become an increasingly public event. Social observers either celebrate the expansion of "eating out" as the emergence of a new food culture or mourn the end of "eating in." Evidence suggests that consumers, especially those with higher income and located in the Global North, allocate an increasing share of their food budget, consume a greater proportion of calories, and spend more time eating out than in the past. Some are concerned about this trend because they view it as the cause of rising obesity and declining family values—a topic linked to issues of domesticity, which we will explore further in the next chapter. Others interpret it as a new form of sociality that is less rigid and more fluid than what was traditionally afforded when most people ate at home and restaurants were restricted to wealthy patrons or special occasions. Eating out has become a prominent way to interact with others and express identities in urban environments, with the restaurant playing a pivotal role. A vast and growing industry has been built on providing urbanites with opportunities to eat in public.

Eating out is not a new phenomenon. Taverns and street markets were essential spaces of social life in ancient Greece and Rome, as they have been in most of the urbanized world. People ate out with others for a variety of reasons: because they were traveling away from home, meeting business or political partners, or celebrating a special occasion. Although the significance of public eating places declined in Europe in the Middle Ages when most people lived in rural areas and ate at home, the urbanization unleashed by the industrial revolution led to their revival. Taverns, inns, and guesthouses became important places for travelers to get a meal, a drink, and some rest. They were often located in people's homes where a single meal, typically consisting of bread and soup or stew, was served at communal tables—also known as *tables d'hôtes*. As cities grew, men of all occupations gathered in alehouses, pubs, and other drinking establishments to quench their thirst, eat a simple meal, socialize, and discuss politics (see textbox 11.1). Women, unless they were servers, sex workers, or undaunted by social expectations, typically did not frequent such establishments. Wealthy individuals ate and entertained at home where private chefs and servants prepared and served meals for them.

In modern times, as more aspects of daily life became commoditized, a wider range of establishments emerged, including the restaurant where elaborate dishes selected from an expanding menu were served by professional staff at separate tables in a decorous atmosphere (Appelbaum 2011). Many historians trace the first *restaurants*—literally meaning things that restore—to the late 1700s in Paris when the French revolution ended the rigidity of professional guilds, allowing chefs, some of whom had lost their employment with aristocratic families or came to Paris in search of opportunities, to create a new form of hospitality, including establishments specializing in "restorative" bouillon for refined consumers (Spang and Gopnik 2019). In the 1800s, Paris experienced a culinary revolution and an explosion of restaurants, brasseries, bistros, and cafés where increasingly sophisticated dishes attracted the expanding bourgeoisie. This vibrant foodscape caught the attention of visitors

The British Pub

The pub, which is the modern version of the ale-house, is recognized as a uniquely British institution. In 1700, following the growth of commercial beer brewing with hops, there were presumably fifty-eight thousand alehouses in England alone. Until recently, every neighborhood, town, and village in England, Ireland, and Scotland had at least one "public house" where beer was served, usually by the pint or fraction thereof (the state-mandated measurement unit for draught beer). In the United States, people may be familiar with the Irish pub—a caricatured version with dark wood panels, a brass bar, and vintage beer signs that became ubiquitous in the 1980s but has since then lost most of its appeal for lacking in authenticity.

Many view pubs as spaces of conviviality that generate a sense of community and solidarity as people of all classes, genders, sexualities, and races mingle and socialize, with drunkenness temporarily removing inhibitions and sociocultural barriers (Jayne, Valentine, Holloway 2016). Pubs are often perceived and portrayed in film and writing as places of belonging, where social misfits and people down on their luck might find sympathy and companionship. For some, pubs are also essential spaces of civil society where news and opinions could be exchanged and debated. As one of the few gathering places open to the working class, pubs have also played a role in labor and political organizing. Often, though, pubs are seen as disorderly places, where people engage in irresponsible, unhealthy, immoral, and risky behavior, including sexual debauchery and violence.

Although sameness is commonly assumed among pub patrons who perform similar drinking activities, this does not mean equality. People come to the pub from different places and return to these places afterward. Furthermore, people experience the pub differently, with drinking in public constructed as entertainment and pleasure for some and problematized for others—mainly women and young people—whose presence is often monitored and regulated.

Today, pubs are under threat of extinction, as they have been replaced by other gathering places such as coffee shops, clubs, and restaurants where different groups may develop and express their identities and socialize in segregated ways. Remaining pubs are primarily the domain of working-class men or tourists in search of a traditional English experience. This trend has been decried as symbolic of the disappearance of public space, community, and Englishness.

and foreign travelers who described it as uniquely Parisian and modern. Indeed, by the mid-1800s, Paris had allegedly more restaurants than any city in the world. These establishments allowed the growing upper-middle class to revel in the luxuries and pleasures afforded by modern society and to distinguish itself from the working class who purportedly ate unrefined and tasteless foods at home or in disreputable and insalubrious places. To this day, French people consider gastronomic restaurants a symbol of French culture and lifestyle.

This brief history is clearly Eurocentric and ignores the fact that restaurants existed in other parts of the world before the French culinary revolution (Rawson and Shore 2019). For example, large Chinese cities boosted restaurants as early as the twelfth century. Concentrated in commercial districts, these establishments catered primarily to business travelers and offered elaborate meals consisting of numerous dishes served at private tables. In Japan, a unique restaurant culture emerged in the sixteenth century with multicourse meals crafted to highlight local and seasonal foods. Other places around the world also had commercial public establishments serving individualized meals, setting them apart from street stands and taverns as restaurants.

In many ways, the modern restaurant as it emerged in Paris in the nineteenth century symbolizes the democratic values of the enlightenment and the expansion of capitalism and consumerism: it was allegedly open to anyone who could afford to pay for a meal that had become a commoditized service and experience. Meals that had once been provided by family members or servants were now sold on the market. Yet, while presumably "public," restaurants were highly segregated by class and gender; different places served different social groups and only a few were considered acceptable for women, especially if unaccompanied by a man.

Speaking of the restaurant as if it were a singular type of establishment is misleading considering the diversity of commercial places serving food. One of the purposes of restaurants is convenience; it provides a service that saves consumers the labor of planning, shopping for, preparing, and cleaning up after a meal at home. This has become more important in a society where people appear to be increasingly busy, including the many women who joined the labor force (see chapter 12).

Of course, restaurants are much more than a commercial operation selling convenience. Eating out is an experience that includes food, but also service, décor, music, scenery, the wait staff, and other eaters. As we will see in the next chapter, the labor involved in creating this experience is rarely acknowledge and purposely kept hidden in the back of the house. Every restaurant, whether a fast-food chain, casual eatery, or Michelin-starred establishment, straddles that line between providing convenience and experience, with the meaning and importance of both aspects varying immensely from place to place. A geographic perspective that conceptualizes restaurants as relational places contingent upon local circumstances and cultures and shaped by larger scale economic processes, political forces, and social norms, helps to understand their significance in urban life as providers of both convenient services and social experiences.

In the past several decades, eating out has been democratized, with more people eating away from home on a regular basis. However, class, race, and gender continue to distinguish eating places and shape the act of eating out. A wide range of restaurants are now serving a highly segmented market, with most consumers frequenting a limited subset of businesses. In the United States, family restaurants, diners, and delis have long occupied the foodscapes of large cities and smaller towns. With rarely changing menus, they serve consistent, filling, and relatively affordable food in a family-friendly setting. Ethnic restaurants also serve a key role in meeting the needs of immigrants and attracting a young and cosmopolitan clientele charmed by their exoticism (see chapter 10). Fast food, especially chains like McDonalds, KFC, and others, has been one of the most transformative forces in contemporary foodscapes, changing the way we eat, homogenizing food culture, displacing older establishments, and extending corporate control of our foodways. As such, it has received much negative attention form cultural elites and public health experts. Meanwhile, high-end restaurants continue to draw the admiration of food critics, even though consumers increasingly reject their snobbery in favor of casual eateries, including revamped diners and family restaurants, ethnic eateries, farm-to-table cafés, and temporary "popup" restaurants in interesting settings. Many other food providers, such as street vendors, farmers markets, public halls, and food trucks, operate alongside these "proper" restaurants.

The expansion of dining out across socioeconomic strata, however, hides trends that put into question the notion of democratization and raises social justice concerns. First, the uneven distribution of restaurants within cities means that some neighborhoods offer a variety of dining options while others only have a few fast-food shops, limiting the ability of residents to enjoy food in the same way. Second, within restaurants, there is a division between those who sit at the tables and those who wait on those tables or work in the kitchen, many of whom could not afford to eat the meals they prepare and serve. Despite the friendliness and conversational ease of most waiting staff, particularly in American restaurants, their relationships with customers remain transactional and shaped by power differentials.

11.3. Gastro-Development and the Urban Food Machine

Because food experiences have become a key feature of urban lifestyles and identities, public officials, business owners, developers, and urban elites have been bolstering their city's unique "food scene" as a way to stimulate consumption and attract investors, high-income residents, and tourists. This strategy is part of a larger urban agenda built on the notion, popularized by Richard Florida (2002), that urban growth and economic vitality depend on young, educated, and creative people (e.g., designers, researchers, software developers, journalists, scientists, venture capitalists, artists) who are attracted to cities that offer an exciting urban lifestyle through cultural amenities such as parks, bike lanes, museums, art districts, independent boutiques, music venues, bars, and restaurants. Rather than using public resources to create jobs, invest in education, build infrastructure, strengthen regional industries, and provide a social safety net as Keynesian economics promoted in the 1950s and 1960s, urban planners and policy makers switched their focus in the late 1970s to making their cities more attractive, catering to investors, affluent consumers, and homeowners who are expected to bring resources to the city and refill public coffers. This approach to urban governance reflects a *neoliberal* agenda to the extent that it alters the role of the government from direct provider of services to facilitator and supporter of private businesses. It also reflects and encourages a shift of economic activity from production and manufacturing to consumption and service, in which culture plays a pivotal role.

Urban foodscapes around the world have been altered by these efforts, with new cultural spaces of food consumption constantly appearing, shifting the attention of investors and elite consumers from one neighborhood to another and transforming the city. I call this process *gastro-development* (Joassart-Marcelli 2021) because it promotes urban development via food-oriented projects like food halls (see textbox 11.2), restaurant districts, street markets, food truck fairs, and festivals. Alternative projects, such as community gardens and farmers markets, are also increasingly enrolled in urban development and revitalization. Other leisure sites, including shopping malls, museums, theaters, and sport stadiums have also been redesigned to feature food as part of the consumer experience such that most forms of entertainment involve "getting a quick bite."

These gastro-development projects are not just the doing of government officials, they involve many actors who together produce and transform urban foodscapes. For instance, planners and public officials allocate public resources to the development of food spaces by clearing space, managing land, providing infrastructure, and selectively granting building permits and operation licenses. Architects, developers, and investors design, build, and finance spaces where food is a major attraction. The food media, including online reviews, local newspapers, and lifestyle magazines, publicize and advertise urban foodscapes with a bias toward new "hot spots," playing a key role in the discursive production of food spaces. Business and economic development organizations that often operate as public-private partnerships also stimulate the creation of neighborhood districts with banners, clean-up, and other "main street" beautification projects. Because some of the most exciting and authentic food spaces are found in neighborhoods that tend to be perceived as dangerous, police officers and private security guards are often tasked with surveilling these public food areas and ensuring the safety of consumers (often at the expense of longtime residents). Finally, restaurateurs and chefs—especially those with celebrity

Box 11.2   **Food Halls as Urban Revitalization Projects**

Food halls and public markets are experiencing a renaissance in many cities. Old public markets, which lost their purpose as supermarkets expanded in the second half of the twentieth century, are being refurbished and vacant buildings such as factories, train stations, warehouses, churches, and hospitals are being repurposed as fashionable spaces of food consumption. For example, *Chelsea Market* in New York, *Ribeira Hall* in Lisbon, and *Depo* in Moscow, each house over fifty food vendors or restaurants, including local artisans and highly rated restaurant outposts that attract a diverse clientele consisting of young and affluent residents, food connoisseurs, and tourists.

Investors in these projects are typically wealthy developers and corporations, such as Google and Time Out, who see this as an opportunity to diversify their investments and transform cities in which they have a stake. Large architectural firms work on multiple projects, replicating spaces from one city to the next. In some cases, local governments contribute real estate and public funding to stimulate neighborhood revitalization and urban development.

In France, where strict food rituals have influenced the timing of meals and table manners, the popularity of food halls and courts further confirms changing foodways. These revamped spaces offer the convenience of fast food without its negative image and with the bonus of publicity, sociality, and distinction. In the past few years alone, large food halls have opened with great fanfare in many cities around the country, including the *Halle Gourmande* in Nice's old Gare du Sud train station (see figure 11.2), the *Halle Boca* in Bordeaux, and *La Cité de la Gastronomie* in Lyon.

While privileging local purveyors and artisans, these spaces embrace a similar aesthetic that combines industrial urban chic, local heritage, and connection to the countryside. Whether in France, Argentina, Hungary, or elsewhere, many of the food stands have English names that seemingly contradict the emphasis on cultural heritage but are recognizable to tourists and convey a sense of hipness. They also serve similar types of food and dishes, including "gourmet" burgers, tacos, sushi, ramen, pizza, and small shareable plates that continue to evolve with trends. Most halls include bars or stands selling alcohol, preferably local wine, craft beers, or artisanal liquor, inviting visitors to stay a while, have a good time, and engage in the *performance* of contemporary urban lifestyle.

The popularity of public food halls has been paralleled by an expansion of upscale markets organized as a series of stalls mimicking independent vendors specializing in niche products. Eataly, with almost forty locations across the world, is one of the best examples of this new form of thematic food retail that combines shopping, eating, and entertainment.

FIGURE 11.2 La Halle Gourmande in Nice, France. This commercial space opened in 2019 in the former Gare du Sud train station, which closed in 1991. It is home to thirty-three businesses, twenty-eight of which sell food.

Source: https://pixabay.com/photos/nice-south-station-covered-market-4927285/.

status—play a key role in "activating" specific neighborhoods. Opening a business in an "up-and-coming" area might be a bold move that not only positions the restaurant as adventurous and hip, but also transforms the surrounding neighborhood by attracting consumers and other businesses. Finally, in the past decade, social media have provided a megaphone to relatively young, educated, and affluent urbanites whose posts and ratings on Yelp and other platforms amplify these trends. These various actors work together as a sort of *urban food machine* to stimulate growth via gastro-development (Joassart-Marcelli 2021).

Most of the recently developed or revitalized urban food spaces share a similar *aesthetic* that, in the geographic imagination, is associated with visions of community, authenticity, and multiculturalism. The community aspect is visible in the design of most places, including large indoor or outdoor open spaces with communal tables and lounging areas meant to invite consumers to dwell and sit together. Artistic murals, reclaimed materials, exposed bricks, and strung lights accentuate the democratic and urban nature of spaces such as repurposed old factories, warehouses, churches, and train stations (see figure 11.2). Media depictions of food halls and street markets often highlight the mix of customers, with "young tech guys in hoodies" sitting next to "bankers in suits" and "moms in yoga pants" sharing "simple" foods like burgers, tacos, or pizza. Despite the appeal to a sense of community, the consumers described in reviews represent a relatively narrow slice of the population that lives a different lifestyle than those who are not sitting at the communal tables, including the many workers preparing and serving food. Whether meaningful interactions between different groups occur in these shared food spaces is debatable.

Authenticity is another highly marketable characteristics of food that is in high demand among foodies. As noted in chapter 10, taste is socially constructed as a mean of distinction and, in an increasingly homogenous and standardized foodscape, the notion of authenticity enables upper-middle-class consumers to distinguish themselves by their cultural capital, which includes knowledge and experience of "good" food. Authenticity, however, always presumes an outsider's perspective as it is typically the food of other people, places, or times that is being judged, valued, and often appropriated for its originality and genuineness. Here too, the settings and service contribute to creating a sense of authenticity that cannot be found in chain restaurants and suburban food courts. Discovering new "holes in the wall" or "hidden neighborhood gems" is an extremely rewarding process for foodies, who view themselves as pioneers, willing to venture into unusual places to experience authenticity.

A related aspect is the *multicultural* nature of trendy urban food spaces, which are being praised for their ethnic and racial diversity. Like community and authenticity, however, the notion of multiculturalism is socially constructed and based on a partial understanding of difference. In this case, ethnic and racial difference is something to be celebrated, with the collective openness to ethnic food arguably signaling an end to racial prejudice and the emergence of cosmopolitanism. Yet, in promoting colorblindness, such multiculturalism hides the persistence of *systemic racism* and its material consequences on the lives of people of color who continue to suffer disproportionately from food insecurity and are often excluded from multicultural spaces, except in very specific roles as workers or cultural ambassadors. While food may be viewed as a means to bring people together and public food spaces may be conceptualized as *spaces of encounter*, some question the nature of encounters that take place in commercial spaces where inclusion is contingent about purchasing power. In other words, community, authenticity, and diversity are commoditized components of the consumption experience.

Given the spatial nature of gastro-development, urban scholars have begun to analyze its connection to *gentrification*. Not only does the arrival of a new coffee

shop, specialty bakery, or wine bar signal the ongoing transformation of a neighborhood, it contributes to change by attracting newcomers with different lifestyles
and higher incomes. These changes affect long-time food businesses. For example, as
ethnic food purveyors are being "discovered" by foodies, they often struggle to both
meet the expectations of new customers and continue to serve long-time residents,
including generations of immigrants. In many cases, they are taken over or replaced
by new establishments that appropriate their foods and market them in a carefully
curated setting that correspond to foodies' geographic imagination of ethnic places
based on their travels, readings, and media exposure. Bolstered by a newly discovered
and expanding foodscape, older urban neighborhoods attract new residents who are
willing to pay more for food, housing, and other necessities in order to enjoy a bohe-
mian urban lifestyle. As a result of the rising cost of living, many longtime residents,
including immigrants, people of color, and low-income households, cannot afford
to stay in their neighborhood and end up having to move elsewhere. Similarly, older
businesses struggle with competition from newcomers and rising rents. The differential treatment of street vendors and food trucks illustrates the struggle to stay put in
cities that cater to the most affluent consumers (see textbox 11.3).

Beyond the physical effect of displacement that manifest through the real estate market, there are less well-documented cultural, social, and emotional aspects of
gentrification associated with the changing rhythms, networks, and organization of

| Box 11.3 | **The Food Truck Wars** |

Food trucks exemplify the "street food" trend that,
although common in most of the Global South, is
taking cities of the Western world by storm. In recent years, famous chefs have given up the four
walls of their kitchen for a set of wheels and food
entrepreneurs have made a name for themselves
by selling upscale food in parking lots. Chef Roy
Choi is often credited for popularizing this trend
in Los Angeles, having built a food truck empire
selling Korean tacos to foodies, tourists, and
bar-hoppers across the region. Cities like Portland, Oregon, and Austin, Texas, are known for
their food truck culture, attracting similar crowds
of young and hip urbanites.

Food trucks in those cities, however, are not
new. In Los Angeles, *loncheras* have long navigated
the city to sell meals to working people, school children, and families near worksites, schools, and
public parks. Mostly operated by immigrants from
Mexico and Central America, these trucks offer
inexpensive food, such as hamburgers, hot dogs,
sandwiches, and burritos, and a variety of snacks
and sodas. They are joined by many vendors who
cannot afford a truck and sell food from push carts,
stands, and folding tables, often without a permit.

These two groups of food vendors—described
by Agyeman, Matthews and Sobel (2017, 3) as
"hip, entrepreneurial *trucksters*" and "illegitimate,

immigrant *hucksters*"—are regarded very differently by public officials, residents, and food experts. On the one hand, gourmet food trucks are
seen as enriching the foodscape. As such they
receive attention from the food media and their
business is often promoted by public agencies
that facilitate the permitting process, grant them
access to public space, and even sponsor festivals and special events where they are invited to
sell their food. On the other hand, *loncheras* and
immigrant street vendors are generally perceived
as a nuisance and health risk, despite the important role they play in serving working class people
and neighborhoods where other types of prepared
foods may not be readily available. Consequently,
their activities are heavily regulated and often
criminalized. Many vendors receive hefty fines and
have their supplies and equipment confiscated if
they fail to follow certain rules.

This differential treatment by public officials and
uneven enforcement of laws and regulations highlights differences in the *right to the city* whereby
affluent consumers and businesses with greater financial, cultural, and legal capital have more power
to influence the foodscape. The fact that immigrants
are disproportionately penalized and face additional
hurdles points to the role of race in organizing the
city and structuring economic opportunities.

everyday life. As the restaurants, grocery stores, and other food retailers that have sustained foodways and livelihoods for generations disappear, many feel "out of place" in their own neighborhood. This loss of belonging is compounded by a rise in food insecurity caused by the disappearance of familiar sources of food.

11.4. Food Deserts and Food Apartheid

One of the most studied and discussed urban foodscapes is the *food desert*. The desert metaphor appeared a little more than a decade ago in the United States and the United Kingdom to describe areas where food is absent or lacking in affordability, diversity, or nutritional value. It quickly made its way into the policy discourse and began shaping public health initiatives.

The United States Department of Agriculture, which is in charge of implementing federal food assistance programs, defines a food desert as a low-income area in which a significant share of the population does not have access to a supermarket or large grocery store where they could purchase fresh, healthy, and affordable food near their home. Access is conceptualized as a function of distance, with stores located beyond half a mile from a person's place of residence considered inaccessible for urban residents (the threshold goes up to ten miles in rural areas). Although technically difficult to compute, the concept is fairly simple to grasp and can easily be displayed on colorful maps that clearly identify deficient areas (see textbox 11.4).

A large body of work has been built on the notion of food desert. Much of this research involves geographic information science (GIS) and focuses on developing increasingly sophisticated measurements and mapping techniques of food "deficits." The idea of food desert has been most influential in public health, where researchers have used it to explain health disparities by race and income. Their primary assumption is that living in a food desert restricts the food choices of residents—mostly people of color and low-income households—who end up consuming food that is high in calories but low in nutrients, leading to disproportionate levels of obesity, cardiovascular disease, diabetes, and other chronic illnesses (see chapter 13). Numerous studies in the United States and England have documented lower densities of supermarkets in nonwhite and low-income neighborhoods, compared to white and affluent areas, suggesting that class and race influence the food environment.

For example, in the United States, over sixty-five hundred census tracts (small areas of about four thousand people defined by the US Census) are considered food deserts because more than a fifth of their population are in poverty and over a third lack access to a supermarket within half a mile. Together, these tracts, over 80 percent of which are in cities, house almost fifty-five million people or about 18 percent of the US population. Using a more lenient measure of access (based on a one-mile distance) leads to a figure of nineteen million people or 6 percent of the population (USDA 2017). The average number of chain supermarkets available to Blacks and Latinos in their neighborhood is less than half the average available to whites (Powell et al. 2007). Furthermore, evidence from case studies suggests that grocery stores in nonwhite neighborhoods are more likely to be small, charge high prices, carry less variety, and offer low-quality and expired food than in other neighborhoods (Treuhaft et al. 2009). Researchers have found that people residing in low-income and high-minority areas are disproportionately exposed to fast food outlets and convenience stores that presumably sell unhealthy food (Kwate 2008). This includes children whose schools are surrounded by small shops selling "junk" food (Lee 2012).

Box 11.4 **Mapping Food Deserts**

Maps are powerful tools for visualizing spatial patterns of inequality and clustering. A single image can accomplish more than a thousand words in highlighting intra-urban disparities in access to food and pointing out areas with limited access to supermarkets, known as food deserts. When compared or overlayed with maps of income and racial distribution, these food access maps point to the socioeconomic and racial nature of food insecurity. For example, figure 11.3 shows how the location of food deserts in San Diego, California, relates to the nonwhite population by census tract—areas defined by the US Census Bureau with populations of about four thousand people. Except for military bases and universities, which are home to many low-income people and have low access to supermarkets, most food deserts are in neighborhoods where the majority of residents are nonwhite, including City Heights and Southeastern San Diego (Joassart-Marcelli 2021).

As many geographers have noted, maps can also be deceiving and must be used with caution. As any type of representation, they reflect the perspective of the author, with the cartographer making decisions regarding data to map,

categorization of values, scale, unit of analysis, color schemes, etc. For instance, mapping access to food requires decisions about what distance makes a store inaccessible and how to measure this distance (e.g., straight line, street patterns, time traveled, public transit route). One of the most important questions to ask when analyzing a map is not what it shows but what it does not show. A critical observer will always consider what is invisible or left out of the map. This is particularly important when maps are used to inform public policy. In the case of figure 11.3, the map is useful to highlight racial disparities, but it does not say anything about what caused these patterns. It also hides sources of food other than supermarkets and does not provide any information about the everyday food geographies of residents in the different areas displayed. Without understanding the historical political and economic factors underlying the existence of food deserts and the way people navigate these spaces, such a map might lead to misguided policy that simply seeks to fill the "gaps" by attracting supermarkets. It might also contribute to place stigma and thereby reinforce existing inequality.

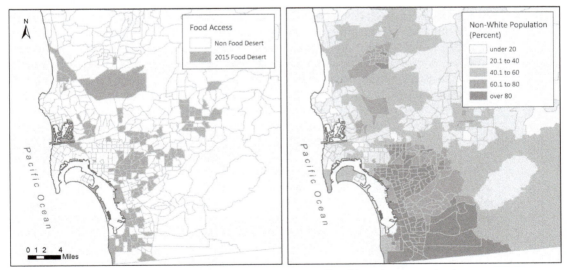

FIGURE 11.3 Maps of food deserts and nonwhite population in San Diego, by census tract.

Source: Author with data from USDA (2017) and US Census (2020).

Public health research on the impact of living in a food desert on health behaviors (e.g., fruit and vegetable consumption) and outcomes (e.g., obesity, chronic disease), however, has been a lot less conclusive. This is partly because it is difficult to distinguish neighborhood effects and to assert the direction of causality (from access to behavior to outcomes) statistically without cross panel data tracking changes in the food environment, behavior, and health over extended periods of time and across different places.

Despite being poorly substantiated, the idea that living in a food desert encourages bad health behavior has gained traction in policy circles and is practically taken for granted. A wide range of policy interventions have been deployed to attract supermarkets in underserved communities through states grants, tax credits, and philanthropy. In many communities, residents have also organized to improve their food environment through urban agriculture and alternative supply chains. However, bringing a supermarket to a low-income neighborhood without addressing deep seated racial and economic inequalities is unlikely to have a huge impact on food security.

Although powerful in attracting attention to the unevenness of urban foodscapes and its relationship to race and ethnicity, the concept of food desert has drawn some criticism from urban geographers for several reasons. First, equating food access with distance to supermarkets reveals a relatively narrow understanding of urban food geographies, which may be augmented by other sources of food, including corner stores, ethnic markets, street vendors, gardens, and other food provisioning spaces and strategies devised by people of color and low-income people to navigate the lack of mainstream retailers. The very notion of desert suggests an absence of food, which is not always an accurate depiction of urban neighborhoods meeting the food desert criteria. Acknowledging existing resources and uplifting ongoing strategies might be more effective in helping residents' meet their food needs and supporting local livelihoods than bringing a big box retailer with no connections to local communities.

A second related criticism is about the stigmatization caused by labeling an area as a food desert. Such labels contribute to the devaluation of place, including its retail sector but also its inhabitants and their foodways. As such, they reproduce stereotypes that Black, Brown, and poor people have unhealthy and dysfunctional relationships to food. Although the research on food desert seems to emphasize environmental factors rather than individual behaviors as explanations for obesity and other health conditions, the portrayal of some neighborhoods as empty, desolate, troubled, and in need of outside intervention may contribute to a lack of investment, a poor sense of place, and feelings of social exclusion, with severe consequences for the well-being of residents, regardless of what they eat.

Thirdly, food deserts are static: they are identified through snapshots of the food environment at a given point in time. These spatial patterns, which are typically represented by maps, are taken as given and the historical dynamics that caused some places to become devoid of large food retailers are rarely examined. Food activists and critical scholars have recently suggested that the term *food apartheid* might better describe the state-sanctioned inequality and segregation processes that characterize urban foodscapes. The term draws attention to the role of systemic racism (see textbox 11.5) in producing food deprivation in some places and food abundance in others. Among these is a process known as "retail redlining" by which supermarkets and chain stores have purposely avoided or withdrawn from neighborhoods of color.

To understand the creation of uneven foodscapes, we need to go back in time and consider how urban areas became segregated. In the 1930s, as the country was recovering from the Great Depression and Black migration from the rural South to northern cities had transformed urban landscapes, the Federal Housing Administration

| Box 11.5 | **Systemic Racism and Urban Spaces** |

Understandings of racism and how people of color face discrimination in many areas of their lives have evolved over time. Traditionally, racism has been conceptualized as *individual racism*: discrimination resulting from individual racist assumptions, beliefs, or behavior. For example, employers may be racist if they refuse to hire or promote people from certain racial groups based on prejudiced assumptions linking job performance to race. This presupposes an individual perpetrator whose actions intentionally hurt people. Today, however, we recognize that some racial biases are unconscious. For instance, doctors and teachers have been shown to have unconscious expectations about patients and students based on race, which in turn influence their interactions with them.

Furthermore, we also recognize that racism does not only manifest through individual actions. The term *institutional racism* refers to discrimination and unfair treatment resulting from prejudiced or biased organizations. For example, the police have drawn the attention of organizations like Black Lives Matter for being an inherently biased institution that does not value Black lives. The media, including the movie industry, has also been accused of racism by failing to allow people of color to represent their own life.

Structural racism describes the society-wide normalization and legitimization of policies, ideologies, and practices that systematically produce cumulative and chronic adverse outcomes for people of color, which translate into lower income, wealth, health, and education. Evidence of structural racism is found in its systemic devaluation of nonwhite lives, including the long-term trauma it creates. Within this framework, it is harder to identify a culprit; instead, racism is understood as the result of structural forces such as capitalism and white supremacy whereby the perpetration of white advantages is premised on the dispossession, displacement, and exploitation of non-white people.

For geographers like Laura Pulido (2010) and Ruth Gilmore (2007), racism is organized and unfolds spatially as people of color are materially and discursively excluded from spaces coded as white. Indeed, Gilmore (2007, 261) defines "racism [as] the state-sanctioned and/or legal production and exploitation of group-differentiated vulnerabilities to premature death, in distinct yet densely connected political geographies." In other words, people of color experience racism as a set of spatially articulated conditions that compromise the quality or longevity of their lives by making them more vulnerable. Highly segregated cities are key constituents of racially differentiated vulnerability, with people of color disproportionately isolated from jobs and livelihood opportunities, deprived of amenities like green space, healthy food, and safe schools, and exposed to pollution and undesirable land uses.

(FHA) was created to address housing needs in cities. It focused primarily on stimulating home ownership for white residents by subsidizing suburban development and lowering the cost of mortgage loans. After World War II, whites moved to suburbs in large numbers—a phenomenon described as "white flight," suggesting that this migration was motivated by a desire to escape an inner city increasingly viewed as nonwhite, dangerous, and unhealthy. Indeed, Blacks and other nonwhites were not allowed to purchase housing in these new developments for fear that property values would decline, as explicitly stated in government documents. Meanwhile, the FHA designated urban areas where nonwhite families were allowed to live as too risky for investment, delineating them with a red line on official maps—hence the term *redlining*. By refusing to insure mortgages in these "risky" areas, the federal government made it virtually impossible for Blacks to get loans, invest in their neighborhoods, and build equity in housing as white people were encouraged to do in the suburbs. Most businesses moved out too, depriving residents of valuable jobs. Discriminatory housing policy was reinforced by transportation, education, land use, tax, zoning, and other policies that systematically favored white and affluent neighborhoods.

Although state-sanctioned racial segregation officially ended in the 1960s, its impacts are still felt in cities today as most redlined neighborhoods struggle to recover from decades of lack of investment and job losses. Today, these areas are characterized by poverty, unemployment, high proportions of nonwhite residents, and low rates of homeownership. They are also much more likely to be food deserts since most retailers have left and supermarket chains are hesitant to open new stores, given the negative image associated with these neighborhoods.

The idea of *food apartheid* acknowledges the role of systemic racism (see textbox 11.5) in shaping access to food, including access to mainstream supermarkets. The history of food retailing in the United States parallels that of white flight (Eisenhauer 2001). Before World War II, grocery stores were clustered in densely populated urban areas. Much smaller than today's supermarkets, they were easily accessible to shoppers who found them within or near their neighborhoods and developed personal relationships with store clerks, making these shops important spaces of social life. Most of them were independent family-owned businesses, with many run by immigrants. When residential suburbanization began, some grocery stores followed, taking advantage of land availability and lower rent to expand their size. Self-service became more common in the 1950s, when the first supermarkets provided customers with shopping carts to gather wrapped products displayed in wide aisles and refrigerated sections. As stores became larger, they benefited from economies of scale that allowed them to lower their costs, capture a larger share of the food retailing market, and outcompete smaller stores that closed one by one. In the following decades, the number and size of suburban supermarkets continued to expand while the central city experienced business closures, causing "leakage" of expenditure outside of the community where such spending could have supported food retailers and attracted other complementary businesses. Following mergers and acquisitions, a few large companies came to dominate the industry. The 1990s witnessed the expansion of megastores and discount stores such as Walmart, Sam's Club, and Costco, which are almost exclusively located in suburban areas, given their very large size, and are inaccessible without a car.

Researchers have documented this process of *retail redlining*—as Eisenhauer (2001) calls it—in cities across the United States, revealing how the closure of food retailers in low-income urban neighborhoods was not an accident but the result of racist policies that systematically privileged white suburbanites while ignoring the needs of central city residents (McClintock 2011, Joassart-Marcelli 2021). Their analysis generates a better understanding of how food geographies emerge and evolve over time, affecting the everyday life of residents who must contend not just with a lack of access to food but with legacies of racism that diminish their life chances. In that sense, food apartheid is much more than the existence of food deserts; it is a process by which food abundance occurs in some places while other places struggle with food insecurity, as witnessed in the lack of supermarkets as well as in widespread poverty, limited employment opportunities, low availability of clean land to grow food, dehumanizing food assistance, stigmatization of local and ethnic businesses as unhealthy, and, more recently, gentrification pressures that exclude long-term residents from improvements in the food environment.

11.5. Urban Agriculture and Alternative Foodscapes

In the past two to three decades, growing food insecurity and discontentment with the global industrial food system have led producers, consumers, and activists to look for alternatives. As described in chapter 5, many turned inward and sought to re-localize food production and consumption by developing and strengthening food supply chains and social connections at the local and regional scales. In cities, this has led to the creation or expansion of community gardens, farmers markets, community supported agriculture (CSA), guerilla gardening, cooperative grocery stores, farm-to-table restaurants, urban animal husbandry, collective kitchens, and other community-based initiatives that provide food locally.

Urban agriculture, which refers to plant cultivation and animal rearing in urban and peri-urban areas, has become the crux of this movement (WinklerPrins 2017). According to its numerous advocates, it offers many benefits such as access to fresh and healthy food, economic opportunities for growers and vendors (many of whom are poor women), physical activity, social cohesion, recycling of urban waste and water, sustainability, urban greening, and climate resiliency. Although it is typically practiced on empty lots, it can also take place on rooftops, in waterways, and in buildings. It may consist of individual gardens, small-scale collective and semi-commercial farms, or large-scale commercial operations.

Individual and collective gardens are common sights in cities of the Global South. According to a frequently cited study from the FAO, eight hundred million people are involved in urban agriculture globally, accounting for 15 to 20 percent of the world's food production. It is estimated that 25 to 30 percent of urban dwellers worldwide are engaged in the agrifood sector (Orsini et al. 2013), with urban agriculture representing one of the fastest growing sources of employment and income in developing countries (Mougeot 1999). However, most urban agriculture is informal and operates outside of the realm of the market economy, making it challenging to gather data on its global scope and effects. Most researchers conduct case studies, highlighting differences between cities and the significance of place in offering unique environmental, social, economic, and political conditions for the expansion of urban agriculture.

For example, in Havana, Cuba, over twenty-six thousand gardens provide food for self-consumption, covering 12 percent of the land and supplying 58 percent of vegetables. Community gardens provide residents with 150 to 300 grams of fresh produce daily, playing a major role in alleviating food insecurity (Cruz and Medina 2003). Urban agriculture was promoted by the government in the 1990s as a solution to food scarcity caused by the loss of assistance due to the collapse of the Soviet Union and the continued United States embargo. An extensive system of urban agriculture was developed, including popular gardens, cooperative production units, and state farms. The government provided training for growers and promoted organic agriculture by banning chemical fertilizers and pesticides and promoting composting and other agroecological methods (Novo and Murphy 2000).

In Dar es Salaam, Tanzania, one of the fastest growing cities in sub-Saharan Africa, urban agriculture has long been a key part of the food system. More than thirty-five thousand farmers grow plants such as leafy greens, tomatoes, amaranths, sweet potatoes, cassavas, cashews, coconuts, and bananas, or raise animals for eggs, milk, or meat (Schmidt 2011). As much as two-thirds of the population is engaged in agriculture to supplement their consumption (Ratta and Nasr 1996). In the 1990s, Dar es Salaam was selected by the United Nations' Habitat and Environment Programme (UNEP) as a site for their Sustainable Cities initiative. The formalization of urban agriculture was advocated to reduce food insecurity, create jobs, and increase environmental sustainability. Yet urban agriculture continues to be perceived by local elites and policy makers as backward and dirty—an embarrassment in a city that wants to compete in the global economy. Consequently, there has been very little change in local policies and planning instruments in support of urban agriculture, which remains highly informal (Mkwela 2013). Most farmers do not have secured legal access to land and their activities are unregulated, if not outlawed. Yet the food they produce feeds thousands of residents, including some of the poorest.

In the past two decades, urban agriculture in developing countries has garnered much attention from international development agencies for its potential role in contributing to food security, generating income, and creating jobs among the poorest urban dwellers, including slum inhabitants, who live in places where adequate food is not available or easily accessible. Yet, with some exceptions (see textbox 6.5 about urban agriculture in Belo Horizonte, Brazil), city planning has lagged behind in incorporating agriculture in urban design and infrastructure.

Although less common, urban agriculture has historically played an important role in the Global North too, especially for immigrants, people of color, and low-income families who have used their gardening knowledge and skills to supplement their diets. During war times, in the United States and Europe, gardening became a patriotic act and a wide range of people planted so-called "victory gardens" (see figure 11.4). In low-income communities of color, there is a long tradition of food activism, including the creation of community gardens and mutual aid societies. For example, Monica White (2018) has documented how Black women have grown food as part of their struggle for self-determination and self-reliance for decades—a story that has been erased from the contemporary food movement. Today, the Detroit Black Community Food Security Network illustrates how gardening forms the basis of resistance against social injustices through self-reliance. By growing and sharing their own food, they confront the industrial corporate food system and regain control over food and other aspects of their lives.

Because of their historical association with immigrants, racial minorities, and working-class people, however, gardening and self-provisioning have often been perceived as backward. In the second half of the twentieth century, as capitalism and

FIGURE 11.4 "The Fruits of Victory," by Leonebel Jacobs, 1918. Poster by the National War Garden Commission promoting gardening as a patriotic act during World War I.

Source: Library of Congress https://www.loc.gov/item/93510433/.

technological innovation brought us processed food and supermarkets, it practically disappeared from cities, except in low-income communities of color where it remained unacknowledged. Even in rural areas, gardening gradually declined as farmers switched to industrial monoculture and stopped selling seasonal fresh produce at farm stands, farmers markets, and urban cooperative stores.

Recently, however, urban agriculture and farmers markets have been rediscovered by planners, policy makers, and community advocates who have promoted them as a path toward food security and what many call *food justice*, which is broadly understood as communities exercising their right to grow, sell, and eat healthy food. What might have started as "alternative" grassroots projects to foster food justice by supporting local growers, improving access to healthy food, caring for land collectively, and reducing food waste have now become "mainstream" in many cities. Aware of consumers' concerns and anxieties surrounding food and their willingness to pay a premium for local products, farmers, business owners, marketers, and developers have embraced the "food movement" and are selling so-called alternative food to the public.

Alternative food spaces have become prominent features of urban foodscapes, mixing comfortably with mainstream supermarkets, grocery stores, and restaurants. Edible gardens have proliferated in both rich and poor neighborhoods and are now adorning school yards, public housing projects, empty lots, churches, farm-to-table restaurants, and new condominiums. A similar trend characterizes farmers markets, which have grown in numbers and are found in a variety of urban areas. The farm-to-table trend has taken over the restaurant world and many stores proudly carry locally grown products. As noted in the previous section, city planners and developers have recognized the importance of food and encouraged these developments.

Food Justice 11.6.

As various groups increasingly invoke the notion of *food justice*, its meaning has become more ambiguous and diluted. Given the numerous injustices caused by the contemporary food system, ranging from labor exploitation, widespread food insecurity and malnutrition, health disparities, uneven environmental impacts, destruction of small farms and local livelihoods, and cultural homogenization, it should come as no surprise that urban food justice initiatives are not part of a unified "food movement" as portrayed in the media. Rather, they are motivated by diverging and often contradictory goals. Thus, it might be appropriate to ask whether "alternative" food spaces are really that different from other spaces of consumption and to consider what sort of urban foodscape would enable food justice.

For many advocates, it is the *local* focus that makes most of these urban food projects alternative. It is assumed that if communities pool their resources and work together, they will become self-reliant, food secure, and sustainable. With support and leadership from philanthropic and nonprofit organizations, volunteers—mostly white and middle-class—have been attempting to build an alternative food system. Yet, most of these efforts aim at filling gaps created by the mainstream food economy rather than challenging its structure. The faith placed on the ability of consumers and nonprofits to reform the food system locally is symptomatic of neoliberalism to the extent that it emphasizes market-based solutions and shifts responsibility onto individual and communities. Such an approach may produce an attractive city, full of exciting places for the creative class to engage in "ethical" consumption, but it fails to address the root causes of food-related social and environmental inequities.

The conflation of "eating local" with food justice is problematic because it assumes that the geographic origin of food is what makes it just, without questioning local relations of production, distribution, and consumption and acknowledging how these are shaped at larger scales by the broader food system. At this juncture, it is helpful to return to the theoretical framework that was developed earlier in this chapter and conceptualizes foodscapes as *produced* relationally out of interactions between people, things, and places and *lived* through everyday experiences, with concrete and material consequences. From that perspective, one might ask how particular foodscapes come to be associated with food justice by focusing on how such landscapes are produced and what they actually do for the people inhabiting them, including how they shape their everyday relationship to food. Doing so reveals the various actors and motives underlying food justice projects in cities. It also highlights the spatial and dynamic nature of such projects, which in their more radical versions are about claiming rights to the city.

Scholars and activists are beginning to recognize the limitations of an alternative food movement that operates within the confines of the market system, privileges the demands of affluent consumers, marginalizes those with the greatest needs, and does not challenge the spatial organization of cities. In response, they have offered different definitions of food justice that emphasize the need to address the structural inequalities underlying food insecurity, labor exploitation, and environmental degradation, which disproportionately affect the most vulnerable members of society (Cadieux and Slocum 2015, Holt-Giménez and Shattuck 2011). For instance, Hislop (2014, 24) defines food justice as "the struggle against racism, exploitation, and oppression taking place within the food system that addresses inequality's root causes both within and beyond the food chain." These more "radical" notions of food justice center anti-racism and the need to dismantle *food apartheid*.

In an article titled "What Does It Mean to Do Food Justice?," Cadieux and Slocum (2015) outline four nodes of organizing and intervention that would create a more equitable food system. Specifically, they emphasize the need to: (1) acknowledge and challenge race, gender, and class inequities and the historical and collective trauma they have brought; (2) build shared economies and systems of exchange that support communal livelihoods and self-determination; (3) reenvision our relationships to land; and (4) value all forms of labor and compensate work fairly. The effects of alternative food projects on food justice could be assessed by their contribution toward these four goals. From a geographic perspective, this is tied to the *right to the city*—an idea developed by Henri Lefebvre to describe a radical vision of cities in which urban space would be controlled and produced by all urban inhabitants rather than by the state and capital. In the context of food justice, it refers to the capacity of all people to participate in the production and transformation of the urban foodscape so that it grants them access and control over healthy and culturally appropriate food and allows them to work in and contribute to the shared food economy with dignity.

According to these criteria, many alternative food projects do not fit well in radical visions of food justice and fail to meet most if any of the goals outlined by Cadieux and Slocum (2015). Indeed, for more than a decade, critical social scientists, including many geographers, have pointed out the shortcomings and risks of not-so-alternative food projects like urban agriculture (Tornaghi, 2014). First, many consumption- and lifestyle-oriented alternative projects fail to acknowledge race and class barriers. Seemingly colorblind and inclusive, they insinuate that everybody is welcome. However, participants are mostly affluent and white urbanites, including gentrifiers, who have the income, time, and presumed moral superiority to engage in this type of activities. Those with lower income and more complicated and traumatic relationships with farming and food, including people of color, are socially and spatially excluded, feeling like they do not belong in the alternative foodscape and resenting how their food practices are being judged as inferior and immoral. These dynamics highlight the *whiteness* of the alternative food movement (Guthman 2008, Slocum 2007, Ramírez 2015).

Second, because volunteers, nonprofit staff, and public agencies think about Blacks and Latinos' place-based relationships to food as deficient, pathological, and in need of intervention, there is very little effort to recognize their practices as assets to the community that could be uplifted to support livelihoods and self-sufficiency. Although many community food projects have their roots in the social justice and civil rights movement of the 1960s (see textbox 11.6), today's movement is dominated by young, college-educated, and white people who hold most of the leadership positions in food justice organizations. Initiatives to attract a new supermarket, create a community garden, or start a farmers market do not take into account and often disrupt the everyday food geographies of low-income people, leading to gentrification and displacement (Joassart-Marcelli 2021).

Third, without secure access to clean land to grow food, space to exchange products, and places to gather and envision a different food future, it is unlikely that the most vulnerable members of society will be able to participate in the alternative economy in a meaningful way. In the United States and around the world, access to land, capital, and public resources remains highly uneven and a major obstacle to minoritized groups' full participation in a more equitable food economy. Land evictions are all too common, as poignantly illustrated by the documentary film *The Garden*, in which dozens of growers are evicted from their community farm in South Central Los Angeles. The market for land in cities is such that there is a scarcity of available and affordable land for urban agriculture. This is tied to the housing crisis that disproportionately affect people of color and curbs their ability to invest in their neighborhood, build equity, and develop a vibrant, fair, and autonomous local food system.

Finally, labor has received very little attention within the alternative food movement. Concerns for the income, exploitation, and alienation of unpaid volunteers, independent farmers and producers trying to make a living, or workers employed in various alternative businesses such as food co-ops and farm-to-table restaurants are rarely voiced, in part because this sort of work has been romanticized as communal and less exploitative. The overrepresentation of women among unpaid alternative food project volunteers raises questions about gender equity. Similarly, the disproportionate representation of people of color among workers and their relative absence from leadership positions point to racial inequities. The fact that few people earn a decent income pursuing alternative food activities without giving up on equity goals suggests that consumers of locally produced food are either unaware of labor issues or unwilling to pay a fair price, thereby reproducing the economic inequalities underlying food apartheid.

These criticisms of seemingly well-intentioned initiatives can be confusing and discouraging. In response, researchers and activists have been trying to move beyond critiques and identify transformative projects. Their work points to the importance of upholding the *right to the city*—that is, the right of all urban citizens to have a say in what their city looks like and to share the responsibilities and returns of caring for the city equitably. For example, Seattle's Clean Green program (Ramírez 2015), San Diego's Project New Village (Bosco and Joassart-Marcelli 2018), Los Angeles' Community Service Unlimited (Garth 2020), Boston's Dudley Street Neighborhood Initiative (Loh and Agyeman 2017), the West Oakland Farmers Market (Alkon 2012), and the Detroit Black Community Food Security Network (White 2018) are noteworthy for their focus on meeting immediate food

| Box 11.6 | **The Black Panther Party's Food Program** |

Although the Black Panther Party (BPP) became known for its armed resistance and violent tactics against racism and police brutality, it is its "survival programs," including community health clinics, neighborhood patrols, legal aid, community schools, senior citizen assistance, and free food programs, that brought it recognition and relevance within Black communities across the United States. These programs were conceived as community-based "self-defense"—an important theme in the BPP agenda which aligns closely with self-reliance and self-determination—against the state's failure to alleviate hunger, provide basic services, and protect the lives of Black Americans.

The Free Breakfast for School Children program began in Oakland in 1969 and quickly spread to over forty cities, serving up to thirty thousand children per day. Thousands of bags of food were also delivered to families as part of the People's Free Food Program (see Figure 11.5). The BPP prided itself on receiving no help from government agencies or philanthropic organization, instead working collectively to collect donations and pack, prepare, distribute, and serve food. Besides providing

FROM THE BLACK PANTHER PARTY'S ANGELA DAVIS PEOPLE'S FREE FOOD PROGRAM

10,000 FREE BAGS OF GROCERIES (WITH CHICKENS IN EVERY BAG)

WILL BE GIVEN AWAY FREE MARCH 29, 30 AND 31, 1972
AT THE BLACK COMMUNITY SURVIVAL CONFERENCE

MARCH 29, 1972 Oakland Auditorium 10th St. OAKLAND, CALIF. (Doors Open at 5:00 P.M.)
MARCH 30, 1972 Greenman Field 66th Ave. (Near East 14th) OAKLAND, CALIF.) (12:00 P.M.)
MARCH 31, 1972 San Pablo Park 2800 Park St. (at Oregon) BERKELEY, CALIF. (12:00 P.M.)

SPEAKERS

FREE ADMISSION ALL THREE DAYS

- BOBBY SEALE,
  CHAIRMAN, BLACK PANTHER PARTY
- SISTER JOHNNIE TILLMAN
  NATIONAL CHAIRMAN,
  WELFARE RIGHTS ORGANIZATION
- ERICKA HUGGINS,
  BLACK PANTHER PARTY
- RONALD V. DELLUMS,
  CALIFORNIA CONGRESSMAN
- JULIAN BOND,
  GEORGIA STATE SENATOR
- D'ARMY BAILEY,
  BERKELEY CITY COUNCILMAN
- IRA SIMMONS,
  BERKELEY CITY COUNCILMAN
- ARTHUR EVE,
  NEW YORK STATE ASSEMBLYMAN
- LLOYD BARBEE,
  WISCONSIN STATE ASSEMBLYMAN
- DONALD WILLIAMS,
  MEDICAL AUTHORITY ON SICKLE CELL ANEMIA
- FATHER EARL NEIL,
  ST. AUGUSTINE EPISCOPAL CHURCH
- MARSHA MARTIN,
  STUDENT BODY PRESIDENT, MILLS COLLEGE
- JODY ALLEN
  CHAIRMAN OF THE B.S.U., LANEY COLLEGE

BOBBY SEALE
CHAIRMAN,
BLACK PANTHER PARTY

SISTER JOHNNIE TILLMAN
NATIONAL CHAIRMAN,
WELFARE RIGHTS ORGANIZATION

RON DELLUMS
CALIFORNIA CONGRESSMAN
FROM THE 7TH
CONGRESSIONAL DISTRICT

ANGELA DAVIS
WILL BE AT THE
BLACK COMMUNITY
SURVIVAL CONFERENCE

ERICKA HUGGINS
BLACK PANTHER PARTY

FOR FURTHER INFORMATION, AT CONFERENCE CONTACT THE BLACK PANTHER PARTY, CENTRAL HEADQUARTERS 1048 PERALTA STREET OAKLAND, CALIFORNIA 94607 CALL (415) 465-5047

FIGURE 11.5 Poster advertising the Black Panther Party's Survival Conference and free food program, Oakland, California, 1972.

Source: Collection of the Smithsonian National Museum of African American History and Culture, https://nmaahc.si.edu/object/nmaahc_2013.46.10.

nutrition, which helped children's school performance, these programs demonstrated the power of organizing the Black community. As Heynen (2009) argues, food became a tool of resistance and liberation—a way to make visible and to confront the contradictions inflicted by capitalism on Black people. The free food program was both an anti-capitalist and *anti-racist* project in the sense that it undermined and rejected a food system oppressive to Black people. Because it engaged men and entire communities in collective food provisioning, it was also an anti-patriarchal project. The BPP built connections with other groups fighting against injustices in the food system, for instance by supporting the United Farm Workers in their boycotts of certain grocery stores (see textbox 4.2).

The Federal Bureau of Investigation (FBI) became increasingly concerned about the prospect of a successful anti-capitalist black liberation movement grounded in community-based food distribution programs, which director J.

Edgar Hoover described as "potentially the greatest threat to efforts by authorities to neutralize the BPP and destroy what it stands for" (Blakemore 2018). The agency led a concerted attack on these programs, spreading misinformation to parents, raiding distribution sites, and destroying supplies. In 1975, the federal government authorized the School Breakfast program, which today feeds fifteen million children before school. The BPP is often credited for demonstrating the need for such a program and putting pressure on the federal government to implement it. Some, however, suggest that the institutionalization of the breakfast program undermined the significance of the BPP and contributed to its demise.

The BPP's food programs provide valuable lessons for today's alternative food movements by highlighting the structural roots of food injustice and demonstrating the radical power of community-led initiatives and self-reliance in challenging oppressive social structures (Hassberg 2020).

needs and increasing food access while simultaneously addressing the systemic racial and economic disparities underlying food apartheid. These projects are led by and for people of color and focus on reclaiming the city and enabling food geographies that embrace difference in food practices.

Conclusion 11.7.

Most of the world population now lives in cities. That share is expected to increase in the coming decades as more people move out of rural areas in search of opportunities. How cities organize to feed growing urban populations is a major economic and logistical challenge that has been met by the expansion of chain supermarkets, megastores, and restaurants that source their products globally. This model, however, is facing significant criticism for contributing to environmental degradation, labor exploitation, and the destruction of local economies and cultures and for neglecting the needs of society's poorest segments.

In this chapter, we used the notion of foodscape to examine how the consumption and distribution of food is spatially organized and socially structured. Despite homogenizing trends driven by global political and economic forces, cities continue to distinguish themselves by their unique foodscapes. Their commercial streets are lined with shops, restaurants, and eating places that reflect local histories, cultures, and conditions. Each city's foodscape is uniquely shaped and reshaped by business decisions, consumer desires, recurring practices, public policy, planning decisions, and community initiatives that are influenced by and respond to global forces.

Urban foodscapes are organized differently across space and evolve over time as social relations of production change and new consumption opportunities emerge. Furthermore, the way foodscapes are lived and experienced varies according to social hierarchies based on gender, race, and class. As sites of consumption, foodscapes are built to stimulate spending and encourage the expression of identities via food behaviors such as shopping and eating. They are designed to meet the desires of the upper and middle classes, which entrepreneurs lure with new and exciting food places that constantly transform the city. In today's cosmopolitan context, these places must appear democratic, authentic, and multicultural to endow patrons with cultural capital.

These aesthetic values, however, hide wide social and spatial inequities in access to food. In the United States, low-income and nonwhite areas tend to have lower access to affordable, healthy, and culturally appropriate food and greater access to fast-food restaurants and stores selling high-calorie and low-nutrient items—a process described here as food apartheid and caused by a long history of racist urban politics linked to white flight and disinvestment.

To address these gaps, many communities have organized to create alternative food spaces and transform their foodscapes to bring about food justice. The so-called alternative food movement has uplifted anti-racist projects and promoted grassroots initiatives, including urban agriculture, through which people who have been marginalized by the mainstream food system can gain greater access to fresh food and reclaim some agency and autonomy in their everyday food practices. However, as the notion of food justice gains popularity, it is increasingly coopted by urban makers more interested in encouraging consumption and beautifying the city than in confronting the structural inequities underlying uneven foodscapes.

Food, Kitchens, and Gender 12

Learning Objectives

- Focus on the kitchen as an economic, social, cultural, political, and emotional space.
- Explore the gendered and racial divisions of labor that characterize home cooking and both paid and unpaid food provisioning work.
- Examine the social relationships and norms surrounding cooking, including patriarchy and domesticity.
- Engage with feminist scholarship to examine cooking as a source of both oppression and power.
- Understand the economic significance of social reproduction and the role of marketing and advertising in shaping it.
- Investigate professional kitchens and the nature of restaurant work, including differences between front and back of the house.

Around the world, the responsibility for food provisioning, including planning, shopping, cooking, and cleaning, disproportionately falls on women. Although unpaid, undervalued, and invisible, this labor is essential for the functioning of the economy and the maintenance of families, homes, and cultures. Without being fed, workers would lack the strength, energy, and concentration needed to be productive and contribute to the economy. Food structures domestic life, with families constituted through practices of care, such as feeding, that both justify their existence and strengthen their cohesion.

From a geographic perspective, one could say that the kitchen, as a relational *place* and *scale*, is connected to other places and scales ranging from the global (e.g., the global economy and contemporary food regime) to the local (e.g., labor market, retail infrastructure, feeding cultures, norms of domesticity) to the body (e.g., the well-fed and cared-for body). Thus, food preparation—like other components of the food system—is intertwined in larger political, economic, social, cultural, and affective relations.

In this chapter, we examine food preparation by focusing on the kitchen as an economic, social, cultural, and emotional space. We pay particular attention to the role of gender and its changing significance. We begin by focusing on home kitchens and examining the social relationships and norms surrounding cooking, including domesticity and patriarchy, which are tangled up with love, care, and responsibility. We then explore mechanisms by which households outsource food preparation by hiring servants, cooks, or private chefs, who are typically of lower social classes and from minoritized groups. This provides us with an opportunity to think about the *intersectionality* of race, class, and gender and the role of the market economy in shaping these relations. A look into professional kitchens helps us draw distinctions and parallels between paid and unpaid labor. We end the chapter by considering the economic realm of home cooking as a large market that is influenced by technological innovations, media, and advertising and shapes the way people cook and relate to food.

12.1. Home Cooking: Domesticity and Patriarchy

Cooking is both a source of oppression and liberation. For many women, who do most of the home cooking, there is a tension between the subservient and tedious aspect of their work in feeding their families and the pleasure they derive from caring for and connecting with them. The field of food studies reflects this ambivalence with scholars conceptualizing domestic responsibilities as a form of oppression and others understanding them as source of power.

12.1.1. Gendered Division of Labor

Almost everywhere around the world and throughout most of history, women learn from an early age that it is primarily their responsibility to feed their household. Even in hunter-gatherer societies, where men were viewed as food providers and women as caregivers, evidence suggests that women contributed as much as 80 percent of all calories through gathering plant-based food, hunting smaller animals, fishing, and eventually growing plants, which were less impressive but more reliable sources of food than hunting. In much of the Global South, especially in the poorest countries, women represent the majority of subsistence farmers and are almost exclusively responsible for domestic food preparation. Because processed food and refrigeration are less common, a lot more work is required to produce even seemingly basic meals. For instance, making tortillas—a staple food in Mexico—can take several hours each day, involving boiling corn, grinding the kernels with a cylinder pestle to a fine masa, shaping the tortillas by hand, and cooking them on a comal griddle over a wood burning fire (see figure 12.1).

As economies diversify, employment opportunities attract people into cities and pull many women out of the domestic sphere of unpaid work into the public realm of waged labor. As women begin earning income, often doing tasks they used to do at home without pay, they have less time for cooking and other domestic responsibilities, slowly eroding gendered domestic divisions of labor. In the nineteenth century United States, poor urban migrants who lived in boarding houses, tenements, rented rooms, cramped apartments, or makeshift housing, often lacked the space and equipment to cook. Instead, they relied on street stands and take-out shops to purchase food one meal at a time. Today, in the Western world, some low-income people turn to fast food chains to get food when a home-cooked meal is not an option. Poor women in Global South cities also spend less time cooking than their rural counterparts, buying bread or other basic food from various vendors and spending the extra time earning wages. For example, most urbanites in Mexico no longer make their own tortillas daily but buy them fresh from a neighborhood tortilleria, processed

FIGURE 12.1 Woman making tortillas by hand, Mexico, 1927.

Source: New York State Archives. http://digitalcollections.archives.nysed.gov/index.php/Detail/objects/1456.

from a supermarket, or stuffed from a taco stand. With urbanization, new markets have emerged to meet the needs of an expanding population and workforce, allowing female workers to allocate less time to cooking and more time to paid work.

Despite these economic and social changes, women continue to spend significantly more time than men performing food-related chores. In the United States, between 2015 and 2019, fully employed married women with at least one child under eighteen years old spent more than twice as much time in food preparation and cleanup (fifty vs. twenty-three minutes) and over three times as much time doing housework (forty-four vs. thirteen minutes) than their male counterparts (Bureau of Labor Statistics 2020). On an average day, less than half of men contributed any time at all to food preparation. In the past fifty years, women have cut their daily food work by about thirty seconds each year, while men have increased their contribution by twenty seconds. At this rate, we might achieve gender parity in domestic food work by about 2050.

Despite these persistent gender inequities, popular narratives often blame the feminization of the labor force—not the lack of men's contribution—for declines in home cooking, family cohesion, and cultural traditions. In these accounts, working women are described as abandoning their domestic responsibilities in favor of professional careers. For conservatives, this phenomenon is alarming because it allegedly imposes high social costs on children, families, and society. Arguably, without the influential family dinner daily ritual, it is becoming difficult to bond with one another, express love, pass on moral values, share traditions, discipline children, and impose a structure to daily life. Thus, divorce, low school achievement, mental illness, drug abuse, obesity, lack of community, and many other social and health problems ensue.

Murcott (2012) describes the decades-old lament over the disappearing family meal as a "moral panic" that is not grounded in factual evidence about meal patterns but in ideas about what a family meal ought to and used to be. Such ideas are common in popular literature, film, and journalism. However, they are more often aspirations than actual events that take place on a regular basis. In upper-class families, until recently, meals were prepared by servants and children typically ate separately.

In poor families, work schedules and lack of space made it difficult for people to eat together. While husbands were fed when they came home from work, kids ate at a different time and wives would often forgo sit-down meals altogether. For families who have been eating together, meals have been frequently a great source of anxiety, tension, and even violence, with mothers fretting to please everybody, parents or siblings bickering, and children misbehaving. Thus, the great family dinner may not be as common and positive as what we think it ought to be.

12.1.2. Patriarchy

Why do we continue to place so much value on the family meal? For feminists, the family meal both symbolizes and reproduces *patriarchy*—the evolving and situated political and social system that assumes that men are superior to women, granting them privileges and giving them licenses to dominate, oppress, and exploit women (hooks 2004). This notion of male superiority is learned through socialization and reproduced through myriads of practices that reinforce gender roles and hierarchies. For instance, patriarchy underlies the ideology that women are natural care givers, which in turn limits their employment opportunities outside the home and reproduces the sort of wage gap that justifies their subordination at home and their disproportionate contributions to domestic work. Patriarchy is intimately tied to capitalism which benefits immensely from women's unpaid contribution to *social reproduction*. By feeding, clothing, and caring for workers and future workers, women help supply labor and keep the wheels of capitalism turning. The idea that this type of work comes naturally to women is what allows it to go on unpaid and what underlies gender wage discrimination in the labor market where feminized service jobs are systematically underpaid. Capitalist patriarchy is not a one size fits all system; it varies over time and space and relies on different ways to maintain women's subordination.

In Western middle-class patriarchal society, women who fail to cook "proper" meals for their loved ones are seen as uncaring and immoral wives and mothers. As we will learn in the next chapter, what constitute a proper meal is highly contextual and shaped by cultural understandings of healthy food. Because food provisioning and emotions are so deeply intertwined and cooking is viewed as an expression of love, women experience immense pressure to feed and serve their families. Although some women are coerced to perform these tasks through domestic violence, in most cases, they put themselves into subordinate positions because of internalized and unspoken beliefs about gender relationships, marriage, motherhood, and domesticity. *Femininity* is performed by serving and caring for others, which is a socially recognized way to be a woman. Watching their grandmothers, mothers, and sisters cook before them; playing with toy stoves and dinette sets; listening to teachers, politicians, and religious leaders praise subordinate women and vilify independent women; and absorbing media images that connect food, womanhood, and domesticity to the ideal family, women continue to be socialized to cook for others.

Of course, this view of femininity and the associated gendered division of labor have been evolving and, in many families, domestic tasks are becoming more evenly distributed. In a foodie culture, where the appreciation of good food and the ability to cook is a source of distinction (see chapter 10), home cooking is becoming less gendered than in the past and more attractive to men. Yet what men and women do in the kitchen still differs and women remain the "default" cook in most heterosexual relationships, particularly when they have children and earn less than their spouse does.

Men who cook are often portrayed as liberated and enlightened heroes, while women's food work is unnoticed and taken for granted. Men are more likely to contribute a special dish, help with a celebratory meal, or take charge of the barbecue, but rarely do they engage in all aspects of food provisioning including planning, shopping, addressing health concerns, balancing food preferences of multiple family members, and cleaning after the meal. Even in families where both partners work in paid employment, women continue to perform most of these physical, mental, and emotional tasks—what Hochschild (2003) described as the "second shift." In her classic ethnographic study, DeVault (1994) shows that married working women spend more time preparing meals than single mothers, partly because of their husband's expectations and sense of entitlement linked to their subjectivities as breadwinners. And when women earn more than their spouse, which would economically justify a decline in their domestic chores in favor of paid employment, their "second shift" rarely diminishes and, in some cases, increases to compensate for their partners' loss of masculinity. Under these conditions, it is not surprising that women are stressed and tired as they attempt to "do it all" and become "supermoms" without much acknowledgment or assistance.

Much of the literature on gender and home cooking is based on heterosexual couples and has an implicit heteronormative bias. Carrington's (2013) study of what he calls "lesbigay" or lesbian, bisexual, and gay couples stands out for his focus on feeding work in queer households. Because preparing and sharing meals is such a symbol of family and home, many queer couples engage in these activities to affirm their status as a family. According to Carrington, affluent lesbian and gay families have more time, money, and resources to invest in domesticity, and therefore are able to achieve a stronger sense of themselves as a family, which they occasionally display at lavish dinner parties. This finding echoes the ability of affluent heterosexual families to use dinner time to reproduce their privilege through the sharing of informal knowledge, class identity, cultural practices, and healthy food. Although more likely to reject cultural gender expectations, both lesbian and gay families still contend with the notion of cooking as woman's work, which presents a bigger conflict for many gay men who resort to various strategies not to be perceived as a "housewife." Like in heterosexual couples, both those who cook and those who do not cook minimize the amount of work it takes, shining light on the difficulty of describing, measuring, and valuing labor that is so intimately connected with love and care.

For many feminists, women's liberation requires a rejection of domestic life or what Betty Friedan calls the "feminine mystique"—the false and debilitating notion that equates womanhood with *domesticity*. To some extent, queer liberation also rests on the elimination of patriarchy and its rigid and binary gender roles. Despite calls to reject the "prison of domesticity" beginning in the late 1960s, women have been mostly left to fend for themselves in their own homes and kitchens, as white middle-class feminists focused their attention on "public" issues such as wage inequality, gender discrimination at work, and reproductive rights. Feminists of color, by drawing attention to the gendered, classed, and raced geographies of home as everyday sites of violence, oppression, resistance, and liberation, have contributed more progressive visions of social reproductions that open homes to the public and promote more communal and collective forms of domesticity that extend into neighborhoods and cities (see textbox 12.1).

12.1.3. Domesticity and Power

For many women, food is a source of power. Surveys conducted in the United States and Western Europe reveal that a significant proportion of women enjoy cooking. By overseeing food provisioning, women are "gatekeepers" who decide what food is being served, how, and when. This gives them some control over domestic finances, family life, and the private sphere of the home. Nowhere is this control more obvious

Box 12.1	**Progressive Homemaking**

Patriarchy and the domestic burden it imposes on women are tied to nuclear and heteronormative families. With the advancement of capitalism and wage labor, production shifted away from the home and the family into factories and offices. Workers moved to cities and settled away from multigenerational kinship networks that had helped feed families in self-sufficient and collaborative ways. Nuclear families consisting of opposite-sex parents and one or more children became the norm, with men earning an income in the public sphere of production and women in charge of maintaining homes, which were increasingly built to house single families. This trend began centuries ago in the Global North and is still unfolding at a rapid pace in the Global South where urbanization and industrialization are splitting kin relationships.

In fighting patriarchy and the oppression of women, manifested in their domestic responsibilities and confinement to the home, radical feminists have called for a rejection of both capitalism and the nuclear family. There are numerous examples of women organizing to socialize domestic work, by taking it into the public sphere (Morrow and Parker 2020). For example, in 1870, a group of women in Cambridge, Massachusetts, created a cooperative bakery and laundry that would serve multiple members. Reformists like Jane Addams also supported cooperative living arrangements for the women and couples who served immigrants at the Settlement House she helped create in Chicago. In Black

communities, mutual aid societies and clubs organized collectively to deliver services, including food (see textbox 11.6 about the People's Free Food Program of the Black Panther Party), and provide safe spaces and housing for women and girls. Other experiments have included various types of communes, cooperative housing, community gardens, and collective kitchens.

In her book *The Grand Domestic Revolution*, Dolores Hayden (1982) borrows from these ideas to propose different spatial designs of cities and homes that would promote a more inclusive, egalitarian, and solidary organization of everyday life. By reorganizing domestic work collectively and simultaneously extending home-like relations of care to the city, Hayden and others seek to blur the distinction between production and social reproduction, thereby challenging both patriarchy and capitalism.

Because the dismantling of the nuclear family threatens capitalism, conservative politicians often adopt "pro-family" agendas that reinstate heteronormative families as morally superior. They do so in part by blaming social problems on nontraditional families, including those with single or same-sex parents, without recognizing the economic contradictions in the notion of nuclear families, such as the fact that lower-class women have always worked outside the home and the challenges that many men face in finding stable employment and earning a decent income.

---

than in the shaping of food habits and disciplining of children. There is also a widely held belief that wives control their husbands through food, as exemplified in popular sayings like "the way to a man's heart is through his stomach." Several scholars have also noted a potent association between food and sex, which has also been illustrated in numerous films and literary works (see textbox 12.2).

One of the rewards of cooking is the sense of connection it provides. Shared meals presumably strengthen ties with family and friends and maintain traditions. Many women feel that by cooking they hold the family together—a sentiment reinforced by idealized images of family dinners discussed in the previous section. Performing food work is an important part of women's identities as caretakers, wives, lovers, mothers, or daughters. In fact, when insufficient resources or unfortunate circumstances prevent women from providing cooked meals for their families, many experience sadness, shame, guilt, and a loss of identity.

Some women use cooking as a form of resistance against various threats in their lives. For example, immigrant women often cook time-consuming dishes to maintain ties to the homeland and resist the dissolution of their culture (see chapter 10).

Box 12.2	**Food and Sex**

There are strong parallels between food and sex. As Elspeth Probyn (2000) argues in her classic book *Carnal Appetites: FoodSexIdentities*, eating and love making both involve similar levels of intimate physicality and sensuality, such as taste, smell, sight, and touch. These parallels have been exploited in art, film, and literature, which often portray lovers feeding each other and depict food in eroticized ways as well as sex in terms of flavors (e.g., hot, sweet, spicy).

Because of this strong connection, feeding work has been sexualized in highly gendered ways. Women who know how to cook are viewed as more desirable, but they may also be more dangerous, since this knowledge gives them power. Having a voracious appetite is seen as an essentially masculine characteristic that applies as much to food as it does to sex. Like cooking, sexuality is often perceived as a site of women's subordination. However, both food and sex can be used to exert

agency and possibly control, manipulate, kill, or ruin men, as when, according to the bible, Eve fed Adam an apple in the Garden of Eden.

Historically, certain types of food have been identified for their aphrodisiac qualities. For example, oysters, chocolate, spices, and wine are believed to stimulate sexual desire and intensify pleasure. They are often served in romantic settings such as the commercialized Valentine's dinner. Some of these foods have occasionally been banned because of their troubling properties.

The very notion of "food porn" also highlights parallels between food and sex. In a context where all sorts of foods are widely available, the aesthetic value of food has become increasingly important. It is not enough for food to be nutritious and flavorful, it should also be visually appealing. This trend is illustrated in the proliferation of images in the food media depicting food in sensationalized ways, making it exciting, unrealistic, and unattainable to most.

Nostalgic foods alleviate homesickness and alienation, granting women a sense of control over their everyday life (Mannur 2007). Some women also cook elaborate meals from scratch to resist the cultural loss, environmental degradation, and social collapse they associate with the modern industrial global food system. Indeed, in recent years, women have played a central role in the *alternative food movement* that advocates a return to local, seasonal, minimally processed, and craftly prepared foods (see chapter 11). Without their commitment to these causes and their unpaid labor as volunteers, gardeners, shoppers, and cooks, it is doubtful that the alternative food movement would have grown as it did. In abusive domestic relationships, women have also used food to appease tensions, ward off domestic violence, and create a semblance of normalcy. And in contexts where employment outside the home is limited or forbidden, women have relied on food to generate income. For example, in *Building Houses Out of Chicken Legs*, Williams-Forson (2006) shows how Black women in Virginia used chicken as a "tool of self-expression, self-actualization, resistance, and even accommodation and power," making a living and showing off their skills by selling fried chicken, hot biscuits, and coffee to train passengers as "waiter carriers."

In these examples, women use food to negotiate, contest, and resist constraints beyond their individual control, be it displacement, segregation, discrimination, poverty, or other structural inequities. Selecting ingredients, preparing unique dishes, serving loved ones, sharing meals with friends and relatives, learning new recipes, working with their hands in creative ways, shaping the rhythm of family life, and earning some income reveal women's *agency* within the confines of patriarchy. Instead of being a domestic prison, the kitchen is a domain of self-expression, ingenuity, creativity, and respite.

Post-feminists argue that it is time to move beyond the polarized image of women as either "feminists" who refuse to cook or "housewives" who cook for others. Instead, they embrace cooking (and eating) as a source of pleasure and well-being and an expression of independence and creativity (Hollows 2003). The notion of "domestic goddess," adopted by Nigella Lawson and other celebrity home cooks, illustrates this new and positive domestic subjectivity.

Women's agency at home, however, does not necessarily translate into greater power outside the home, in the public sphere of work and politics, and may in fact reproduce gender inequality and stereotypes. The ambivalence between cooking as pleasure and chore suggests that, for most women, agency and oppression coexist in various ratios depending on circumstances such as cultural pressure, time, and financial resources.

12.2. Outsourcing Social Reproduction

Households who have the means to do so, often outsource social reproduction activities, including cooking, cleaning, laundering, nursing, elderly care, tutoring, psychological counseling, and other types of care work, partly relieving women of some of the responsibilities they have been socially assigned. They do so by having cooks in their homes or purchasing prepared food in commercial outlets.

12.2.1. Domestic Workers

Historically, wealthy families had cooks and servants who performed the various tasks required for meals to be served, including planning, shopping, cooking, setting the table, serving, and cleaning. Domestic workers are typically paid wages for their labor, turning social reproduction into a commoditized service that can be purchased for a negotiated price. However, for much of history and into the present, cooks have been enslaved people or other unpaid workers tied to families through coercive or paternalistic relationships in which their labor would be exchanged for room and board. Even in the most advanced and regulated capitalist economies, domestic labor arrangements remain highly informal and exploitative, with private homes escaping labor monitoring and regulations. As a result, many domestic workers and personal cooks work long hours, receive low wages, and are subject to verbal, emotional, and physical abuse.

Around the world, poor women have been relegated to lives of servitude, leaving their families behind to care for others and perform tasks that their employers deemed too tedious and demeaning for themselves. Private home cooks, whether paid or unpaid, are mostly lower-class women, including migrants, racial, ethnic, and religious minorities, and Indigenous women. Their social position can be understood as a result of their gender, class, and race, which *intersect* in different ways depending on circumstances, history, and geography. In the US South, the daily meals of white affluent families and their guests would often be prepared by enslaved Black people. After the Civil War, those families kept formerly enslaved people as hired "help" to perform the same tasks (see textbox 12.3). Restrictions on Black people's employment and travel meant that domestic work was one of the few options available for Black women, forcing them to leave their own families to make a living. Those able to leave the South in what is known as the Great Migration often found jobs in northern cities as cooks and maids. This type of work was and continues to be poorly paid and virtually unregulated, untouched by labor unions and left out of federal social protections afforded to other workers—the result of racist concessions to Southern lawmakers dating back to policies adopted during the New Deal era.

In Latin America, it is still common for upper-middle-class families, especially in cities, to have domestic workers, many of whom are poor, rural, migrant, and/

Box 12.3   The Help

In the US South, as in many other parts of the world, domestic workers have been called "the help"—a term suggesting that their work is ancillary, subservient, and not very valuable. At the same time, "the help" is often imagined as a de facto member of the family, with white people portrayed as innocent, paternalistic, generous, and colorblind. The power-laden transaction in which domestic labor is coerced by slavery or obtained in exchange for low wages, food, and/or shelter is masked by an imagined family-like and affectionate relationship.

The best-selling novel *The Help* (Stockett 2009), which became an instant book club favorite and was turned into a movie in 2011, perfectly illustrates this assumption of Black subservience and white innocence that appeals to white audiences (Murphy and Harris 2018). The main protagonist of this story is Skeeter, a young woman who returns home in Jackson, Mississippi, after graduating from college in the early 1960s and struggles to start a career as a writer in a sexist world. Hired by a local newspaper to write a column about housekeeping, she begins questioning the way her childhood friends and her own family treat the Black women who work for them. Over several months, she convinces these domestic workers to share their stories of personal hardship and employer abuse, which she eventually publishes in a book that draws attention to their plight and lands her a job in New York City. In the book and film, Skeeter is portrayed as a brave and caring woman willing to challenge racial injustice, illustrating the idea of "white savior" so heavily criticized by race scholars. In contrast, the black characters are seen as subservient, victimized by poverty, crime, and dysfunctional families, and reluctant to participate in their own liberation, reproducing racial stereotypes and emphasizing their lack of agency. This sort of narrative, echoed in many other popular movies, absolves white people of responsibility in perpetuating racism and white supremacy and erases a long history of activism led by people of color.

Deeply racist, the image of "the help" as embodied in the naturally caring, loving, and happy Black Mammy has endured in the United States for decades and has only recently begun fading, with brands like Quaker Oats dropping their iconic Aunt Jemima label (see figure 12.2) and ConAgra doing the same with Mrs. Butterworth's, amid worldwide protests against racism. Some, however, argue that these brands represented and even promoted a form of integration as Aunt Jemima could be found in the kitchens of many white families, honoring the lives of Black women like Nancy Green who was born into slavery and adorned boxes of Aunt Jemima pancake flour for years, having allegedly created the recipe.

FIGURE 12.2 Advertising for Aunt Jemima's pancake flour, 1909. Depictions of Black cooks are typically essentialist, caricatural, superficial, and therefore racist.

Source: Library of Congress, Chronicling America: Historic American Newspaper. https://chroniclingamerica.loc.gov/lccn/sn83030214/1909-11-07/ed-1/seq-44/.

or Indigenous women with few other employment options. Domestic work is also common in the Middle East, especially in the wealthier Persian Gulf states such as Bahrain, Kuwait, Saudi Arabia, and the United Arab Emirates, where cooks come primarily from Indonesia, Bangladesh, the Philippines, and other countries with large Muslim populations. In many Global South cities, like Mumbai, Singapore, Cape Town, Rio, Mexico City, and Shanghai, the number of domestic workers has increased over the past decade, reflecting deepening income inequality that enables some families to hire people to perform tedious domestic tasks while forcing others to sell their own labor to take care of families not their own.

The rise in migration from the Global South since the 1970s has generated a large supply of labor by women of color whose skills have been systematically devalued and for whom domestic work is one of the few occupations available. Today, in the United States, there are 2.2 million people who work in private homes cooking, cleaning, and caring for children and the elderly. According to a recent report (Wolfe et al. 2020), the vast majority (91.5 percent) are women and well over half (57.1 percent) are Black, Latino, or Asian workers, who make just over a third (36 percent) of the workforce in other occupations. About a third are foreign-born, compared to a sixth in other types of jobs. Perhaps unsurprisingly, only 3.7 percent of all domestic jobs are filled by white males, who represent over a third of the entire labor force. Domestic workers earn about $12 dollars per hour, well below the median hourly wage of $20 for all occupations. They are also more than three times as likely to live in poverty, with fewer than one out of ten covered by a retirement plan.

12.2.2. Restaurants and Commercial Food Preparation

Today, in the Global North, although many middle-class households hire workers for a variety of personal services, home cooks are less common than in the past. This is partly because there is now a much wider range of options to purchase prepared food from grocery stores, restaurants, and catering services—most of which can be ordered online and delivered quickly or picked up by car from a drive-through window, especially in urban settings. Processed foods, epitomized by frozen dinners and boxed meals, are widely available in grocery stores. Supermarket chains compete with each other based on their selection of prepared foods and are dedicating growing space to display a variety of gourmet items freshly packaged or cooked on the spot. Food halls that offer casual, diverse, and fast dining options are expanding (see textbox 11.2). Convenient foods, including prewashed lettuce, cleaned and chopped vegetables, and fresh meal kits, have simplified cooking and reduced the amount of time required to put a "proper" meal together, alleviating the burden of cooking and diminishing the need for hired help. Many people do not even bother cooking at home and instead eat out in restaurants where all sorts of meals and snacks can be found at any time of the day and night at a variety of price points.

Since 2010, people in the United States have been spending more money on food away from home than on food at home, following decades of changing food habits (Saksena et al. 2018). In 2019, they spent $850 billion in various types of restaurants and $790 billion in grocery stores and other food retailers (USDA 2020). Over a third of calories consumed come from food away from home, more than twice what it was in the late 1970s. This trend is mostly attributed to the increase in "quick service" restaurants, including fast food, limited service, and casual restaurants—the fastest growing segment of the market.

In France, where home cooking and family meals remain anchored in the national culture, the proportion of meals eaten outside the home has increased as well. Yet, according to 2019 data from Eurostat (2020), a much larger share of food expenditure continues to be allocated to at-home food (67 percent) than to catering services (33 percent), which includes all types of restaurants, food stands, and cafeterias. For the twenty-seven countries of the European Union, food at home in 2019 represented 63 percent of total food expenditure, slightly down from 65 percent in 2010. Globally, according to Euromonitor International (2012), the average figure was 71 percent, suggesting that very few countries rival the United States in the amount of expenditure allocated to eating out.

The propensity to eat out varies geographically based on factors such as work arrangements, income distribution, eating cultures, gender norms, and diversity, availability, and affordability of restaurants, including fast food options, which are still frowned upon in most European countries despite their growing popularity. In general, however, those who eat out more often tend to earn higher incomes, be younger, live in smaller households, and reside in urban areas. Indeed, class is pivotal to the ability to outsource food preparation to the extent that eating out and purchasing prepared meals, except perhaps at street stands and in fast food establishments, is typically more expensive than cooking at home.

The convenience and success of the food service industry depends on the existence of a large working class willing to fill low-wage jobs such as cooks, dishwashers, and waiters. As discussed in chapter 4, in middle- and high-income countries, there are now many more jobs in food processing, manufacturing, and services than there are in agriculture, which has become heavily industrialized and capital intensive. Like hired domestic cooks, commercial food workers are mostly people of color and immigrants who tend to be exploited, as demonstrated by the very low wages and benefits and the high incidence of poverty, abuse, and work-related injury reported in chapter 4. Thus, in addition to class and gender, race plays a key role in structuring the outsourcing of food preparation.

Unlike in private homes where food work is mostly performed by women, men dominate the commercial food industry, particularly when it comes to slaughtering and butchering animals, processing food, cooking with industrial and dangerous equipment, and storing and transporting food. Women are more likely to be employed in food service jobs, as cashiers, hosts, waiters, and servers, which are seen as more feminine and appropriate. In fact, these jobs represent an extension of the caring work that women are presumably naturally predisposed to do and are therefore devalued as "low-skill" jobs.

The distinction between women's and men's jobs is highlighted by the comparison between domestic and professional kitchens. The former is seen as a private space, historically relegated to the back of the house or even outside. In wealthy families, it is the domain of servants, but in lower- and middle-class households, it is often viewed as a feminine space within the home and a "women's place." Housewives presumably rule over "their" kitchen; they determine what food comes in, know the place of every pot and pan, decorate it according to their taste, and restrict access by children and spouses. Regardless of how knowledgeable and skilled they are, homemakers will never be chefs. Some rely on traditions—recipes transmitted across generations of women—and others turn to cookbooks, magazines, websites, and television shows to learn new tricks and expand their repertoire, but they will arguably never gain the sophistication, precision, creativity, and strength that professional chefs have acquired through culinary school and arduous apprenticeships. In contrast to home kitchens, restaurant and commercial kitchens are seen as a man's world where women could not "take the heat." Indeed, they are extremely hot, contain heavy equipment, advanced technology, and dangerously sharp knives, and buzz with masculine energy and potentially toxic banter. Full of stainless steel, they are nothing like home kitchens.

In the past two or three decades, chefs have become popular icons and professional kitchens have opened their doors to the public. The overwhelming majority of celebrity chefs are white men (e.g., Gordon Ramsey, Jamie Oliver, Bobby Flay, Mario Batali, Anthony Bourdain), and although they each have their own style, they embody similar versions of masculinity that emphasize expertise, discipline, hard work, leadership, and toughness. To be sure, women have also gained celebrity status for their cooking skills, but they have done so as traditional homemakers, loving wives and mothers, or "domestic goddesses"—not as chefs in their own right. In an article entitled "Why Are There No Great Women Chefs?," Druckman (2010) argues that women are pigeonholed into a very limited set of personae that are always contrasted to and measured against the yardstick of professionally trained male chefs (except when it comes to their physical bodies). When they succeed, women are described as exceptions—defying norms and bucking the system—and as such they lack the power to change the sexist dynamics of restaurant kitchens.

Despite popular images of celebrity chefs owning dozens of restaurants in multiple cities and controlling commercial and media empires of kitchenware, spices, condiments, cookbooks, and television shows, it is important to remember that the majority of restaurant workers earn very low income and, perhaps ironically, struggle to feed their own families. As we learned in chapter 4, most food service jobs, including cooks and waiters, are poorly paid and offer little security. In *Behind the Kitchen Door*, Jayaraman (2013) reveals the working conditions, experiences, and circumstances of food workers. She specifically points to the irony of the foodie culture and the so-called alternative food movement in which consumers want to know everything about their food but ignore how kitchen workers and wait staff are treated. Relying on interviews of restaurant workers in eight US cities, including New York, Los Angeles, Miami, and Chicago, Jayaraman paints a dire picture of restaurant work through the stories of tipped workers who went home without pay after breaking dishes or when their customers failed to pay for their meals; bussers who were denied promotion to waiter after years of service because of the color of their skin; dishwashers who were repetitively harassed in the kitchen; hostesses whose bodies were objectified by customers who are "always right"; line cooks who went to work with tuberculosis because they had no sick-leave benefits; chefs who verbally and physically abused their staff; and servers who earned an hourly minimum wage of $2.13 because of antiquated federal laws. Throughout these stories, she highlights the racism and sexism that prevail within restaurants and lead to occupational segregation and significant wage disparity.

Such disparities follow the spatial organization of the restaurant, with a clear separation between the "front of the house" and the "back of the house." The former represents the public face of the restaurant, including the host station, the bar, and the dining room. Most workers in this section are white and embody a certain racialized aesthetic that signifies hospitality in their friendly yet respectful disposition and proper attire. When people of color or immigrants are hired as waiters, hosts, sommeliers, and bartenders, they are often objectified and exoticized to convey a certain image and create a cosmopolitan experience. For instance, many restaurants in US cities hire European waiters with thick French or Italian accents and favor "charming" aspiring actors and artists, rather than promote experienced but less glamorous long-term servers and bussers. Several studies describe the taxing and emotional nature of this type of caring work that consists of smiling, engaging in superficial conversations, and displaying constant attention to others' needs (see textbox 12.4).

The "back of the house" or the kitchen, which may be subdivided into storage room, preparation area, cooking line, holding area, and dishwashing area, is hidden from the public. This is where people of color and the lowest-paid jobs are found and

| Box 12.4 | **Emotional Labor** |

Emotional labor is a term coined by Arlie Hochschild (1983) to describe the management of emotion as part of one's job, typically in service and hospitality. For instance, servers, flight attendants, hairdressers, nurses, and customer care and sales representatives all engage in some type of acting in performing their job. This involves both hiding negative emotions and faking positive ones. This includes waiters who put on a smile, engage in cheerful conversation, joke, and sometimes flirt with customers to make sure they order a lot and tip well. They also address complaints and put up with rude behavior without getting visibly upset. Their job consists of being attentive to other people's emotions, while repressing or modifying theirs. While some of this acting might seem superficial, it often runs deeper and involves the alteration of one's own emotions.

As Crang (1994) argues, restaurant work is defined by unique workplace *geographies of display* that consist of displaying attention and care while hiding the management and surveillance regulating those practices in order to create a seamless and pleasurable experience for consumers. Waiters are trained and monitored to perform scripted

and choreographed tasks. However, they must also go beyond what is expected, especially if their earnings depend on tips, allowing them to change the script and negotiate the various *social encounters* that constitute their job. Such encounters do not happen in the vacuum of the restaurant, but are shaped by wider social relations of sexuality, gender, age, race, and class and the power associated with them. For instance, restaurant work is highly sexualized, with bodies, including physical appearance, clothing, and gestures, central to the performance as suggested by the widely held belief that better looking and flirtatious waiters get more tips.

At the end of the day, when servers go home, they are often physically but also emotionally exhausted. Over time, this can have a deleterious effect on mental health by making one doubt and question their own emotions. Indeed, people engaged in emotional labor are more likely to experience stress and burnout. Crang argues that dealing with people, rather than with things as one would in a factory, is a central feature of work in contemporary postindustrial society.

where most of the abuse and exploitation take place—away from the eyes and ears of unsuspecting dinners.

Several studies document the high level of *informality* in the food service industry (see Joassart-Marcelli 2021). Informality refers to the lack of enforcement of social and labor protections. This is true in the Global South, where street vendors and small unregistered businesses dominate, as well as in the Global North, where immigrants and other marginalized workers are disproportionately engaged in jobs related to food preparation either as employees or as entrepreneurs. As seen in chapter 10, restaurants and food stores are some of the most common avenues for ethnic entrepreneurship. Often a result of necessity in the face of labor market restrictions, these small and informal businesses depend on self-exploitation, support from networks of relatives and friends, and low wage labor to stay afloat. Food is one of the industries with the highest proportion of undocumented immigrants, along with agriculture and construction. In the United States, undocumented immigrants represent about a fifth of restaurant cooks and dishwashers—well above their 5 percent share of the labor force (Passel 2015). In cities like Los Angeles, the restaurant industry virtually depends on Latino workers, especially those who are undocumented, to fill back-of-the-house jobs like cooks, expeditors, and dishwashers. This exploitation has led to calls for serious reforms and even the "abolition" of tipping and restaurants.

In short, the convenience to which many of us have grown accustomed comes at a high cost for the many workers who toil in kitchens, assembly lines, refrigerated rooms, grocery store aisles, and warehouses to process and prepare our food.

12.3. Marketing Domesticity

In consumer societies, expectation about what and how to cook and who should be doing it are deeply influenced by the media and more specifically advertising. Much of what people eat today, whether it is jarred baby food, canned haricots, frozen pizza, microwavable popcorn, or coffee brewed from individual capsules, has been sold to us by savvy marketing campaigns that mold our relationship to food. Convenience has always been a highly marketable feature. The idea that processed foods, new appliances, and various gadgets could save time and tedium has fueled a large industry and led to changes in food preparation. Taste, affordability, health, and more recently sustainability and equity are also important selling points that often seem to compete with convenience.

Advertising and food media in general reflect, influence, and normalize gendered, classed, and raced relationships to food and contribute to the commoditization of feeding work. Although similar trends can be observed regarding the types of products advertised around the world, commercials are often adapted to local societies and cultures and their messages evolve over time.

In the United States, the 1950s witnessed an explosion of advertising for domestic appliances and convenience foods. The economic boom that followed the end of World War II led to an increase in mass production and mass consumption. As many middle-class (and mostly white) families moved to suburban homes, they bought refrigerators, electric stoves, and dishwashers to appoint their larger kitchens. They also bought mixers, blenders, and toasters that crowded their shiny counters or filled their built-in cabinets. This was also the time, as we discussed in chapter 11, when the size of grocery stores grew exponentially, turning them into supermarkets with wide aisles filled with expanding selections of packaged and processed food.

Advertisements at the time were primarily directed at women, convincing them that these modern products would help them fulfill their domestic duties more easily. Rather than questioning the gendered division of labor by advertising products that could be prepared or used by men, women, and children, they reinscribed the existing gender division of labor by portraying these innovations as beneficial to women. They also put the family meal on a pedestal and raised expectations about the diversity and quality of meals that women were supposed to prepare and serve. Now that cake could be made from a box, it had to be served more regularly. And since the dishwasher took care of doing the dishes, women had no reasons to complain about the extra mess. In addition to raising expectations, these technological innovations also contributed to deskilling domestic labor and making it even less visible.

Ads from the 1950s turned cooking into a form of consumption that was more about shopping for new products and acquiring the right kind of appliances than about physical food work. These ads invariably included depictions of white heterosexual nuclear families where husbands were portrayed as hardworking suit-wearing breadwinners, children were well behaved, and fancily clad wives were happy to serve them, even if they had to resort to a few tricks to please everyone (see figure 12.3). According to Inness (2001), these images contribute to a "kitchen culture" that normalizes the kitchen as a woman's place.

Over the subsequent decades, food innovations and gender roles continued to shape each other. As more women joined the labor force and began resisting domestic expectations, convenience became even more relevant. Still, rather than freeing women from their responsibilities, inventions like the microwave simply lightened their duties. Some deplore how these technologies took the joy and the creativity out of cooking, turning it into a monotonous and mechanized activity that simply

FIGURE 12.3 Advertising for vitamin-fortified cereals. Ads often reveal gendered notions of femininity associated with domesticity, subservience, and physical appearances and suggest that food innovations ease domestic work.

Source: https://pixy.org/1155263/.

required that the freezer be stocked. With modernization, kitchens lost their warmth and smells, women forgot how to cook, and recipes stopped being passed on to the next generation. As we will learn in chapter 13, this perspective, which simultaneously blames women and the food industry for the disappearance of family meals and the loss of food culture, has become a dominant narrative in explaining the rise of obesity and other social ills.

Since the 1980s, foodie culture has changed the way we think about cooking. Authenticity, diversity, purity, freshness, and health now compete with convenience in influencing consumer purchases, at least among higher income consumers. A whole new set of appliances have appeared to meet and promote this "turn to quality," including espresso machines, pasta makers, sous-vide ovens, dehydrators, wine refrigerators, water carbonators, and waste composters that allow modern cooks—still primarily women—to showcase their skills and knowledge and care for their families in presumably more enlightened ways. Chefs, turned into celebrities by the media, have been teaching us new skills and have introduced new ingredients into our everyday food. Architecture and lifestyle magazines began devoting more space to kitchen design, and kitchens became once again the heart of the home, with well-appointed large and open kitchens adding real estate value to any property.

The evolution of mainstream "kitchen cultures" reveals changing consumer preferences and larger social dynamics regarding work and domesticity. While very few people are unswayed by advertising and the media, not everyone follows mainstream trends with the same enthusiasm. First, many low-income people cannot afford to purchase appliances and remodel their kitchens according to the latest trends, particularly if they are renters. Second, for many working-class, immigrant, and ethnic families, food practices seem to be more heavily influenced by tradition than by popular trends. Predictability and familiarity of meals are prioritized over innovation and diversity. As with many other consumer trends, mainstream kitchen cultures align more closely with the white middle class, for whom "keeping up with the Joneses" is important. Third, a growing number of people appear to be questioning and resisting consumerism and the urge to buy new technology and more heavily processed food. Such resistance is often grounded in nostalgia and romanticization of the past, as illustrated by the Slow Food movement. Yet, although they consume differently and perhaps with greater scrutiny, these consumers are still influenced by advertising and marketing.

12.4. Conclusion

Part haven and part prison, kitchens are significant places in the organization of familial, social, and work lives. Cooking may provide joy, including the pleasures of eating, caring for loved ones, crafting something with one's hands, and passing on traditions. However, for many people, cooking is also an immense burden that is exacerbated by a lack of time and resources. Because patriarchy continues to dictate that women should primarily be responsible for home cooking and social reproduction in general, the domestic division of labor remains heavily gendered, despite recent advances made by women in the workplace. Today, in heterosexual families, unpaid and devalued food work continues to be carried out primarily by women.

Wealthier families can outsource cooking and avoid domestic tensions by hiring workers to perform these tasks, purchasing prepared food, or eating out more often. The food service industry has been one of the fastest growing segments of the economy, but its profitability disproportionately rests on the low-wage labor of women, immigrants, and people of color. Consumers who prioritize convenience, affordability, health, and quality seem to have little concerns for the exploitation of food workers, which is hidden in plain sight. The way we cook and eat is reinforced by a commercial culture that masks food work and encourages consumers to buy more without challenging gendered inequities or labor exploitation.

Food, Bodies, Health, and Nutrition 13

Learning Objectives

- Define the body relationally, as produced materially and discursively through interactions with the social, cultural, economic, political, and physical environments.
- Use the notion of embodiment to examine fatness and its gendered, raced, and classed aspects.
- Examine different explanations of obesity, drawing attention to the global food regime and parallels with food insecurity.
- Question the moral panic surrounding the obesity "epidemic" and the pathologization of fatness in public discourse and popular media.
- Critically assess mainstream nutrition and consider what "healthy eating" truly means.
- Consider diets and everyday food practices as social and emotional resistance.

Health has emerged as one of the biggest concerns about the contemporary food system. Fears and anxieties regarding food-related diseases, illnesses, and conditions, which include obesity or fatness, have figured high on the social consciousness, with many pundits blaming the food system for making us fat and sick. Unlike the problems of environmental degradation and labor exploitation that may only affect us remotely or indirectly at this time, unhealthy food touches us directly, individually, and intimately, entering our bodies and putting our own lives at risk.

Food contaminated with bacteria, viruses, parasites, or toxic chemicals could kill us. Globally, according to the World Health Organization (WHO 2020), six hundred million people—one in ten—are estimated to get sick each year from eating unsafe food. This results in 420,000 annual deaths that disproportionately affect children under five and people living in places where food safety is compromised by the lack of sanitation and proper food cleaning, storage, and cooking facilities. Still, in the United States, there are 48 million food poisoning cases diagnosed each year, leading to over 3,000 deaths. Increasingly complex and transnational food supply chains have exacerbated the risk of contamination and made it difficult to track.

Although not technically unsafe to eat, much of the food available through mainstream grocery stores and restaurants is unhealthy and negatively impacts health in slow and chronic ways, by clogging our arteries, weakening our immune system, causing inflammation, and destabilizing our endocrine system. Countless popular books and documentaries in the past two decades have shed light on the deleterious health effects of industrial food that are high in calories but low in nutrients and include numerous toxins linked to fertilizers, pesticides, additives, and packaging. In particular, there has been a collective concern—some would say a *moral panic*—regarding the rise in average human body weight around the world (see chapter 6).

By definition, health concerns are about *the body* and the viruses, bacteria, toxins, calories, and nutrients that enter it. Critical geographers have approached the body relationally as a scale interconnected to others and shaped by these interconnections. The body is not an independent and isolated unit, it is in constant interaction with its environment, socially, materially, and emotionally. This idea has important ramifications for how we think about food, diets, nutrition, and health in general.

In this chapter, we examine the relationship between food and health through a geographically informed understanding of the body. We begin with a theoretical discussion of the body and introduce the notion of embodiment and other key concepts. Equipped with these, we then turn to the so-called "obesity epidemic" and explore fatness and its gendered, classed, and raced aspects, paying attention to moral questions. We examine different explanations of increasing fatness, including the role of environmental, political, economic, and cultural factors. We end the chapter with a discussion of healthy eating and an investigation of how mainstream nutrition and ideas about good food emerge and influence food practices. We consider how power relations shape diets and everyday eating and how food practices can reproduce but also resist social norms regarding health, bodies, and nutrition.

13.1. The Body

The recent attention given to the body in social theory is part of a movement toward the corporeality and materiality of social relations and culture. It stems from a rejection of Cartesian *dualisms* between mind and body, rationality and feelings, and abstraction and practice and is an attempt to bring together body, mind, and spirit. Feminist, poststructuralist, queer, and critical race theorists, including many cultural geographers, argue that "we do not *have* bodies, we *are* our bodies" (Trinh 1999, 258). Focusing on the body has several implications for the study of health and nutrition.

First, our bodies are not merely physical containers for our minds, they are the primary ground for experience and knowledge—what Haraway (1988) calls *situated knowledges*. Our cognitive and reflexive capacities are only possible because our embodied beings interact affectively with what surrounds them. Our own *subjectivities*—the way we think about ourselves—are produced and expressed through these embodied experiences. *Affect* plays a central role in this process, suggesting that emotions and bodily sensations are inseparable from the mind.

Second, the body is not simply a given biological entity but a work in progress, constituted and shaped through contextualized interactions. As Butler (1997, 404) puts it, "The body is always an embodying of possibilities both conditioned and circumscribed by historical convention." There is a metabolic and affective interchange between the body and the environment that subverts distinctions between human bodies, landscapes, and self. The notion of *embodiment* highlights how social difference is metabolized by and inscribed on the body. The permeability of the body to the broader environment is both material and symbolic.

Food provides one of the most poignant examples to illustrate this concept. Because food is ingested by the body, it connects us *viscerally*—through the gut—to what is around us. Environmental pollutants and toxins become part of foodstuffs that enter the body and transform it. The built environment, including access to food retailers or land, influences what people eat. Economic inequality and poverty, which are linked to food insecurity, also affect the body by depriving it of nutrients. In other words, the body is materially constituted within political, economic, and social structures, suggesting that there is a *political ecology of the body* (Guthman 2012, Hayes-Conroy and Hayes-Conroy 2015).

In addition to being materially shaped by their surroundings, bodies are also symbolically and discursively shaped by prevailing ideologies. Social norms about bodies, health, and beauty have profound impacts on eating practices and embodied subjectivities. As a society, we categorize bodies, including our own, according to moral values we attach to specific types of bodies and ways of eating. For example, in the Global North, discipline, hard work, and high socioeconomic status are discursively embodied in thin bodies. In contrast, fat bodies have come to be associated with laziness, lack of will power, and low socioeconomic status. Such associations are not universal and vary across time and space. Race and gender also influence the way bodies and bodily practices are being judged and stigmatized. In that sense, social difference is symbolically "inscribed" on the body.

The concepts of biopolitics and biopower have been influential in this scholarship on the body, with important ramifications for health. According to Michel Foucault (1978), *biopolitics* refers to the administration of human life. In that framework, the body and life itself are politicized and governed by regulations, interventions, ideologies, and self-imposed disciplining. *Biopower* refers to the technologies and techniques that monitor and control human social and biological processes. There are many state-sponsored and institutionally enforced rules about what people are allowed to do to their bodies, including the regulation of smoking, alcohol consumption, and diets for children, pregnant women, and hospital patients. However, in modern society, most of these "rules" are self-imposed: we regulate and monitor our own behavior according to socially and morally prescribed norms of health. Thus, according to Foucault, power in modern society is no longer the sort of sovereign power that coerces people by force, including the threat of death, but biopower or the ability to influence life by making people do things on their own will.

For instance, the notion of "health consciousness" suggests that health is something that can be achieved through self-discipline and individual responsibility—an idea fostered by "self-help" books, videos, and magazines. To the extent that health is increasingly associated with appearances, being healthy means having a thin, fit, and toned body, like that of actors and celebrities omnipresent in the media. In that context, the body becomes an individual *project* of self-discipline. Those who fail to get the ideal body often blame themselves and are seen by others as lacking awareness, discipline, and moral values.

Bodies, however, do not just materially absorb and symbolically reflect social difference, they also actively engage in negotiation, contestation, or resistance. We exercise some agency in the way we carry, use, care for, and shape our body. As diets, physical exercise regimen, plastic surgeries, choice of clothing, makeup, tattoos, and piercings show, bodies can be modified through individual actions. Some people modify their bodies and bodily practices to fit social norms, but others use their bodies as a form of *resistance* to confront expectations. These actions, along with our bodily emotions, shape our identities and social worlds. Yet this agency is always contingent on the larger political, economic, social, and cultural contexts that make such actions conceivable, possible, and desired.

The process of embodiment pivots on the places and spaces where bodies are materially and discursively constituted through unique political ecologies and biopolitics, which is one of the main reasons this theoretical approach has been so appealing to geographers. The environment-food-body nexus has prompted important scholarly work among geographers on obesity, fatness, and nutrition, with particular attention to gender, sexuality, race, and ethnicity.

13.2. The Obesity Epidemic and the Pathologization of Fatness

In chapter 6, we learned that obesity and overweight have been on the rise globally, with about 39 percent of the world's adult population considered overweight and 13 percent obese in 2016—more than twice what it was forty years earlier. These trends have been alarming public health experts because of their human, social, and economic costs, leading the World Health Organization to declare an "obesity epidemic" in the late 1990s. Concerns are especially high when it comes to children who are now predicted to live shorter lives than previous generations because of the higher body weight they carry and its negative impacts on their health. Overweight and obesity primarily affect people in rich countries like the United States where two-thirds of adults are now considered overweight or obese. However, it appears to be growing rapidly in middle-income countries like Argentina, Egypt, and Brazil (see figure 6.2)—a phenomenon that some observers describe as "globesity." For instance, in India, a country with widespread undernutrition, obesity is increasing faster than the global average, worrying public health officials about the spread of noncommunicable diseases such as diabetes, heart disease, hypertension, and several types of cancer. As noted in chapter 6, obesity and hunger are not mutually exclusive; they are both forms of malnutrition that can manifest themselves simultaneously. In the Global North, excessive weight is more likely to burden low- and middle-income people, while in the Global South, it is usually upper-income people who are bigger and heavier. As a result, corpulence is interpreted differently across time and space.

The phrase "obesity epidemic" seems odd considering that obesity is not technically a disease and is certainly not communicable in the traditional sense of an epidemic, which implies that it is biologically transmitted from one person to another at a rapid rate. However, in light of concepts such as biopolitics and biopower, the phrase makes sense as a way to underscore urgency, mobilize resources, and legitimize various types of interventions. It engenders a *moral panic* around body weight that paints fatness as the cause of numerous social problems, ranging from low labor productivity and lack of military readiness to the high economic burden of caring for the health of overweight people, all of which threaten capitalism and require to be curbed via drastic measures. The resulting fat-phobic culture blames individuals for these problems (Guthman 2011, LeBesco 2010).

One of the key assumptions in the mainstream public health *rhetoric* on obesity, exemplified in the epidemic discourse, is the notion that thinness is healthier than fatness. Obesity refers to an excess of body fat, which studies have linked to several health problems such as coronary disease, diabetes, and cancer. However, rather than diagnosing and measuring these problems directly, they are being predicted based on fatness. Specifically, the body mass index (BMI), which is computed based on a ratio of weight to height, has been used extensively by medical professionals and public health scholars to identify people and populations at risk. As physical appearances become predictors of health, large body sizes carry increasingly negative stigma. Fatness is *pathologized*, becoming a condition or disease in need of medical treatment or intervention.

Although there is a correlation between BMI and certain diseases and illnesses, some scholars question the notion that fatness automatically means poor health. Many big bodies are overall healthier than thin bodies, especially when considering the physical and emotional harm caused by efforts to remain or become thin, including a rise in eating disorders. If health is about physical, emotional, and spiritual well-being, it cannot be assessed simply by looking at the shape or size of someone's body.

Fat studies has emerged as a new field of scholarship concerned with the various meanings of fatness and experiences of inhabiting fat bodies. Geographers such as Colls and Evans (2009), Guthman (2011), and Longhurst (2005) have made important contributions to this critical literature by questioning and challenging the stigma associated with corpulence and considering the role of place and space in shaping what it means to be fat.

What society considers to be a healthy body or deems to be pathologically fat has changed over time and varies across space, suggesting that there is no universal agreed-upon standard of health but rather socially constructed ideals. For instance, it is typically assumed that fatness is a desired symbol of wealth in societies where food is scarce and many people do not have enough to eat. During the Renaissance, women with large and soft bodies were considered more beautiful, as suggested in notorious paintings by Rubens and others.

Feminists have raised concerns that, over the past decades, the ideal female body has shrunk significantly, particularly in the Western world. They have shown that the average size of fashion models and celebrities is much smaller today than in the past and practically unattainable for most women, two-thirds of whom wear a US size 14 or larger. In her classic book *Unbearable Weight*, feminist scholar Susan Bordo (1993) argues that eating disorders such as anorexia nervosa and bulimia are logical manifestations—embodiments—of a culture in which women's identities are tied to their bodies and bodies are viewed as malleable and controllable. She understands women's bodies as a site of struggle. Women are constantly at war with their bodies: on the one hand they are encouraged to eat by consumerism, but on the other hand they are expected to stay fit and thin. Bordo's work has been foundational in the interdisciplinary field of body and fat studies and influential in questioning the connection between thinness and health.

In response to these concerns, *body positivity* has emerged as a social movement seeking to challenge social standards of beauty and health and promote the acceptance of all bodies regardless of physical appearance, age, gender, sexuality, race, or physical capacity. It has influenced advertising, with products increasingly marketed by models of all sizes, and infiltrated social media, where "real" photos are accompanied by hashtags celebrating beauty in its various forms. Some have criticized the body positivity movement for normalizing fatness and discouraging people to take initiatives to improve their health. Others have attacked it for failing to address the systemic causes of both obesity and fatphobia, instead putting responsibility on individuals again—in this case to feel good about their bodies regardless of their shape. Such a positive, almost cheerful, discourse masks the internal struggles and self-hatred that often comes with living in a fat body.

Explaining Obesity 13.3.

13.3.1. Energy Imbalance

There are several competing explanations for why some people gain weight and why obesity is occurring globally today. The dominant narrative focuses on "the energy imbalance that occurs when a person consumes more calories than their body burns," which the U.S. Center for Disease Control claims is the cause of obesity (CDC 2020). Influenced by this assumption, numerous studies have reported changing diets, rising caloric intake, and declining levels of physical activity around the world. In high-income regions, where food is readily available and lifestyles have become more sedentary, people have been consuming more calories than they presumably need. For example, in the United States, the average person consumed 3,766 calories per day in 2017, well above the 2,500 and 2,000 calories recommended for relatively sedentary men and women, respectively. In most of the Global North, more calories are consumed than necessary for a healthy life. The composition of these calories is also problematic, with a large share coming from sugar and saturated and trans fats. As processed food has become more common, people have been consuming fewer fresh fruits and vegetables, with many eating less than the recommended number of servings. A decline in physical activity is also blamed for causing an increase in body size. This is typically attributed to changes in the nature of work, growing dependence on the automobile, increasing use of television, internet, and video games for leisure, and declining access to safe outdoor spaces due to crime and urbanization patterns.

For some scholars, there is a biological explanation for our tendency to consume more calories than we need and for the body's predisposition to store fat. Throughout much of human history, food was scarce, so humans evolved to eat whenever they could and store fat reserves to prepare for future penuries. Known as the "thrifty gene hypothesis," this explanation suggests that our ancestors developed a genetic predisposition to store fat to ensure survival of our species. Genetics might also explain why we crave sweet, fat, and salty food, although many people argue that these are "acquired" tastes, learned through food practices in the earlier years of life. Today, in an environment where food is plentiful and lifestyles more sedentary, such genes no longer serve a useful role. Instead, they encourage us to eat more than we need, gravitate toward unhealthy foods, and store excessive and unnecessary reserves of fat. While there is little doubt that genetic factors influence the likelihood of accumulating fat and becoming obese, large variations in body weight among people with similar genetic makeup suggest that other factors are at play in activating these genes or effectively resisting their influence. Furthermore, these genes presumably evolve slowly and have existed for thousands of years, thus they cannot explain the sudden rise in obesity without considering the changing environment.

Energy imbalance explanations that emphasize caloric intake and energy expenditure have been criticized for blaming individuals for being overweight and failing to acknowledge the role of extra-individual factors in influencing behavior as well as non-behavioral explanations.

13.3.2. Environmental Factors: The Ecological Model

Looking beyond individual behaviors and flaws, scholars have turned to environmental factors as explanations for expanding body sizes. Influenced by environmental justice research demonstrating how exposure to health risks (e.g., pollution) and access to health amenities (e.g., public parks) are related to race and class, they argue that differences in obesity rates might be explained through an *ecological model* that takes into account how neighborhood, school, home, and work environments shape health behavior and outcomes (Egger and Swinburn 1997). For instance, the notion of food desert (see chapter 11) emerged out of a desire to understand how neighborhood access to affordable healthy food might influence eating behavior and contribute to health disparities, including the fact that in the United States low-income, Black,

Latino, and Native people are more likely to be overweight and suffer from related diseases such as diabetes than more affluent, white, and Asian people are. The suggestion that physical and social environments could be *obesogenic*, meaning that they could promote weight gain and prevent weight loss, resonates with many health geographers, such as Smith and Cummins (2008) and Pearce and Witten (2010). This behavioral ecological model is also extremely important in the field of public health.

Over the past two decades, numerous studies have documented, mapped, and analyzed the distribution of environmental factors such as neighborhood safety, air quality, and access to parks, bike lanes, supermarkets, convenience stores, and fast-food restaurants and estimated their potential impacts on diets and health outcomes. Many have found some statistical associations between class, race, and food access as well as between food access and eating behavior at the neighborhood-level (see chapter 11). As Guthman (2011) argues, causality has been more difficult to substantiate because the spatial relationship between the absence of healthy food in a neighborhood and the prevalence of obesity may not be reduced to behavioral dynamics linked to energy imbalance. For example, low-income people (who are more likely to be overweight) might move to a neighborhood with less desirable food environments because it is one of the few places where they can afford to live, reversing the presumed direction of causality. Similarly, retailers may choose to locate in areas based on the demographic characteristics of residents and assumptions about their diets. There is evidence that environmental factors, including chemicals used in food production and packaging, could also influence obesity physiologically through toxic exposure and disruption of the endocrine system, rather than just through caloric intake and individual decision making.

Under these circumstances, grasping how obesogenic environments, including food deserts, are produced politically and economically in ways that reinforce class and race inequality is a critical step toward understanding the rise in fatness and its greater incidence among socioeconomically and racially marginalized people. It is also a necessary step to move away from explanations that stigmatize unhealthy places and ultimately blame their residents by suggesting that they lack the knowledge, motivation, and strength of character to organize and improve their own communities.

This has prompted some scholars to broaden their perspective on place and turn to structural explanations of obesity, expanding the scalar focus of inquiry. Instead of asking whether living in a food desert causes obesity, they are asking deeper questions about how food deserts and toxic environments come into existence and why they are concentrated in certain places and populated by certain groups of people. So-called obesogenic environments are not fixed or given, they are produced by and evolve with political, economic, and social relations, including private investment, public policy, and city planning decisions that spatially reproduce social divisions and urban segregation along the lines of race and class (see chapter 11). They are also embedded within the larger food regime. These concerns with political economic forces distinguish the *political ecology of the body* approach adopted by critical geographers from the so-called *ecological model* which dominates mainstream research on obesity.

13.3.3. The Food System

Considering political economic forces beyond the neighborhood and the city takes us to the national and global scales of the contemporary capitalist food system, which we studied in chapter 3. Critical scholars have examined the role of agrifood corporations and the state in inundating markets with cheap food products high in refined carbohydrates and unhealthy fats that promote obesity. Some have focused on the consumption side, investigating how corporations encourage over-eating through

FIGURE 13.1 Carl's Jr. Super Bowl XLIX advertising, featuring Kate Upton, 2015.

Source: YouTube, https://www.youtube.com/watch?v=BpWlYZzJ9iY (fair use allowance).

marketing and advertising. Others have looked more closely into production, including farming, trade, and regulatory policy, showing how the state has promoted a system in which the production of unhealthy food prevails.

Advertising is often blamed for stimulating consumption and hiding the potential effects of food on weight and health (Nestle 2013). For instance, commercials for fast food typically show young, thin, and seemingly healthy people eating. In the United States, a series of ads featuring barely dressed and highly sexualized models and celebrities eating "all natural" burgers drew much criticism from public health advocates and feminists alike (see figure 13.1). Having conveyed the unrealistic idea that you could indulge in fast food and stay thin to millions of viewers, the company eventually withdrew its ads.

Food corporations allocate a large share of their budgets to advertising, with about half of it directed at children who are often unable to distinguish between ads and regular programing such as cartoons. In 2016, in the United States, children saw an average of ten food ads per day on television and many more through product placement, on social media, and even at school (Rudd Center 2017). These ads focus primarily on fast food, followed by sugary drinks and breakfast cereal—all of which are high in calories and low in nutrients. Black and Latino young people are disproportionately targeted by these ads. Research indicates that food ads are effective in influencing taste, creating brand loyalty, triggering cravings, and increasing consumption of advertised foods among children, leading some experts to blame advertising for the increase in childhood obesity. Some public health advocates argue that such ads should be banned or regulated, like cigarette ads. Although some countries have moved in that direction, in the United States the decision is left to food companies who self-regulate advertising to children under twelve years old through a voluntary program. In the past decade, faced with significant criticism, corporations have begun positioning themselves as part of the solution by engaging in healthy lifestyle campaigns. Such campaigns encourage consumers to eat diverse diets, consume with moderation, and engage in regular exercise, leaving room for the occasional consumption of fast-food, sweets, and alcohol. As such, they reinforce the idea that health is an individual and moral choice, while simultaneously encouraging consumption. The burden of obesity is on individuals, including the parents of young children, not corporations.

As political economists argue, corporations operate with backing from the state, whose primary role under neoliberalism is to support free markets and private enterprise to sustain economic growth and capital accumulation. Several popular and influential books such as *Fast Food Nation* and *The Omnivore's Dilemma* and films such as *Food Inc.* came out in the early 2000s, drawing the public's attention to the power of agrifood industries in driving what we eat by prioritizing profits over the health of consumers with the complicity of governments. Focusing primarily on the United States, these books and films document how food corporations, working with lawyers, lobbyists, advertisers, marketers, and experts, undermine dietary advice, influence government policy in their favor, and ultimately manipulate eating habits by limiting consumer choice. Empowered by huge profits, the food industry is well organized and works on multiple fronts to shape the food system in ways that threaten our health and contribute to obesity, while simultaneously causing food insecurity.

First, they financially support political candidates involved in food and agriculture policy, including chairs and members of congressional agricultural committees, who receive hundreds of thousands of dollars in campaign contributions and soft money. This likely influences their positions on matters related to farm policy, trade, food assistance, and food industry regulations.

Second, they influence policy through thousands of lobbyists who feed policy makers partial and biased information. Policy briefs based on research funded by agrifood corporations and their philanthropic arms shape the content of public health messages, impact food regulations such as label requirements and the prohibition of certain chemicals and ingredients, and influence taxes, subsidies, and trade protections of specific products, with dramatic consequences on the price and availability of food.

Third, the food industry infiltrates government agencies through the existence of "revolving doors" that put industry leaders in important decision-making positions within public agencies and turn former politicians with insider knowledge into lobbyists for food corporations. For example, Tom Vilsack was selected by President Biden as secretary of the USDA. Having previously served in that position for the Obama administration, he subsequently became executive vice president of Dairy Management Inc. and president and CEO of its subsidiary U.S. Dairy Export Council before returning to the USDA. These connections between public agencies and the food industry are likely to influence policy decisions.

Fourth, food corporations file lawsuits against unfavorable laws and regulations. For example, various attempts by state and city governments to tax soda and sugary drinks have been challenged in court. The court system is also used to intimidate and silence dissenting voices, accusing them of libel. Several states have passed so-called "food disparagement" laws that make it easier for food producers to sue their critics, including advocates for animal rights, the environment, and public health. Texas is one of those states where in 1997 a group of cattle-industry executives sued television show host Oprah Winfrey for allegedly disparaging the beef industry by suggesting that beef might be dangerous to eat in a conversation about the so-called mad cow disease. Although Oprah won her legal case, such high-profile lawsuits create a climate of intimidation where scientists and advocates are afraid to raise public concern about the dangers of eating meat, dairy, sugar, and highly processed food.

Through these different mechanisms, agrifood corporations have been able to influence policy and regulations in many domains, with severe consequences for how food is produced, marketed, advertised, regulated, and distributed. For example,

Nestle (2013) discusses how required food labels and public health educational materials, including the food pyramid and "my plate" infographics used by the USDA, are the result of intense negotiations between scientists, government officials, and corporate lobbyists. The latter fight hard to ensure that their products, including red meat and dairy, continue to be recommended as parts of a healthy diet and are not burdened by labels that could suggest a potential risk to consumer's health. As a result, Americans continue to consume foods that scientific evidence indicate might negatively affect their health.

The political influence of corporations has been critical in allowing food manufacturers to avoid regulations and monitoring of their activities. This is illustrated by the lack of authority that the Food and Drug Administration (FDA) and U.S. Department of Agriculture (USDA) have in overseeing food processing plants and recalling unsafe food products. Indeed, these federal agencies need permissions to visit facilities and collect microbial samples. In most cases, they can only recommend recalls of contaminated food, leaving the decision to food companies to voluntarily recall their products and incur a large monetary loss. In response to numerous food-borne illnesses outbreaks, several bills have been introduced over the past decades to increase the oversight of federal agencies, but most have been rejected by congress—likely a result of intense lobbying and pressure from food companies.

Big agrifood corporations have also been meddling with the Farm Bill (see textbox 3.2), which must be reapproved every four years to ensure the continuation of subsidies and protections for the sectors it represents, including corn, soy, dairy, meat, and sugar. As a result of these subsidies and protections, the cost of unhealthy food is kept artificially low, giving it an unfair advantage over healthier options such as organic produce grown on small and biodiverse farms that do not benefit from the same level of government support. The widespread use of corn syrup in processed food illustrates the deleterious effects of public policy shaped by special interests (see textbox 13.1).

Box 13.1   **Corn Syrup**

Corn syrup, especially in the refined form of high-fructose corn syrup (HFCS), has become a symbol of everything that is wrong with our food system. Chemically produced from corn starch, it is found in a wide variety of industrially manufactured food products including soft drinks, juices, breads, salad dressings, sweetened yogurt, breakfast cereal, granola bars, and frozen pizza. Similar to sugar in taste, HFCS is used for its texture, flavor, stability, and long shelf life, but perhaps more importantly because of its affordability, which is a direct result of corn subsidies and trade restrictions on the import of sugar cane.

In the early 1970s, the Farm Bill changed the way it guaranteed minimum target prices for commodities like corn. Instead of buying potential surpluses, the government began paying farmers deficiency payments to cover the difference between market and target prices on their crops. This encouraged production of corn and led to a decline

in its market price, providing an incentive for food manufacturers to substitute corn syrup for cane sugar and add it to their products, making them increasingly sweeter. High-fructose corn syrup became more widespread in the 1970s following these changes in corn subsidies.

HFCS became targeted in the popular media as a negative symbol of industrial food production and a major cause of obesity. In response, to avoid HFCS, many foodies have turned to "traditional" sodas made with "real sugar," including vintage brands that have become popular again in hip restaurants. Manufacturers of products containing HFCS turn to science to defend themselves and argue that sugar and high-fructose corn syrup have relatively similar effects on health and obesity. This ignores the fact that by lowering the cost of sweetened food, HFCS has increased the consumption of empty calories, particularly among people with tight budgets.

As figure 13.2 illustrates, corn syrup—the ancestor of HFS—has long been promoted as a delicious and even patriotic alternative to sugar that is unique to America. Thanks to corn syrup, Americans do not need to "change their habits," even in war time; they can keep sweetening their pancakes and enjoying good candy (which the ad describes as a necessity). Today, ads might be less explicit in encouraging the consumption of sweeteners and avoid language that directly puts the responsibility of feeding families on the "housewife."

FIGURE 13.2 Advertising for Karo Corn Syrup, 1917.

Source: https://commons.wikimedia.org/wiki/File:Karoadvert-1917.jpg.

As we discussed in chapter 3, economic and trade policy, whether imposed unilaterally by national governments or multilaterally by international organizations like the World Bank and the International Monetary Fund, have also played a central role in opening markets, fostering foreign investment, removing subsidies and regulations, and devaluing currencies, with significant impact on food production and consumption. For instance, the expansion of GMO corn, soy, and wheat and its growing presence in our food supply can be traced to economic policies that encouraged large scale monoculture and exports around the world.

In this political and economic context where governments seem to be more concerned with the economic well-being of the food industry than with the health of their own people, corporate power remains unchecked and continues to grow, producing unhealthy yet highly profitable food and promoting its consumption around the world. This is particularly true in industrial societies like the United States where people are almost completely dependent on corporations to obtain food. In areas where subsistence farming remains a significant part of the economy and people grow their own food, corporations hold less power over consumers. However, as we learn in chapter 3, the global capitalist industrial food regime is threatening the livelihoods of small farmers, decreasing their ability to feed themselves, and increasing their dependence on imported and processed food. This contributes to both increased food insecurity and rising obesity in the Global South, where diets are changing rapidly as illustrated by the increase in processed food consumption and obesity in Mexico (see textbox 3.4) and in many Pacific islands (see textbox 6.3), following the implementation of structural adjustments programs and free trade policies. In other words, obesity and hunger are caused by the same structural mechanisms that grant agrifood corporations the power to control how food is produced and distributed.

13.3.4. Culture

As we learned in chapter 10, cultural factors play a role in shaping our relationship to food, including composition and timing of meals. American food culture has an especially bad reputation when it comes to promoting unhealthy habits. Around the world, the United States is known and derided for its fast-food drive-through restaurants where overweight people purchase meals, snacks, or drinks to be consumed hurriedly in their car at any time of day or night. Although this perception might be exaggerated and is problematic in the way it homogenizes and vilifies eating practices in the United States, it nevertheless points to consumerism and the importance of convenience in the American eating culture. In an overworked society, the ability to eat quickly, without too much preparation, and whenever desired is highly valued and has fueled a whole industry of fast food, take out, deliveries, frozen meals, and packaged food. As we learned in chapter 12, cooking is a major source of stress and tension in many households for whom convenient food provides a way to avoid conflict. As a result, many Americans no longer eat three meals a day following a relatively stable schedule of breakfast, lunch, and dinner. Instead, they eat "all the time" and "snack" throughout the day—a phenomenon linked to increased obesity.

Outside of the United States, many people, especially cultural elites and public health experts, fear the globalization of this trend and the threats it poses to their food cultures. For example, the *Slow Food* movement grew out of this fear of fast food and its putative negative impacts on health, society, and culture. Founded in Italy in protest to the opening of a McDonald's restaurant in one of Rome's most iconic public spaces, the Slow Food organization quickly expanded to other countries in Europe and beyond. By advocating a return to traditions, including "simple" food prepared at home with ingredients grown locally by small farmers and crafted with care, Slow Food spearheaded a cultural shift from convenience to quality. Today, its network comprises producers in 160 countries, including many food communities in the United States. While rarely focusing on health and obesity directly, the underlying assumption of the Slow Food movement is that "food habits, food quality, and lifestyle determine our health and influence our life expectancy." According to the organization's website, "eating can easily be healthy and enjoyable at the same time. . . . All we have to do is look to our rich gastronomic traditions" (Slow Food 2020). This type of message emphasizes the cultural aspect of healthy eating, while reinforcing the idea that it is a choice.

Evidence suggests that countries where rigid food traditions dictate the timing, location, composition, and size of meals might be more successful at keeping their population thin. School lunches provide an opportunity to explore these cross-cultural differences in food practices (see figure 13.2). The use of solid plates, glasses, and silverware outside of the United States suggests a certain attention to detail and level of care. The separate serving of appetizers (usually a vegetable-based soup or salad), main courses (including protein, vegetable, and starch), and desserts (typically a piece of fruit, cheese, or dairy product) indicates that meals ought to be structured and that balanced diets are important. In many countries, not only do school lunches provide children with nutritious food, but they are part of a curriculum to teach nutrition, culture, and taste.

Attributing these differences entirely to culture however is misleading. Pictures like the ones presented in figure 13.3, which typically depict meals in countries like France, Japan, Korea, Italy, Greece, Spain, and Israel, whose cuisines enjoy a healthy reputation, may be giving a false if not fabricated image of what school lunches are, reinforcing ethnic and racial stereotypes. In reality, there are significant variations in school lunches from day to day and across schools, particularly between public and private schools and between rich and poor districts. Countries differ in budgets allocated toward school lunches, guidelines, and regulations regarding what food to include and exclude, and competition from home-packed lunches, outside options, and vending machines.

For example, in France, children are typically required to eat the lunch offered at school as part of nutritional and cultural education. Although the state spends about $2 on food per child per day—more than twice what is spent on average in the United States—most parents are also required to contribute to the cost of lunch according to a pay scale, ranging from $0.20 for the lowest-income families to $7 for the wealthiest families (Le Billon 2012). This helps fund nutritious and varied school lunches.

In the United States, the School Lunch Program was created in 1946 and feeds more than thirty million children. Government subsidies vary depending on parents' income and range from about $0.30 for "paid" lunches to almost $3 for "free" lunches. Unless their children qualify for free lunches, parents are expected to pay the extra cost beyond what is covered by the government. As a result, lunches in Mill Valley—one of the wealthiest districts in the nation located north of San Francisco in a region known for its foodie culture—cost parents an average of $6.80 per day—a result of parents' insistence that their children be fed healthy lunches (Macarow 2016). Across the bay, in Oakland, most schools qualify for free lunches for their entire student body, meaning that parents do not have to pay and that meals must be prepared within the budget set by the $2.82 federal reimbursement rate, most of which is spent on labor costs. In elite private elementary schools, healthy and gourmet meals prepared by trained chefs or catering services are included in tuition and used by schools to advertise how much they care about the health and well-being of students. Parents who cannot pay the fees or high tuitions might feel that they are failing to provide the best for their children.

FIGURE 13.3 School lunches around the world.
  a. Japan: tofu curry with rice, side salad, quarter apple, with milk
  b. Finland: Chicken in cream sauce, carrots, cabbage, steamed potatoes, and lingonberry sauce, with milk and water
  c. USA: hamburger, tater tots, and ketchup, with chocolate milk
  d. France: cheese omelet, carrots, side salad, apple, bread, and butter, with water

Source: Finland: https://commons.wikimedia.org/wiki/File:Finnish_school_lunch.jpg.
    Japan: https://commons.wikimedia.org/wiki/File:Japanese_tofu_school_lunch.jpg.
    USA: https://flic.kr/p/7XPdUm.
    France: https://nutrition.org/french-brown-bag/.

In recent years, images bashing school lunches have proliferated on social media, generating yet another moral panic about food and blaming cafeteria workers and parents for the rise in childhood obesity. Advocates, including public figures and celebrities such as Michelle Obama, Alice Waters, and the "Renegade Lunch Lady," have argued for reforms of the federally funded program. The 2010 Healthy, Hunger-Free Kids Act established new nutrition standards (reducing portion sizes and setting up minimums for fruit, vegetables, and whole-grain servings and maximums for sodium, sugar, and fat content), provided funding to encourage the use of local produce and gardens, and restricted the sale of competitive food from vending machines. However, such programs must be regularly reapproved and are subject to budget cuts. In 2019, the Trump administration rolled back some of the regulations put in place, arguing that it made school lunches too costly.

In a context where poor nutrition is blamed on cultural deficiencies rather than political and economic factors, the burden to provide healthy alternatives, maintain traditions, and eat "slow food" has primarily fallen on women (see chapter 12). Indeed, women's relationship to food is almost always one of responsibility, whether it is about scarcity, inappropriateness, or excess. Grassroots efforts to improve school lunches and nutritional education, including maintaining school gardens, often rest on the free labor of parent and teacher volunteers. Those who have the time, resources, and inclination to get involved in school projects or pack healthy

Box 13.2   **Bento Boxes and the Mommy Wars**

Bentos or *obentōs* are boxed lunches popular in Japan and East Asia. The simplest versions include cooked rice or noodles, a piece of fish, meat, or tofu, and a side of vegetables. However, bentos can be very elaborate, with a variety of miniature servings of various dishes aesthetically arranged in compartmentalized boxes. Although bentos can be purchased in convenience stores, mothers of young children are expected to prepare nutritious, balanced, and enticing bentos for their children. Because of strict school rules, children are required to eat everything packed in their bento boxes, putting pressure on mothers to prepare perfect meals every day—a skill that many internalize as a symbol of being a good mother and that Allison (1991) interpreted as a form of ideological control by the state.

Interestingly, bento boxes have become popular in US preschools, where parents (especially mothers) face similar pressure to produce healthy and appetizing lunches for their young children. Upper-class mothers seemingly compete with each other in creating sophisticated lunches and ensuring that their children acquire discerning palates that will set them on an individual path to lifelong health. This sort of "mommy war" is in full display at birthday parties and on sports fields where a choice of homemade healthy snacks and fresh fruits will convey social distinction to the mothers who supply them. In contrast, those who bring store-bought cookies and sugary drinks might make children happy but will be judged by presumably more enlightened mothers. While these food rules may occasionally be enforced by schools, parent associations, or sports clubs, they are often unspoken and self-imposed, reflecting social fears about obesity and understandings of health.

meals for their children are viewed as good citizens, doing their share to promote health and sustainability. This has led to "mommy wars" in which mothers try to outdo each other in preparing healthy lunch and distinguish themselves through their children's healthy food habits (see textbox 13.2). This reinforces the idea that bodies can be managed through individual behavior and reinscribes differences based on class and race.

Anti-Fat Politics and Its Consequences 13.4.

Critical scholars argue that obesity is an embodiment of *neoliberal capitalism* through the "neoliberal diet" of industrial, unregulated, highly processed, calorie-intensive, nutrient-poor, and toxic food that transform bodies from the inside. Furthermore, they posit that the contemporary biopolitics linked to the management and regulation of fat bodies are shaped by that system as well. The pathologizing of fatness and the associated panic surrounding fat have become the primary mechanisms that *govern* bodies under neoliberal capitalism. Scientists, experts, marketers, pundits, celebrities, and anxious consumers all contribute to portraying fatness as a disease and instilling fear and disgust of fat in the collective psyche, with the media playing an instrumental role (see textbox 13.3). Statistics, accompanied by terms like epidemic and crisis, contribute to the sense of urgency and exacerbate the haunting notion that obesity is spreading uncontrollably. Such fatphobia pushes people to self-discipline, reinforcing the idea that obesity can be avoided or eliminated through individual behavior, not government intervention or systemic change.

The concept of *bio-citizenship* denotes the idea that individuals have a responsibility to take care of themselves and control their weight for the social good (Halse 2009). The general assumption, particularly within the dominant energy imbalance paradigm, is that people become overweight because they fail to do what is good for them (and for society) due to some individual weakness such as a lack of will power,

FIGURE 13.4 "U.S. Needs Us Strong: Eat Nutritional Food," public health poster from World War II era.

Source: Government and Geographic Information Collection, Northwestern University Libraries. https://dc.library.northwestern.edu/items/822521c7-646f -40b8-88db-f0e6425a0055.

knowledge, or moral values. Those whose lifestyles seemingly contradict the cultural norm of a well-managed body become socially excluded and morally lose their right to belonging and citizenship. As a result, they are ridiculed, shamed, blamed, and discriminated against (see textbox 13.3). In contrast, those who eat well not only improve their chances of being (or at least looking) healthy, but also display the appearance of self-control and moral superiority. During war times, eating well has been portrayed as a patriotic act (see figure 13.4)—an idea that remains true today. This notion of bio-citizenship is indicative of neoliberalism to the extent that it extols personal responsibility and self-discipline through individual choice.

To the extent that people of color are more likely to be overweight for several reasons we discussed above, the conflation between thinness, health, and personal responsibility tends to reinforce race differentiations. In fact, in the popular imagination, the obesity epidemic is typically associated with Black and Brown people—an image generated by repeated statistics and ubiquitous images of dark fat bodies in the media. Black women's bodies have historically been marked as unruly, unfit, lazy, in excess, and in danger, compared to the normalized white body (Strings 2019). In

Box 13.3 **Fatphobia and Fat Shaming on Television**

Television shows are replete with instances of *fat-phobia* (fear of weight gain in oneself or in society), which often lead to *fat shaming* (ridiculing and belittling someone for being fat). This is especially noticeable in reality television shows such as *My 600-Pound Life* or *The Biggest Loser* that portray obesity as a major cause of physical and mental health problems affecting a rapidly growing number of people. People with fat bodies are stigmatized as lazy, weak, depressed, uneducated, and unloved. Their bodies are often filmed in ways that dehumanize them, zooming in on fat parts of their bodies and cutting out their faces. While the occasional success story reinforces the notion that becoming thin is an individual effort, the fact that most participants lose a few pounds before reverting to bad habits suggests that many people lack the required strength of character. Studies have shown that exposure to this type of show increases anti-fat attitudes (Domoff et al. 2012).

In fictional shows, overweight characters are also represented as two-dimensional people whose entire existence and personality is defined by their body size. Comedic films allow the audience to laugh at and make fun of fat characters, who may be funny on the surface but are presumably miserable. The recent Netflix film *Insatiable* was highly criticized for stigmatizing fat people and reinforcing the notion that losing weight is the only path to happiness. It tells the story of Patty, a fat teen (played by a thin actor in costume), who is constantly bullied at school until she gets punched in the face. As a result of this incident, her jaw must be immobilized and she becomes unable to eat, causing her to lose weight. Her ensuing sudden attractiveness allows her to take revenge on those who had bullied her.

Biases against fatness are also revealed in the bodily appearances of television news anchors, reporters, and show hosts, most of whom are thin and attractive by mainstream standards. It is extremely rare to see a big person in a leadership position on television. Together these different images reinforce a culture in which being fat is something to be afraid or ashamed of.

---

many ways, the obesity epidemic discourse extends this narrative, justifying discriminatory practices and reproducing racial inequality.

In that social context, many people go to great lengths and take drastic measures to manage their weight and engage in preventive care. Dieting, which puts people through cycles of starvation and highly restricted eating, is extremely common. In a 2018 survey, almost half of Americans listed "losing weight or getting in shape" as a New Year resolution—the most common resolution behind saving money, well above "buying a house" or "finding love" (Statista 2020a). Indeed, many people believe that being thin will bring happiness and are quick to try the newest diet fads, whether low-carb, paleo, ketogenic, or intermittent fasting. In 2020, just over one hundred million Americans—more than a third of adults—claimed to be watching their diet, with the primary reason being to lose weight. Dissatisfaction with one's body size and fitness were particularly high among millennials (born between 1981 and 1996), with about 80 percent saying that they could be healthier. These figures illustrate a general concern with health that pivots on body size.

Even among those who do not follow strict diet plans, many are paying attention to the type of food they eat. The success of stores like Whole Foods and the growing popularity of "natural" and organic "superfoods," free of additive, fat, sugar, gluten, MSG, and GMO, indicate that people are willing to pay more for what they consider to be healthier food. In a capitalist system, it is not surprising that producers have capitalized on people's fears, earning profits by supplying products that consumers have come to believe will make them thin and healthy. This has often been interpreted as a sign of consumer sovereignty, suggesting that, in a market economy, consumers can vote with their forks and force producers to supply healthier products through their purchases. More critical interpretations, however, emphasize that corporations remain in control of what and how food is produced and marketed.

Through diversification and market segmentation, they benefit directly from the price premium that high income consumers are willing and able to pay, while continuing to market cheaper and presumably less healthy food to lower-income consumers. This market solution has created a bifurcated system in which healthy food is reserved for those who can afford it. Retailers locate their food outlets accordingly, cementing these inequalities spatially in the foodscape. There is also evidence that the expansion of the health food market has done little to curb consumption and might instead stimulate it by assuaging consumers' guilt. Artisanal organic potato chips that have been kettle-cooked in avocado oil and lightly sprinkled with Himalayan pink salt still have the same caloric content as mainstream versions. And although fruit-sweetened candy contains less added sugar than traditional candy, it is still devoid of most nutrients and encourages children to develop a "sweet tooth."

Dietary supplements such as vitamins, minerals, herbal supplements, amino and fatty acids, and probiotics have witnessed a similar upward trend. Despite their mostly unproven benefits, consumers believe that such supplements will improve their health and help them compensate for the lack of nutrients that modern convenient food supplies. Seventy-seven percent of US adults take at least one dietary supplement, contributing to an estimated market of $46 billion in 2020 (Mordor Intelligence 2020). The global market is valued at $170 billion and growing rapidly, with the fastest growth occurring in China and Asian countries. At the fringe of this market are dangerous, expensive, and unregulated medications that suppress appetite, speed up metabolism, and purge the body. Young women are particularly vulnerable to empty promises of rapid results bolstered by "before and after" pictures and testimonials on social media. The medicalization of obesity has also fueled more drastic body mutilation, ranging from liposuction to bariatric surgery.

Public health interventions to promote healthy food habits and physical activity have increased dramatically in the past several decades and often contribute to the social pressure that many people experience to stay or become thin. For example, many schools and workplaces have programs in place to support, encourage, and monitor healthy eating. While well intentioned considering the severity of weight-related diseases, these programs create an atmosphere of surveillance and social control that many find oppressive and might ultimately be counterproductive. Because they often target places with high proportions of Black and Latino people (e.g., schools in urban districts), these programs also contribute to the racialization of obesity.

This sort of monitoring is particularly intense when it comes to children. For example, Dempsey and Gibson (2018) show how the child's body became a scale of intervention in public schools through the US School Lunch Program, which was originally justified to ensure national security by raising healthy, productive, and militarily ready future citizens. As awareness of growing obesity among children led to panic over a possible epidemic, the school lunch program became the site of heightened scrutiny and multiple interventions, with administrators, public officials, health advocates, teachers, and parents debating their fruit, vegetable, fat, sodium, whole grains, and sugar content and children becoming increasingly self-disciplined in "making good choices." Some school districts include children's BMI on their report card and send warning notes to parents who pack lunches deemed unhealthy, suggesting healthy alternatives. Birthday celebrations have been banned or become heavily regulated shunning cake in favor of smoothies or fruit kebabs.

Smart phone technology, including the growing number of applications that allow people to track how many calories they eat and burn, steps they take, hours they sleep, and grams they lose or gain daily, also contribute to a sense of surveillance. This is especially true when these apps are being promoted by employers, health insurers, and online retailers.

Despite the constant angst and stigma associated with fatness and efforts to lose weight, diets often fail to generate sustainable long-term weight loss. This might be because the current biopolitics is causing *resistance*. Most people respond better to carrots (positive rewards) than sticks (punishment). Fat shaming is not an effective way to motivate behavioral change. Similarly, providing people with a long list of dos and don'ts, as most diet plans and public health campaigns do, takes much of the pleasure out of eating, reducing food to its biological function. It also infantilizes eaters and limits their *agency*. As a result, people are more likely to openly rebel or secretly cheat on their diet, as children tend to do. Choosing a bacon burger when everybody else is eating a garden salad is a way to reassert agency by rejecting cultural expectations. Men, who have been socialized to think that "real men don't diet" (Gough 2007), are more likely to resist social pressure to restrict their food intake. Although this might reveal a positive relationship to food and a confident body image, taken to an extreme, this sort of response can also be detrimental to one's health.

In trying to understand why diets so often fail, it is important to remember that food provides more than biological sustenance. Eating is an act of *pleasure*; it pleases our taste buds and makes us feel good, at least temporarily. The popularity of "comfort food"—food with nostalgic and sentimental value—illustrates the emotional aspect of eating. In most cultures, there are specific dishes that people associate with good feelings, often related to childhood, home, family, and identity. For many Americans, dishes like macaroni and cheese, fried chicken, meatloaf, and mashed potatoes signify comfort. The pleasure also comes from the sociality of eating and the capacity of food to bring people together, express identities, and sustain cultures (see chapter 10).

Berlant (2006) argues that obesity reveals the lack of optimism and balance that stressed, fatigued, overwhelmed, and alienated workers experience under capitalism. The pressure of always working toward a better future—getting an education, earning an income, keeping a job, buying a house, staying on top—is so intense, especially among working-class people who can barely stay economically afloat, that food provides a sort of sideway escape. Food, like sports, sex, television, and alcohol, is a form of self-medication that allows people an "interruptive episode [that] suspends the desire to be building towards the good life." Rather than a conscious and autonomous form of resistance, over-eating episodes are "small vacation from the will itself . . . interrupting the liberal and capitalist subject called to consciousness, intentionality, and effective will" (Berlant 2006, 35). In that context, legislating, educating, and shaming will not change people's appetites. Unless everyday life is transformed through radical systemic change, we will continue to eat as an escape.

Healthy Eating 13.5.

What can we learn from these different explanations of obesity and the realization that current biopolitics are more conducive to shaming and dividing people than cultivating well-being? More specifically, what does healthy eating look like? The field of *critical nutrition* offers some useful answers, including valuable contributions by geographers (Guthman 2014a, Hayes-Conroy and Hayes-Conroy 2013).

Critical nutrition emerged in reaction against mainstream nutrition, which Hayes Conroy and Hayes-Conroy (2013) call *hegemonic nutrition* and describe as practices and discourses that: (1) assume that the food-body relationship can be standardized through the energy balance model and measured quantitatively through body mass indices and calories; (2) reduce nutrition to calories and nutrients;

(3) decontextualize food by ignoring the importance of place, culture, society, and history; and (4) privilege expert knowledge over other forms of embodied knowledge about food and health. Such discourse, which is promulgated by experts, as well as celebrity chefs, journalists, bloggers, and people with limited scientific credibility, has clearly been ineffective in reducing Americans' body size. In fact, the proliferation of science-based dietary advice in the past three or four decades seems to have been paradoxically accompanied by a worsening of our health, suggesting major limitations in mainstream nutrition.

Instead, critical nutrition scholars and advocates recommend an approach to eating that acknowledges the centrality of the material body in our relationship to food, that is, the *visceral* and *embodied* reaction that most of us experience when thinking about, smelling, seeing, and ingesting food. They suggest that we listen to our bodies, practice reflexivity, be attuned to what feels right, and question our judgments and prejudices regarding certain foods and more importantly our own bodies. This reflexive, gentle, caring, and forgiving approach to eating is part of what they describe as *body literacy*—the ability to "read" our bodies and learn through them.

A key aspect of critical nutrition is recognizing that food has multiple meanings beyond nutrition and that these meanings are geographically, historically, and socially contingent. Within mainstream nutrition, emotional connections to food and *embodied knowledge* have been ignored if not outright dismissed in favor of quantifiable data such as calorie counts. People are often urged to give up foods deemed unhealthy and replace them with ingredients and dishes recommended by experts. People of color and immigrants are regularly told that the food they love is not good for them. Even if this message is not conveyed explicitly, it is communicated implicitly by the absence of ethnic food in nutritional education materials that normalize the mainstream white diet.

Critical nutrition encourages people to reclaim foods that make them feel good in a broad sense that includes physical, emotional, and spiritual well-being. For instance, Calvo and Esquibel (2016) advocate for Mexican American and Latino people to "decolonize" their diets and return to the healthier foods of their ancestors. This includes a focus on heritage crops such as corn, squash, beans, herbs, and seeds that people were consuming long before colonialism and the growth of industrial agriculture. In her book *Sistah Vegan* (2010), Breeze Harper gives voice to Black vegan women, showing that veganism is not limited to affluent white people and can be a transformative political project to resist racial injustices and reclaim health. These and other similar works have begun to unravel the *whiteness* of nutritional advice and help create a more contextualized and individualized form of nutrition.

Individualized nutrition in this context does not mean a diet plan tailored to one's genetic makeup, like a drug regimen. Instead, it is an acknowledgment that there are many different paths to acquire essential nutrients and cultivate well-being and that, therefore, nutritional advice should not be based on strict and predefined "one-size-fits-all" guidelines but vary according to individual circumstances and preferences (Hayes-Conroy, Hite, Klein, Biltekoff and Kimura 2014). In other words, nutrition ought to be *contextualized*.

Critical nutrition recognizes that health as an individual *project* is neither a realistic nor a healthy endeavor. Fighting obesity is a collective project that goes well beyond providing nutritional advice. It includes redressing power imbalances in the food system such that all people have access to healthy food, the food industry provides decent and well remunerated livelihoods, food is grown and processed in ways that sustain and regenerate healthy environments, products are marketed and distributed in equitable and transparent ways, and corporations and institutions are held accountable for the quality of our food.

The idea of nutrition as a collective project also points to the role of communities in creating a supportive environment conducive to healthy relationships to food. From urban gardens to collective kitchens, community initiatives can play a role in improving access to healthy food, supporting economic solidarity, encouraging body literacy, and producing new visions of what it means to be healthy. Of course, as noted in chapter 11, community-based and alternative food initiatives might also play into the hands of neoliberalism by requiring communities to step up and take responsibility for problems caused well beyond their borders. Perhaps this is a reason they have been increasingly championed in mainstream nutrition and have come to be associated with whiteness. Yet, grassroot food projects can also be radical attempts at promoting food sovereignty.

Conclusion 13.6.

Obesity is one of the major concerns with the contemporary food system. Evidence suggest that it is increasing around the world and is especially prevalent in the United States, which has received most of the research attention and is the primary focus of this chapter. Much can be learned by investigating the material and discursive relationships between food and body within this context.

The notion of embodiment is particularly useful to consider how bodies are the product and reflection of food discourses and ideologies as well as the broader food environment which they inhabit. Energy imbalance, which focuses on calorie intake and expenditure, is the dominant explanation for increasing average body sizes. However, there are other aspects to the body-food-environment interaction that warrant attention. In this chapter, we highlighted the need to consider urban food environments, the larger food system, including the role of corporations and the state, cultural differences, and the punitive nature of mainstream nutrition to better understand the so-called "obesity epidemic." We argued for a more gentle and critical approach to nutrition that acknowledges the multifaceted and situated meanings of food and its entanglement with race, class, and power relations.

New Food Geographies 14

Learning Objectives

- Summarize the major problems facing the contemporary food system.
- Contrast different plausible food futures.
- Synthesize the benefit of a geographic approach to inform place-based solutions.
- Recall the notions of right to food, food sovereignty, and food justice to outline steps toward a more sustainable and equitable food system.
- Provide suggestions on how to get involved in changing the food system through individual behavior, collective organizing, and political engagement.
- Highlight hopeful directions and examples of successful initiatives that create new food geographies.

Food has shaped the history of humankind, the organization of society, and the surface of the earth. The future of humanity depends on food. Our ability to live healthy, peaceful, and dignified lives, to express ourselves, to sustain our cultures, and to care for our communities now and in the future requires that food be produced, processed, distributed, and consumed sustainably and equitably.

A Broken Food System? 14.1.

Unfortunately, the current food system appears not to be working for the vast majority of the world's population. First, those who are working along increasingly globalized food supply chains (chapter 5) are suffering from poverty and labor exploitation (chapter 4). This includes millions of men, women, and children engaged in subsistence farming (chapter 2), working in factory farms and food processing plants, selling food in street markets and megastores, and serving customers in various types of restaurants (chapter 12). Crippling debt and desperation have raised suicide rates among farmers, who, in the United States and around the world, are at the mercy of the transnational seed, chemical, and agrifood technology companies that control input markets and the food distributors, manufacturers, and retailers that dominate output markets on the other end of the supply chain (chapters 3 and 4).

Second, despite seemingly expanding food choices, millions of people are hungry, malnourished, or both (chapter 6). Food insecurity is not only affecting people in low-income countries, but it is also common in high-income nations where inequality is on the rise and social safety nets have been dismantled. Although we produce more than enough food to feed the world's population, many people are denied the means to acquire adequate food. Around the globe, the monoculture of cash crops for export is threatening food security and sovereignty, weakening people's ability to grow and access diverse, culturally appropriate, and environmentally sound crops (chapters 3 and 7). Industrially produced food, which often lacks nutrients but packs calories, is making people sick with diabetes, cancer, and other diseases, if not from the food they eat (chapter 13), then from the fertilizers and pesticides that poison the environment (chapters 7 and 8) or the pathogens bred through factory farming.

Third, global yields are declining despite increased efforts and inputs, revealing a planet in danger (chapters 7 through 9). Calls to reduce our dependence on oil and preserve soil, water, and biodiversity have been mostly ignored under the mantra of feeding the world and promoting economic development. Most of the animals we eat are treated inhumanely, and many species are threatened by pollution, habitat destruction, and overexploitation, including insects, birds, and fish.

Fourth, rich food cultures and traditions are being lost under similar pressures to grow and globalize (chapter 3), making people long for food that gives them joy, creates a sense of community, and values their identity (chapter 11). For too many, cooking is a chore and eating is fraught with anxiety and shame (chapters 12 and 13), denying people one of the greatest sources of pleasure.

The global *COVID-19 pandemic* has exposed the shortcomings of a food system that depends heavily on global supply chains, relies on cheap unprotected and migratory labor, promotes the spread of zoonotic diseases through land use change, and counts on the good will of corporations and charities to feed the poor. Even in countries of mass consumption, supermarket shelves were empty for weeks. Around the world, nutritional challenges have been exacerbated by disruptions along the supply chain that have further reduced access to food and a global economic recession resulting in the loss of food livelihoods and rising poverty. As a result, the number of hungry people in the world increased by almost 120 million from 2019 to 2020, erasing any progress accomplished in the past fifteen years and making the UN sustainable development goal of zero hunger by 2030 practically unattainable. So-called "essential workers" continue to be exposed to greater risks of contracting the virus in most types of food-related jobs, whether farming, processing, or retailing. Yet they must accept these dangerous working conditions and low wages, given the lack of legal and social protections and limited alternative employment options. Lockdowns have resulted in job losses that primarily affect unprotected workers, including informal laborers, small business owners, women, immigrants, and racial/ethnic minorities. These economic challenges have increased vulnerability to weather events and raised tensions in conflict-prone areas. While some argue for the need to "build back" the food system, others point to this crisis as an opportunity to "build better." If anything, the COVID-19 pandemic has raised awareness of the disparities that characterize people's multifaceted relationships and experiences with food.

Who, you might ask, benefits from the current food system? Throughout this book, we have learned that a handful of corporations control the production and distribution of food and reap enormous profits (chapter 3). This is because the global food regime is primarily governed by the logic of capitalism, which measures success in terms of profit and economic growth, not the well-being of people or the health of the planet. In fact, in the economic framework that informs corporate behavior and public policy, malnutrition, diseases, and environmental destruction are *externalities*—meaning costs that are not "internalized" into the price tag of food. And agrifood businesses intend to keep it this way, spending millions in research that questions scientific evidence and lobbying to shape policy to their advantage. Consequently, governments around the world have focused on uplifting agrifood corporations through agricultural and trade policies, deregulation, and selective subsidies, instead of demanding that they produce healthy food, treat people and animals with respect, and preserve the earth's finite resources. Under the fabricated threat of a looming global food crisis, states have bought into the idea that we must ramp up production at all costs, giving corporations a license to ignore so-called externalities and reap large profits. For this reason, we might argue that the current system is far from broken, as some have suggested, but instead is working exactly as intended.

Aside from powerful corporations, affluent consumers, mostly in the Global North, might also benefit from a widening range of food products that provide seemingly endless eating options and experiences. Indeed, on any given night, consumers with enough disposable income can choose between fast or slow, exotic or comfort, organic or conventional, healthy or decadent, local or global, seasonal or year-round, and artisanal or industrial foods—a dizzying array of choices allowing them to experience food in its multifaceted nutritional, social, and cultural capacities while distancing themselves from the problems facing the food system. The rest of the world, whether low-income people in rich countries or most of the Global South's population, have much more limited choices.

Food Futures 14.2.

We currently face multiple plausible food futures. We could stay on the same path, trying to ramp up production while ignoring the social inequality and environmental degradation that ultimately cause greater food insecurity. Given the overwhelming evidence against exerting more stress on the environment, very few people are advocating this "business as usual" approach, except for those who profit from it.

Alternatively, we could change how we grow, process, distribute, and consume food to create a more sustainable and equitable food system that would ensure *food security* in the broadest sense of the term. Some argue for market incentives and regulations that would promote sustainability through technological innovation and ethical consumerism. This is illustrated in the expansion of organic and other labels, genetically engineered seeds, lab-grown and plant-based meat, and increasingly sophisticated computer applications that manage irrigation, pest, soil, transportation, storage, and waste.

Others, however, question the usefulness of the market framework in promoting food security and advocate for more radical change that would reconfigure power relations and restore people's control over knowledge, seeds, land, water, biodiversity, and culture. Considering the many injustices documented here, such radical visions emerge as the only way to truly address global food insecurity and engender more just, healthy, and ecologically sound relationships between people, food, and the earth.

The related notions of right to food, food sovereignty, and food justice, which were called upon repeatedly throughout this book, are useful in delineating specific goals and identifying paths to achieving them. The *right to food* is defined as physical and economic access, either directly or through exchange, to healthy, diverse, and culturally adequate food that enables individual and collective flourishing. It calls for a much broader and comprehensive understanding of food security than what is typically provided by government agencies and international organizations. The idea of *food sovereignty* is helpful in specifying how we might achieve this goal by ensuring that the people who produce and consume food participate in the decision-making process regarding how land, water, and biodiversity are used to produce food that meets people's nutritional, social, and spiritual needs now and in the future. This suggests that realizing the right to food requires *food justice*; that is, the removal of racism, sexism, exploitation, and oppression that prevent farmers and food entrepreneurs from participating in the food system on an equal footing and having their contributions and traditions recognized.

For proponents of these ideas, charity or pro-market economic policies fail to address the power imbalances that characterize the food system and underlie food insecurity. Instead, they advocate for structural changes that confront the deeply entrenched social inequalities that stand in the way of ensuring that all people—regardless of race, ethnicity, gender, class, religious affiliation, or any other characteristics—not only have access to food, but also the ability to define their own food system through economic opportunities and to shape food and agricultural policies through participatory democratic processes. They endorse a comprehensive response to the multifaceted shortcomings of the food system that simultaneously focus on ending hunger and promoting healthy diets, supporting small farmers and food livelihoods, and restoring balance in agricultural ecosystems by prioritizing community needs and removing barriers to economic opportunity and political participation, whether based on gender, class, religion, race, or other forms of social difference. To achieve this broad and radical vision of food security, which aligns with the principles of the right to food, food sovereignty, and food justice, it is paramount that we acknowledge and confront the disproportionate economic and political power granted to large corporate agribusinesses and take steps to restore people's control over knowledge, seeds, land, water, biodiversity, and culture.

While this is a tall order, there is hope in the various food movements and initiatives unfolding around the world. As we have learned throughout this book, today's food system consists of myriad actors, places, and practices connected through global capital flows, economic policies, population movements, cultural trends, knowledge claims, ecological relations, and a changing climate. Because these dynamic networks are always evolving, they are unstable and therefore mutable. Although it is easy to dismiss small-scale efforts as inconsequential, it is important to view them as potentially reconfiguring power relations in ways that might lead to broader change both within and beyond the food system.

14.3. New Food Geographies

This book has emphasized a geographic perspective in analyzing the social and environmental concerns facing the food system and understanding the multiple meanings of food in the human experience. The main contributions of geography to the study of food consist of: (1) integrating a variety of disciplines from the social sciences, environmental studies, and humanities to better understand the full complexity of food; (2) facilitating systemic thinking by focusing on the interdependencies between politics, society, economy, culture, and the environment; (3) drawing attention to the spatial patterns of food production, distribution, and consumption and their unevenness across the globe; (4) articulating how food both shapes and is shaped by the places where it is produced, transformed, sold, cooked, eaten, and discarded; (5) highlighting the networks, chains, and circuits connecting producers, consumers, and the environment across many scales, ranging from the body to the globe; and (6) exposing and challenging the relations of power, exploitation, and oppression underlying these connections.

This knowledge can be mobilized to create new *food geographies*—socio-environmental spatial relations that produce more equitable and sustainable food systems for all. Specifically, a geographic perspective that views food and place as relational can help us envision how localized food initiatives may form the basis of systemic transformation. These new food geographies can reconfigure power relations through the many networks connecting people and places. These include *horizontal networks* that enable productive connections with actors engaged in similar projects elsewhere or in fighting overlapping social and environmental injustices and *vertical networks* that cross scales of governance and bring about change at the regional, national, or global level.

Acknowledging interdependencies between places and scales reveals how we may engender new food geographies by encouraging alternative local food practices, reclaiming traditional and local knowledge, creating more democratic institutions, supporting individual farmers, fishers, artisans, grocers, vendors, and other food workers, including women and people from marginalized communities, and strengthening local and global connections between them.

Get Involved 14.4.

Although the evidence examined in this text paints an alarming and perhaps discouraging picture, it is not too late to act. You and I can play a role in challenging the current system and take steps toward building a better food future—one that affirms the right to food and achieves true food security by redistributing power and prioritizing people and the earth. Through individual behavior, collective organizing, and political engagement, we can improve material circumstances and well-being, redistribute power, nudge public policy, and change narratives around food.

14.4.1. Learning and Reflecting

Throughout this book, you have gained an appreciation of the complex social, economic, political, cultural, and ecological relations embedded in food. You know about the diversity of food experiences around the world and the disparities that race, gender, ethnicity, class, religion, and other social factors produce. You understand the role of place in shaping access to food and the ability to control food materially and discursively. You have also learned to reflect on your own assumptions and practices. This educational process is necessary to initiate meaningful change. Here are some ways you can continue your learning journey:

- Learn about your city or region's food landscape: local government agencies, food policy councils, and universities might have reports available regarding the most important issues facing the area, resources available, and ongoing initiatives.
- Investigate food sourcing, working conditions, and waste issues on your campus.
- Conduct your own research: do an independent study, volunteer as an intern with a local organization, join an ongoing research project at your university.
- Listen to people talk about their experiences: pay particular attention to marginalized groups whose perspectives are often absent from mainstream discourse and whose contributions in creating a more just and sustainable food system have not been properly recognized.

- Ask questions when you shop or eat out: where does the food come from? what is local or in season? what chemicals were used? are workers earning livable wages? how are tips being shared?
- Reflect on your own behavior and relationship to food.
- Adopt a nonjudgmental and compassionate attitude about fatness and corporeal difference.
- When eating foods from cultures different than your own, particularly when you travel, be open minded and cautious not to exoticize, pigeonhole, or appropriate for your own benefit.
- Be aware of privileges and biases you may have in accessing and knowing food.
- Question dominant economic, environmental, political, and public health narratives about food, including the way they might hide injustices, sustain power relations, and prevent structural change.
- Do not take advertising and certification labels at face value.
- Read and follow critical food media such as Civil Eats, Foodtank, The Counter, Real Food Media, FERN News (Food and Environment Reporting Network), NPR's The Salt, and others.

14.4.2. Eating

Over the past two decades, calls to eat more sustainably have become louder. You could go beyond this call, by committing to the environmental as well as the social and economic dimensions of sustainable food. Eat food that is grown in harmony with the environment, generates small ecological footprints, provides producers with well-paid, meaningful, and dignified livelihoods, and helps create a more inclusive society where food is an individual and collective source of health and pleasure. Unfortunately, the information needed to identify such food is not readily available to most consumers, which is why knowledge and accountability are so important. However, there are a few basic rules that typically increase the likelihood that what you eat meets these ethical criteria:

- Eat locally and seasonally.
- Avoid food that is heavily processed and contains numerous additives.
- Use your own bags and containers to shop and stay away from food packaged or served in single-use plastics.
- Reduce your consumption of animal products, especially factory-raised beef, industrial dairy products, and endangered fish species.
- Support farmers who grow food agroecologically and adopt regenerative practices.
- Shop at locally owned businesses, including grocery stores and restaurants that pay living wages and provide economic opportunities to underrepresented groups.
- Buy directly from the producer, for example through Community Supported Agriculture (CSA), farmers markets, cooperative stores, and farm stands.
- Grow and cook your own food.
- Eliminate consumer waste by planning your meals and creatively repurposing leftovers.
- Compost food waste and recycle packaging.
- Broaden your tastes: try new food and recipes that support biodiversity.
- Eat slowly and intentionally.
- Enjoy meaningful food and, if you can, share both the food and the stories surrounding it with others.

- Get to know the people who grow, sell, and prepare your food and thank them for their work.

Of course, these suggestions for "making good choices" place the responsibility on you. As critical scholars have pointed out, this is one way that the negative externalities of the global/industrial/corporate/neoliberal food system are transferred onto consumers. While you may be able to make a difference "with your forks" and influence what is produced and how, these actions do little to take power away from large agrifood businesses who pass on additional costs to you and capitalize on your willingness to pay premiums for products labeled as fair, organic, local, or otherwise ethical, while workers and nature continue to be treated as expendable and exploitable. Some of us might find this route difficult and feel excluded from a movement that elides or erases the historical and contemporary contributions and struggles of people within the food system.

14.4.3. Volunteering, Advocating, and Organizing

To reconfigure power relations, collective action is needed. While much of the so-called food movement emphasizes individual consumption practices like the ones described above, it has also spurred a host of community-based initiatives that seek to reclaim control over local food systems by building tighter relationships between producers and consumers through urban agriculture, community farms, seed banks, cooperative retail, farmers markets, etcetera. Such projects typically aim at addressing ongoing social and environmental injustices. For example, many urban agriculture initiatives in the United States take place in areas struggling with environmental racism and food apartheid. Growing food together is a way to increase food security, foster self-reliance, create economic opportunities, green neighborhoods, strengthen communities, and encourage physical and social activities. To be successful, these projects require the establishment of new forms of decision making and democratic governance that encourage participation, value different forms of knowledge, and promote collective accountability and community ownership. You can play a role by helping create new organizations or supporting existing ones:

- Join a community garden or simply help on volunteer days.
- Grow food with friends, family, and/or neighbors: you can contribute by providing land, building raised beds, building soil, setting up irrigation, planting, weeding, harvesting, sharing produce, or even cooking.
- Volunteer with food assistance organizations and mutual aid groups.
- Intern with your local food policy council or other public agencies or nonprofits working to improve the food system.
- Collaborate with fellow students, staff, faculty, and administrators to bring about change on your campus, such as providing resources for students struggling with food insecurity, creating a food pantry, sourcing local, seasonal, and organic produce for the cafeteria, composting and recycling waste, banning single-use plastics, establishing a farmers market, and starting a community garden that could become a learning, food provisioning, and healing space.
- Help challenge dominant and harmful narratives through productive conversations: ask questions and speak up when myths and stereotypes are being spread, organize events on campus, advocate for a better food system, invite interesting and thought-provoking speakers, and make room at the table for different voices.
- Demand accountability from grocery stores, restaurants, and other food providers.

- Remember that meeting immediate needs is important and necessary but will not eliminate the causes of ongoing crises unless we address the systemic inequities that distort our relationships to food.
- Do not automatically seek a leadership role but observe and listen to find out where your labor, skills, and/or resources are most needed, especially when working with marginalized communities who often have firsthand experience with food injustices and a long (yet unrecognized) history of food activism.

14.4.4. Political Participation

Food injustices and insecurities are deeply rooted in political and economic systems that limit people's ability to control inputs and access nutritious, sustainable, and culturally meaningful food. Such vulnerability results from systemic inequalities, including the ideology, practices, and policies that result in different food experiences according to one's geographic location, social position, and race. Policies on issues such as labor, immigration, education, economic development, housing, food assistance, environmental health and safety, corporate mergers, and so on have a direct or indirect impact on people's right to food. As voters, residents, citizens, and taxpayers, we have a right to participate in the decision-making process that results in policies influencing subsidies to farmers, spending on school lunches, living wages, worker protections, food labels, regulations of dangerous chemicals, and other issues affecting our well-being and that of the planet and future generations.

- If you can vote, do not waste it: vote in federal, state, and local elections, including special ballot initiatives that relate to the food system.
- Seek information about candidates' positions on various food issues and share what you learn with others.
- Participate in organizing efforts to raise awareness about important issues: help striking workers, boycott negligent companies, join protests, share information. Remember that social pressure creates political will.
- Support democratic reforms that reduce the influence of agrifood corporations on elected officials via lobbying and campaign financing.
- Protect and enhance voting rights.
- Encourage voters in your family, among your friends, on campus, and in your community to cast their vote on important issues. You can do that informally or by joining organized "get-out-the-vote" campaigns.
- Remember that local politics matter too: learn about your municipal government and your state legislature.
- Know who your federal, state, and local government officials are. Write or call them about issues that you care about. Petitions and letters signed by multiple stakeholders are especially influential.
- Use social media purposefully to share information, raise awareness, and increase social pressure.

14.5. Hope

There are numerous ways to get involved in pushing back against the social inequality and environmental destructiveness of the current food system and contributing to the creation a better food future. Your interest in geographies of food brings me hope that such future is possible. As more people, like you, learn about the contemporary food system and question its ability to sustain us in the future, I am hopeful that they will join the growing number of food activists who are working to create new food geographies and transform the ways we grow, share, and enjoy food.

For example, in New York, the Seneca Nation established Gakwi:yo:h (i.e., Good Food) Farms in an effort to increase food security and sovereignty (Pietrorazio 2021). The farm and its new cannery are engaging community members in traditional agricultural practices, cultivating culturally meaningful crops like corn, potatoes, and sunflowers; tapping trees for maple syrup; raising buffalo; and processing their own food. In collaboration with agricultural and conservation agencies, the Nation trained new farmers and established a small herd of buffalo that was recently released onto the reservation to be hunted seasonally. In the United States and around the world, Indigenous people are taking back ancestral lands, reclaiming foodways, promoting sustainable practices, enhancing self-sufficiency, and rebuilding food sovereignty (Grey and Patel 2014).

In Detroit, Black women have been transforming vacant land to create a community-based food system that increases access to healthy food, provides safe and green space for the community, and nurtures activism (White 2018). Unlike social enterprises that bring outside investors to capitalize on cheap land to make profit, the Detroit Black Community Food Security Network shows that food justice is a matter of racial justice; it brings about meaningful social and political change to abandoned and polluted urban Black communities. Similar community-led urban agriculture projects are transforming urban neighborhoods in cities across the globe, engaging and empowering their poorest residents, including women, people of color, and immigrants (WinklerPrins 2017).

In Kitale, Kenya, hundreds of farmers are learning about agroecology and low-input farming. Through its training and certificate program, local nonprofit Manor House Agricultural Center is helping farmers to adopt techniques that improve the soil, reduce the need for scarce water and expensive chemical inputs, and increase the quantity, nutritional value, and diversity of crops (Oakland Institute 2015). With yields two to four times higher than in conventional agriculture, farmers can feed their families three healthy meals a day, sell their surplus at the market, and afford to send their children to school. Kenya's 2018–2021 national climate action plan emphasized sustainable agriculture (e.g., crop diversification, agroforestry, water harvesting, and soil management) as a path to ensuring the right to food, which is enshrined in the country's constitution. Over fifty grassroots organizations are currently working together to pressure the government and international organizations to take further action toward strengthening agroecology and supporting small farmers. Throughout the Global South, subsistence farmers are weaving productive networks enabling them to grow food in harmony with the environment and feed their communities with dignity (Wise 2019).

In Manila, the Philippines, hundreds of community pantries have flourished in the city's poorest neighborhoods as a response to the COVID-19 pandemic (Diaz 2021). Alongside grassroots organizations, nonprofits, and churches, groups of private citizens are filling up carts with canned goods, rice, fruits, and vegetables, making them available to anyone who might need them, with the principle of taking what you need and giving what you can. Local farmers and fishers also donated food. Although they have been criticized by the government as breeding "communist rebels," these community pantries have provided invaluable food aid at a time when the government was slow to act. Mutual aid and solidarity groups have emerged in many other places to make food available to those in need without relying on state assistance or bureaucratic food banks. In Paris, Berlin, and Madrid, community fridges open to anyone who wants to drop off or pick up free food have become a common sight. These grassroots initiatives reveal the power of spontaneous community organizing and solidarity economies, while challenging the shame and stigma associated with most food aid.

In Los Angeles, street vendors won an important victory in 2018. Years of organizing led to the city passing a new ordinance to streamline the permitting process and decriminalize street vending. Although it is still unclear how permits will be allocated, particularly under the restrictions imposed by the COVID-19 pandemic, they will presumably allow vendors to sell food without fear of police harassment, eviction, or confiscation. In a city where more than fifty thousand vendors operate without permit, this should help formalize the informal businesses that generate jobs and income, increase access to food, and enhance community life in neighborhoods where jobs are scarce and mainstream retailers are practically absent (Rosales 2020). Similar successes have been achieved recently in New York City and other places where the positive roles of street vendors are being recognized and efforts put in place to formalize and protect their livelihoods.

And in my city, growers, fishers, restaurant workers, consumers, policymakers, food activists, philanthropists, researchers, and others have come together under the leadership of the San Diego Food System Alliance (2021) to develop the San Diego 2030 Food Vision, laying out goals, objectives, and strategies for a more just, healthy, and sustainable regional food system. Input gathered through interviews, community circles, and broad outreach activities draw attention to the need to address equity issues and "cultivate justice" if we want to confront climate change and build resilience. In other words, creating a strong food system requires that we simultaneously fight against hunger, racism, poverty, environmental injustices, and other forms of oppression and discrimination in and out of the food system. Civil society organizations, food policy councils, and planning bodies elsewhere are engaged in similar conversations and creating new forms of governance to guide strategies for a better food system (Andrée et al. 2019).

These are just a few examples that illustrate how to enact more democratic and participatory forms of governance, fight stigma and shame, respect and share different forms of knowledge, do things in harmony with nature, redress past and current injustices, change narratives around food, strengthen food livelihoods, and promote self-determination—in other words, create new *food geographies*. Along with many other ongoing projects, they give me hope that together we can build a food system that will nourish us all.

References

Abebe, Tatek, and Sharon Bessell. 2011. Dominant Discourses, Debates and Silences on Child Labour in Africa and Asia. *Third World Quarterly* 32(4): 765–86.

Adger, W. Neil. 2006. Vulnerability. *Global Environmental Change* 16(3): 268–81.

Afelt, Aneta, Roger Frutos, and Christian Devaux. 2018. Bats, Coronaviruses, and Deforestation: Toward the Emergence of Novel Infectious Diseases? *Frontiers in Microbiology* 9: 702.

The Agrifood Atlas. 2017. *The Agrifood Atlas*. Berlin, Germany: Heinrich Böll Foundation and Rosa Luxemburg Foundation, and Brussels, Belgium: Friends of the Earth Europe. https:// cn.boell.org/sites/default/files/agrifoodatlas2017_facts-and-figures-about-the-corporations -that-control-what-we-eat.pdf.

Agyeman, Julian, Caitlin Matthews, and Hannah Sobel. 2017. *Food Trucks, Cultural Identity, and Social Justice: From Loncheras to Lobsta Love*. Cambridge, MA: MIT Press.

Agyeman, J., and J. McEntee. 2014. Moving the Field of Food Justice Forward through the Lens of Urban Political Ecology. *Geography Compass* 8(3): 211–20.

Albala, K. 2010. The Tomato Queen of San Joaquin. *Gastronomica* 10(2): 55–63.

Alkon, Alison Hope. 2012. *Black, White, and Green: Farmers Markets, Race, and the Green Economy*. Athens: University of Georgia Press.

Alkon, A., and J. Guthman. 2017. *The New Food Activism: Opposition, Cooperation, and Collective Action*. Oakland: University of California Press.

Allen, Patricia. 2010. Realizing Justice in Local Food Systems. *Cambridge Journal of Regions, Economy and Society* 3(2): 295–308.

Allison, Anne. 1991. Japanese Mothers and Obentōs: The Lunch-Box as Ideological State Apparatus. *Anthropological Quarterly* 64(4): 195–208.

Altieri, Miguel A. 2018. *Agroecology: The Science of Sustainable Agriculture*. Boca Raton, FL: CRC Press.

Anderson, Brett. 2020. The Thanksgiving Myth Gets a Deeper Look This Year. *New York Times*, November 17. https://www.nytimes.com/2020/11/17/dining/thanksgiving-native-ameri cans.html.

Andrée, Peter, Jill K. Clark, Charles Z. Levkoe, and Kristen Lowitt. 2019. *Civil Society and Social Movements in Food System Governance*. New York: Routledge.

Appadurai, Arjun. 1988. How to Make a National Cuisine: Cookbooks in Contemporary India. *Comparative Studies in Society and History* 30(1): 3–24.

———. 1986. *The Social Life of Things: Commodities in Cultural Perspective*. Cambridge: Cambridge University Press.

Appelbaum, Robert. 2011. *Dishing It Out: In Search of the Restaurant Experience*. London: Reaktion Books.

Arezki, Rabah, Klaus W. Deininger, and Harris Selod. 2012. The Global Land Rush: Foreign Investors are Buying up Farmland in Developing Countries. *Finance & Development* 49(1): 46–49.

Atkins, Peter, and Ian Bowler. 2001. *Food in Society: Economy, Culture, Geography*. London: Arnold.

Bales, Kevin. 2012. *Disposable People: New Slavery in the Global Economy, Updated with a New Preface*. Berkeley: University of California Press.

BananaLink. 2020a. Our Work in Guatemala. https://www.bananalink.org.uk/partners /guatemala/.

———. 2020b. Banana Value Chain. https://www.bananalink.org.uk/wp-content/up loads/2019/05/Banana-value-breakdown-between-main-supplying-countries-and-the-EU .png.

Banerjee, S. B. 2011. Voices of the Governed: Towards a Theory of the Translocal, *Organization* 18(3): 323–44.

Barndt, Deborah. 2007. *Tangled Routes: Women, Work, and Globalization on the Tomato Trail.* Lanham, MD: Rowman & Littlefield Publishers.

Bauer, Shane. 2019. *American Prison: A Reporter's Undercover Journey into the Business of Punishment.* New York: Penguin Books.

Beaumont, Nicola J., Margrethe Aanesen, Melanie C. Austen, Tobias Börger, James R. Clark, Matthew Cole, Tara Hooper, Penelope K. Lindeque, Christine Pascoe, and Kayleigh J. Wyles. 2019. Global Ecological, Social and Economic Impacts of Marine Plastic. *Marine Pollution Bulletin* 142: 189–95.

Belasco, Warren. 2002. Food Matters: Perspectives on an Emerging Field. In *Food Nations: Selling Taste in Consumer Societies*, edited by Warren Belasco and Philip Scranton, 2–23. New York: Routledge.

Belhabib, Dyhia, and Philippe Le Billon. 2020. Illegal Fishing as a Trans-National Crime. *Frontiers in Marine Science* 7: 162.

Bell, David, and Gill Valentine. 1997. *Consuming Geographies: We Are Where We Eat.* London: Routledge.

Benns, Whitney. 2015. American Slavery, Reinvented: The Thirteenth Amendment Forbade Slavery and Involuntary Servitude, "except as punishment for crime whereof the party shall have been duly convicted." *The Atlantic.* September 21. https://www.theatlantic.com/business/archive/2015/09/prison-labor-in-america/406177/

Berlant, Lauren (2006). Cruel Optimism. *Differences: A Journal of Feminist Cultural Studies* 17(3): 20–36.

Berry, Wendell. 2009. *Bringing It to the Table: On Farming and Food.* Berkeley, CA: Counterpoint.

Blakemore, Erin. 2018. How the Black Panthers' Breakfast Program Both Inspired and Threatened the Government. *History.com.* https://www.history.com/news/free-school-breakfast-black-panther-party.

Bordo, Susan. 1993. *Unbearable Weight: Feminism, Western Culture and the Body.* Berkeley, CA: University of California Press.

Borunda, Alejandra. 2019. See How Much of the Amazon Is Burning, How It Compares to Other Years. *National Geographic.* August 29. https://www.nationalgeographic.com/environment/2019/08/amazon-fires-cause-deforestation-graphic-map/

Bosco, Fernando J. 2006. Actor-Network Theory, Networks, and Relational Geographies. In *Approaches to Human Geography*, edited by Stuart Aitken and Gill Valentine, 136–46. Thousand Oak, CA: Sage.

Bosco, Fernando J., and Pascale Joassart-Marcelli. 2018. Spaces of Alternative Food: Urban Agriculture, Community Gardens and Farmers Markets. In *Food and Place: A Critical Exploration*, edited by Pascale Joassart-Marcelli and Fernando Bosco, 187–207. Lanham, MD: Rowman & Littlefield.

Boserup, E., 2017. *The Conditions of Agricultural Growth: The Economics of Agrarian Change under Population Pressure.* New York: Routledge.

Boucher, Doug, Pipa Elias, Katherine Lininger, Calen May-Tobin, Sarah Roquemore, and Earl Saxon. 2011. *The Root of the Problem: What's Driving Tropical Deforestation Today?* Cambridge, MA: Union of Concerned Scientists.

Bourdieu, Pierre. 1984. *Distinction: A Social Critique of Taste.* Translated by R. Nice. Cambridge, MA: Harvard University Press.

Braxton Little, J. 2018. Fire and Agroforestry Are Reviving Traditional Native Foods and Communities. *Civil Eats.* October 11. https://civileats.com/2018/10/11/fire-and-agroforestry-are-reviving-traditional-native-foods-and-communities/.

Brillat-Savarin, Anthelme. 1826. *Physiologie du Gout, ou Meditations de Gastronomie Transcendante.* Paris: Hachette.

Brown, Robert. 1892. *The Story of Africa and Its Explorers.* London: Cassell & Co.

Bryant, R. L., and M. K. Goodman. 2004. Consuming Narratives: The Political Ecology of "Alternative" Consumption. *Transactions of the Institute of British Geographers* 29(3): 344–66.

Bureau of Labor Statistics. 2020. *American Time Use Survey.* https://www.bls.gov/tus/.

Busch, Lawrence, and Arunas Juska. 1997. Beyond Political Economy: Actor Networks and the Globalization of Agriculture. *Review of International Political Economy* 4(4): 688–708.

Butler, Judith. 1997. Performative Acts and Gender Constitution: An Essay in Phenomenology and Feminist Theory, in *Writing on the Body: Female Embodiment and Feminist Theory*, edited by K. Conboy, N. Medina, and S. Stanbury, 401–17. New York: Columbia University Press.

Butler, Rhett A. 2020. *Calculating Deforestation Figures for the Amazon.* https://rainforests.mongabay.com/amazon/deforestation_calculations.html.

Cadieux, K. Valentine, and Rachel Slocum. 2015. What Does It Mean to Do Food Justice? *Journal of Political Ecology* 22: 1.

California Rural Legal Assistance, Esperanza: The Immigrant Women's Legal Initiative of the Southern Poverty Law Center, Lideres Campesinas, and Victim Rights Law Center. 2010. *Sexual Violence against Farmworkers: A Guidebook for Legal Providers.* https://www.victimrights.org/sites/default/files/Farmworkers%20Legal%20Providers_0.pdf.

Calvo, Luz, and Catriona Rueda Esquibel. 2016. *Decolonize Your Diet: Plant-Based Mexican-American Recipes for Health and Healing.* Vancouver, BC: Arsenal Pulp Press.

Carr, E. R. 2008. Men's Crops and Women's Crops: The Importance of Gender to the Understanding of Agricultural and Development Outcomes in Ghana's Central Region. *World Development* 36(5): 900–915.

Carrington, Christopher. 2013. Feeding Lesbigay Families. In *Food and Culture: A Reader* (third edition), edited by Carole Counihan and Penny Van Esterik, 187–210. New York: Routledge.

Carus, Michael, and Lara Dammer. 2013. Food or Non-Food: Which Agricultural Feedstocks Are Best for Industrial Uses? *Industrial Biotechnology* 9(4): 171–76.

Castree, Noel. 2008. Neoliberalising Nature: Processes, Effects, and Evaluations. *Environment and Planning A* 40(1): 153–73.

Castree, Noel, Neil Coe, Kevin Ward, and Michael Samers. 2004. *Spaces of Work: Global Capitalism and Geographies of Labor.* London: Sage.

CDC. 2020. Behavior, Environment, and Genetic Factors All Have a Role in Causing People to Be Overweight and Obese. Centers for Disease Control. Genomics and Precision Health. https://www.cdc.gov/genomics/resources/diseases/obesity/index.htm.

Chapman, Peter. 2014. *Bananas: How the United Fruit Company Shaped the World.* New York: Canongate.

Chappell, Jahi. 2018. *Beginning to End Hunger: Food and the Environment in Belo Horizonte, Brazil, and Beyond.* Berkeley: University of California Press.

Chisholm, G. G. 1889. *Handbook on Commercial Geography.* London: Longman, Green, and Co.

Cisneros-Montemayor, Andrés M., Daniel Pauly, Lauren V. Weatherdon, and Yoshitaka Ota. 2016. A Global Estimate of Seafood Consumption by Coastal Indigenous Peoples. *PLoS ONE* 11(12): 1–16.

CIW. 2020. *Organization Website.* Immokalee, FL: Coalition of Immokalee Workers. https://ciw-online.org.

Clapp, Jennifer. 2012. *Hunger in The Balance: The New Politics of International Food Aid.* Ithaca, NY: Cornell University Press, 2015.

Clapp, Jennifer. 2016. *Food.* 2nd edition. Cambridge: Polity Press.

Cloke, Paul, Philip Crang, and Mark Goodwin. 2005. *Introducing Human Geographies.* New York: Routledge.

Colls R., and B. Evans. 2009. Questioning Obesity Politics: Introduction to a Special Issue on Critical Geographies of Fat/Bigness/Corpulence. *Antipode* 41(5):1011–20.

Connor, M., and C. J. Armitage. 2002. The Social Psychology of Food. Maidenhead, UK: Open University Press.

Conrow, Joan. 2018. Developing Nations Lead Growth of GMO Crops. Cornell Alliance for Science. https://allianceforscience.cornell.edu/blog/2018/06/developing-nations-lead-growth-gmo-crops/.

Cook, Ian. 2004. Follow the Thing: Papaya. *Antipode* 36(4): 642–64.

———. 2006. Geographies of Food: Following. *Progress in Human Geography* 30(5): 655–66.

Cook, Ian, and Peter Crang. 1996. The World on a Plate: Culinary Culture, Displacement and Geographical Knowledges. *Journal of Material Culture* 1(2): 131–53.

Cook, Ian, and Michelle Harrison. 2007. Follow the Thing: "West Indian Hot Pepper Sauce." *Space and Culture* 10(1): 40–63.

CPS. 2020. Industries at a Glance: Food Services and Drinking Places. U.S. Bureau of Labor Statistics. Current Population Survey. https://www.bls.gov/iag/tgs/iag722.htm.

Crang, Philip. 1994. It's Showtime: On the Workplace Geographies of Display in a Restaurant in Southeast England. *Environment and Planning D: Society and Space* 12(6): 675–704.

Crippa, M., E. Solazzo, D. Guizzardi, F. Monforti-Ferrario, F. N. Tubiello, and A. Leip. 2021 Food Systems Are Responsible for a Third of Global Anthropogenic GHG Emissions. *Nature Food* 2: 198–209.

Cruz, María Caridad, and Roberto Sánchez Medina. 2003. *Agriculture in the City: A Key to Sustainability in Havana, Cuba.* Kingston, Jamaica: Ian Randle Publishers.

Curtin, D. W., and L. M. Heldke, L. M. (Eds.). 1992. *Cooking, Eating, Thinking: Transformative Philosophies of Food.* Bloomington: Indiana University Press.

Dahl-Bredine, Phil, Jesús León Santos, Judith Cooper Haden, Susana Trilling, and Miguel A. Altieri. 2015. *Milpa: From Seed to Salsa: Ancient Ingredients for a Sustainable Future.* Judith Haden Photography.

Dance, Amber. 2019. These Corals Could Survive Climate Change—and Help Save the World's Reefs. *Nature*, November 27.

Daneshkhu, Scheherazade, Lindsay Whipp, and James Fontanella-Khan. 2017. The Lean and Mean Approach of 3G Capital. *Financial Times*, May 7. https://www.ft.com/content/268f73e6-31a3-11e7-9555-23ef563ecf9a.

Davis, Mike. 2002. *Late Victorian Holocausts: El Niño Famines and the Making of the Third World.* New York: Verso Books.

de Blij, Harm. 2008. *The Power of Place: Geography, Destiny, and Globalization's Rough Landscape.* Oxford: Oxford University Press.

Dempsey, Sarah E., and Kristina E. Gibson. 2018. Food, Biopower, and the Child's Body as a Scale of Intervention." In *Food & Place: A Critical Exploration*, edited by Pascale Joassart-Marcelli and Fernando J. Bosco, 253–67. Lanham, MD: Rowman & Littlefield.

Denton, Ginger, and Jonathan Harris. 2019. The Impact of Illegal Fishing on Maritime Piracy: Evidence from West Africa. Studies in Conflict and Terrorism, online: 1–20. https://doi.org/10.1080/1057610X.2019.1594660.

De Schutter. 2010. *Report on Agroecology and the Right to Food. Submitted to the United Nations' General Assembly Human Rights Council.* Special Rapporteur on Right to Food. http://www.srfood.org/images/stories/pdf/officialreports/20110308_a-hrc-16-49_agroecology_en.pdf.

de Souza, Rebecca T. 2019. *Feeding the Other: Whiteness, Privilege, and Neoliberal Stigma in Food Pantries.* Cambridge, MA: MIT Press.

DeVault, Marjorie L. 1994. *Feeding the Family: The Social Organization of Caring as Gendered Work.* Chicago: University of Chicago Press.

de Waal, Alex. 2018. The End of Famine? Prospects for the Elimination of Mass Starvation by Political Action. *Political Geography* 62: 184–95.

———. 1997. *Famine Crimes: Politics & the Disaster Relief Industry in Africa.* Bloomington, IN: Indiana University Press.

Diamond, Jared. 1997. *Guns, Germs, and Steel: The Fates of Human Societies.* New York: W.W. Norton.

Diaz, Glenn. 2021. Pandemic Pantries in the Street? You Communist! *New York Times*, May 7. https://www.nytimes.com/2021/05/07/opinion/philippines-covid-pantries.html.

Díaz, Steffan Igor Ayora. 2012. *Foodscapes, Foodfields, and Identities in Yucatan.* New York: Berghahn Books.

Diener, Alexander C., and Joshua Hagen. 2009. Theorizing Borders in a "Borderless World": Globalization, Territory and Identity. *Geography Compass* 3(3): 1196–216.

Diner, Hasia R. 2009. *Hungering for America.* Cambridge, MA: Harvard University Press.

d'Odorico, Paolo, Joel Carr, Carole Dalin, Jampel Dell'Angelo, Megan Konar, Francesco Laio, Luca Ridolfi, et al. 2019. Global Virtual Water Trade and the Hydrological Cycle: Patterns, Drivers, and Socio-Environmental Impacts. *Environmental Research Letters* 14(5): 053001.

Dolan, Catherine. 2001. The "Good Wife": Struggles over Resources in the Kenyan Horticultural Sector. *Journal of Development Studies* 37(3): 39–70.

Domoff, Sarah E., Nova G. Hinman, Afton M. Koball, Amy Storfer-Isser, Victoria L. Carhart, Kyoung D. Baik, and Robert A. Carels. 2012. The Effects of Reality Television on Weight Bias: An Examination of *The Biggest Loser*. *Obesity* 20(5): 993–98.

Douglas, M. 1984. *Food in the Social Order: Studies of Food and Festivities in Three American Communities*. New York: Russell Sage Foundation.

Doumbouya, Alkaly, Ousmane T. Camara, Josephus Mamie, Jeremias F. Intchama, Abdoulie Jarra, Salifu Ceesay, Assane Guèye, Diène Ndiaye, Ely Beibou, Alan Padilla, and Dyhia Belhabib. 2017. Assessing the Effectiveness of Monitoring Control and Surveillance of Illegal Fishing: The Case of West Africa. *Frontiers in Marine Science* 4(50): 106–15.

Druckman, Charlotte. 2010. Why Are There No Great Women Chefs? *Gastronomica* 10(1).

Dumond, D. 1987. *The Eskimos and Aleuts*. Second edition. London: Thames and Hudson.

Dunn, Elizabeth. 2011. The Food of Sorrow: Humanitarian Aid to Displaced People. In *Food: Ethnographic Encounters*, edited by Leo Coleman, 139–49. New York: Bloomsbury.

DuPuis, E. Melanie. 2002. *Nature's Perfect Food: How Milk Became America's Drink*. New York: New York University Press.

Earle, Rebecca. 2012. The Columbian Exchange. In *The Oxford Handbook of Food History*, edited by Jeffrey Pilcher, 341–57. Oxford: Oxford University Press.

ECA. 2019. Biodiversity in Farming: Audit Preview. European Court of Auditors. https://www.eca.europa.eu/en/Pages/DocItem.aspx?did=50151.

Edwards, Chris. 2018. Agricultural Subsidies. *Downsizing the Federal Government*. Washington, DC: Cato Institute. https://www.downsizinggovernment.org/agriculture/subsidies#_edn28.

Egger, Garry, and Boyd Swinburn. 1997. An "Ecological" Approach to the Obesity Pandemic. *BMJ* 315(7106): 477–80.

Eisenhauer, Elizabeth. 2001. In Poor Health: Supermarket Redlining and Urban Nutrition. *GeoJournal* 53(2): 125–33.

EJF. 2019. *Blood and Water: Human Rights Abuse in the Global Seafood Industry*. London: Environmental Justice Foundation.

Essex, Jamey. 2012. Idle Hands Are the Devil's Tools: The Geopolitics and Geoeconomics of Hunger. *Annals of the Association of American Geographers* 102(1): 191–207.

Estabrook, B., 2012. *Tomatoland: How Modern Industrial Agriculture Destroyed Our Most Alluring Fruit*. Kansas City, MO: Andrews McMeel Publishing.

Euromonitor International. 2012. Home Cooking and Eating Habits: Global Survey Strategic Analysis. *Market Research Blog*. https://blog.euromonitor.com/home-cooking-and-eating-habits-global-survey-strategic-analysis/.

Eurostat. 2020. Final Consumption Expenditure of Households by Consumption Purpose. (COICOP 3 digit). https://ec.europa.eu/eurostat/databrowser/view/NAMA_10_CO3_P3__custom_431578/default/table?lang=en.

Evans, Hannah, and Pascale Joassart-Marcelli. 2018. Ethical Food and Global Commodity Chains. In *Food and Place: A Critical Exploration*, edited by Pascale Joassart-Marcelli and Fernando Bosco, 87–105. Lanham, MD: Rowman & Littlefield.

Fair Trade USA. 2019. *1998–2018: Celebrating 20 Years*. Annual Report. https://www.fairtradecertified.org/sites/default/files/filemanager/documents/Annual%20Reports/DEV_AnnualReport18_200108.pdf.

Fairtrade International. 2011. *Monitoring the Scope and Benefits of Fairtrade*. Third edition. Fairtrade Labelling Organization International. Bonn: Germany. https://files.fairtrade.net/publications/2011_MonitoringReport_3rdEd.pdf.

———. 2013. *Monitoring the Scope and Benefits of Fairtrade*. Fifth edition. Fairtrade International. Bonn: Germany. https://files.fairtrade.net/publications/2013_MonitoringReport_5thEd.pdf.

———. 2014. *Monitoring the Scope and Benefits of Fairtrade: Overall*. Sixth edition. Fairtrade International. Bonn: Germany. https://files.fairtrade.net/publications/2014_MonitoringReport_6thEd.pdf

———. 2020. *Monitoring the Scope and Benefits of Fairtrade: Overall*. 10th edition. Fairtrade International. Bonn: Germany. https://files.fairtrade.net/publications/2019_Monitoring_summary_10thEd.pdf.

FAO. 2020a. Food and Agriculture Data: Trade of Crops and Livestock products. FAOSTAT. http://www.fao.org/faostat/en/#home.

———. 2020b. Suite of Food Security Indicators. FAOSTAT. http://www.fao.org/faostat/en/?#data/FS.

———. 2020c. Land Use. FAOSTAT. http://www.fao.org/faostat/en/#data/RL.

———. 2020d. Crops. FAOSTAT. http://www.fao.org/faostat/en/#data/QC.

———. 2020e. Fertilizer Archive, FAOSTAT. http://www.fao.org/faostat/en/#data/RA.

———. 2020f. *The State of World Fisheries and Aquaculture 2020: Sustainability in Action.* Rome: United Nations' Food and Agriculture Organization. https://doi.org/10.4060/ca9229en.

———. 2020g. Food Balance Sheet. United Nations' Food and Agriculture Organization. FAOSTAT. http://www.fao.org/faostat/en/#data/FBS.

———. 2019a. Rankings: Countries by Commodity. FAOSTAT. Food and Agriculture Organization of the United Nations. Food and Agriculture Data. http://www.fao.org/faostat/en/#rankings/countries_by_commodity.

———. 2019b. *The State of Food Security and Nutrition in the World: Safeguarding against Economic Slowdowns and Downturns.* Rome: Food and Agriculture Organization of the United Nations. http://www.fao.org/3/ca5162en/ca5162en.pdf.

———. 2019c. *The State of the World's Biodiversity for Food and Agriculture.* Rome: Food and Agriculture Organization of the United Nations. http://www.fao.org/3/CA3129EN/CA3129EN.pdf.

———. 2019d. *Fisheries and Aquaculture.* Rome: Food and Agriculture Organization of the United Nations. http://www.fao.org/rural-employment/agricultural-sub-sectors/fisheries-and-aquaculture/en/.

———. 2018. *TAPE: Tool for Agroecology Performance Evaluation. Process for Development and Guidelines for Application.* Rome: Food and Agriculture Organization of the United Nations. http://www.fao.org/3/ca7407en/ca7407en.pdf.

———. 2017a. *Global Soil Partnership Endorses Guidelines on Sustainable Soil Management.* Food and Agriculture Organization of the United Nations. http://www.fao.org/global-soil-partnership/resources/highlights/detail/en/c/416516/.

———. 2017b. *Migration, Agriculture, and Climate Change: Reducing Vulnerabilities and Enhancing Resilience.* Food and Agriculture Organization of the United Nations. http://www.fao.org/3/I8297EN/i8297en.pdf.

———. 2015. *National Investment Profile: Malawi. Water for Agriculture and Energy.* Food and Agriculture Organization United Nations and AGWA Partnership for Agricultural water for Africa. http://www.fao.org/fileadmin/user_upload/agwa/docs/NIP_Malawi_Final.pdf.

———. 2014. *Final Report for the International Symposium on Agroecology for Food Security and Nutrition.* Rome: The Food and Agriculture Organization of the United Nations. http://www.fao.org/3/a-i4327e.pdf.

———. 2012. Percentage of Area Equipped for Irrigation. FAO GeoNetwork. http://www.fao.org/geonetwork/srv/en/metadata.show?id=38006&currTab=simple

———. 2011a. *Women in Agriculture: Closing the Gender Gap for Development.* The State of Food and Agriculture 2010–11. Rome: Food and Agriculture Organization of the United Nations. http://www.fao.org/3/i2050e/i2050e.pdf.

———. 2011b. Gender and Land Rights Database. Rome: Food and Agricultural Organization of the United Nations. https://ourworldindata.org/employment-in-agriculture.

———. 2011c. *The State of the World's Land and Water Resources for Food and Agriculture: Managing Systems at Risk.* Earthscan and the Food and Agriculture Organization of the United Nations. http://www.fao.org/3/i1688e/i1688e.pdf.

———. 2010. *Second Report on the State of the World's Genetic Resources.* Rome: Food and Agriculture Organization of the United Nations. http://www.fao.org/agriculture/crops/core-themes/theme/seeds-pgr/sow/sow2/en/.

Farmer, Paul. 2011. *Haiti after the Earthquake.* New York: Public Affairs.

Farmworker Justice. 2011. *No Way to Treat a Guest.* https://www.farmworkerjustice.org/sites/default/files/documents/7.2.a.6%20fwj.pdf.

Fine, B., 1994. Towards a Political Economy of Food. *Review of International Political Economy* 1: 519–45.

Fine, B., and E. Leopold. 1993. *The World of Consumption*. London: Routledge.

Florida, Richard L. 2002. *The Rise of the Creative Class: And How It's Transforming Work, Leisure, Community and Everyday Life*. New York: Basic Books.

Food and Land Use Coalition. 2019. *Growing Better: Ten Critical Transitions to Transform Food and Land Use*. The Global Consultation Report of the Food and Land Use Coalition. https://www.foodandlandusecoalition.org/wp-content/uploads/2019/09/FOLU-GrowingBetter-GlobalReport.pdf.

Food and Water Watch. 2015. *Factory Farm Nation*. Washington, DC: Food and Water Watch. https://www.factoryfarmmap.org/wp-content/uploads/2015/05/FoodandWaterWatchFactoryFarmFinalReportNationMay2015.pdf.

Forbes. 2018. Global 2000: The World's Largest Public Companies 2018. https://www.forbes.com/sites/kristinstoller/2018/06/06/the-worlds-largest-public-companies-2018/?sh=24f262b4769f.

Foster, John Bellamy. 1999. Marx's Theory of Metabolic Rift: Classical Foundations for Environmental Sociology. *American Journal of Sociology* 105(2): 366–405.

Foucault, Michel. 1978. *An Introduction*. Vol. 1 of *The History of Sexuality*. Translated by Robert Hurley. New York: Vintage.

Francis, C., G. Lieblein, S. Gliessman, T. A. Breland, N. Creamer, R. Harwood, L. Salomonsson, J. Helenius, D. Rickerl, R. Salvador, M. Wiedenhoeft, S. Simmons, P. Allen, M. Altieri, C. Flora, and R. Poincelot. 2003. Agroecology: The Ecology of Food Systems. *Journal of Sustainable Agriculture* 22(3): 99–118.

Friedberg, Susanne. 2004. *French Beans and Food Scares*. New York: Oxford University Press.

———. 2003. Cleaning up Down South: Supermarkets, Ethical Trade and African Horticulture. *Social and Cultural Geography* 4(1): 27–43.

Friedlander, Alan M., Yimnang Golbuu, Enric Ballesteros, Jennifer E. Caselle, Marine Gouezo, Dawnette Olsudong, and Enric Sala. 2017. Size, Age, and Habitat Determine Effectiveness of Palau's Marine Protected Areas. *PLoS One* 12(3): e0174787.

Friedman, Thomas L. 2000. *The Lexus and the Olive Tree: Understanding Globalization*. New York: Farrar, Straus and Giroux.

Friedmann, Harriet. 1987. International Regimes of Food and Agriculture since 1870. *Peasants and Peasant Societies* 2: 247–58.

Fry, Jillian P, Nicholas A. Mailloux, David C. Love, Michael C. Milli, and Ling Cao. 2018. Feed Conversion Efficiency in Aquaculture: Do We Measure It Correctly? *Environmental Research Letters* 13: 024017.

Gabaccia, Donna R. 2000. *We Are What We Eat: Ethnic Food and the Making of Americans*. Cambridge, MA: Harvard University Press.

Gálvez, Alyshia. 2018. *Eating NAFTA: Trade, Food Policies, and the Destruction of Mexico*. Berkeley: University of California Press.

GAO. 2018. *Working Children: Federal Injury Data and Compliance Strategies Could Be Strengthened. A Report to Congressional Requesters*. Washington, DC: United States Government Accountability Office. https://www.gao.gov/assets/700/695209.pdf.

Garnett, Tara. 2011. Where Are the Best Opportunities for Reducing Greenhouse Gas Emissions in the Food System (Including the Food Chain)? *Food Policy* 36: S23–S32.

Garth, Hannah. 2020. Blackness and "Justice" in the Los Angeles Food Justice Movement. In *Black Food Matters: Racial Justice in the Wake of Food Justice*, edited by Hannah Garth and Ashanté Reese, 107–30. Minneapolis: University of Minnesota Press.

Gassert, F., M. Landis, M. Luck, P. Reig, and T. Shiao. 2014. Aqueduct Global Maps 2.1. Washington, DC: World Resources Institute. https://datasets.wri.org/dataset/aqueduct-global-maps-21-data.

Gereffi, Gary, and Miguel Korzeniewicz (Eds.). 1994. *Commodity Chains and Global Capitalism*. Westport, CT: Praeger.

Gibbs, H. K., and J. Megham Salmon. 2015. Mapping the World's Degraded Lands. *Applied Geography* 57: 12–21.

Gilmore, Ruth. 2007. *Golden Gulag: Prisons, Surplus, Crisis, and Opposition in Globalizing California*. Berkeley: University of California Press.

Goodman, D., and M. Watts (Eds.). 1997. *Globalising Food: Agrarian Questions and Global Restructuring.* Routledge, London

Goss, Jon. 2004. Geography of Consumption I. *Progress in Human Geography* 28(3): 369–80.

Gottlieb, Robert, and Anupama Joshi. 2013. *Food Justice.* Cambridge, MA: MIT Press.

Gough, Brendan. 2007. "Real Men Don't Diet": An Analysis of Contemporary Newspaper Representations of Men, Food and Health. *Social Science & Medicine* 64(2): 326–37.

GRAIN. 2018. Daewoo's Overseas Agribusiness Expansion. Against the Grain, May 2018. https://grain.org/en/article/5951-daewoo-s-overseas-agribusiness-expansion.

Greaves, Russell D., and Karen L. Kramer. 2014. Hunter–Gatherer Use of Wild Plants and Domesticates: Archaeological Implications for Mixed Economies before Agricultural Intensification. *Journal of Archaeological Science* 41: 263–71.

Greenberg, Miriam. 2013. What on Earth Is Sustainable? Toward Critical Sustainability Studies. *Boom: A Journal of California* 3(4): 54–66.

Greenpeace. 2017. Hope in West Africa Ship Tour. Summary of Findings. https://www.greenpeace.org/static/planet4-africa-stateless/2018/10/7cf01664-7cf01664-hopeinwestafricashiptour_finalbriefing.pdf

Greshko, Michael. 2018. Window to Save World's Coral Reefs Closing Rapidly. *National Geographic*, January 4.

Grey, Sam, and Raj Patel. 2015. Food Sovereignty as Decolonization: Some Contributions from Indigenous Movements to Food System and Development Politics. *Agriculture and Human Values* 32(3): 431–44.

Gros, Jean-Germain. 2010. Indigestible Recipe: Rice, Chicken Wings, and International Financial Institutions: or Hunger Politics in Haiti. *Journal of Black Studies* 40(5): 974–86.

Gruby, Rebecca L., and Xavier Basurto. 2014. Multi-level Governance for Large Marine Commons: Politics and Polycentricity in Palau's Protected Area Network. *Environmental Science & Policy* 36: 48–60.

Guptill, Amy E., Denise A. Copelton, and Betsy Lucal. 2013. *Food and Society: Principles and Paradoxes.* Malden, MA: Polity Press.

Guthman, Julie. 2014a. Introducing Critical Nutrition: A Special Issue on Dietary Advice and Its Discontents. *Gastronomica: The Journal of Food and Culture* 14(3): 1–4.

———. 2014b. *Agrarian Dreams.* Berkeley: University of California Press.

———. 2012. Opening up the Black Box of the Body in Geographical Obesity Research: Toward a Critical Political Ecology of Fat. *Annals of the Association of American Geographers* 102(5):951–57.

———. 2011. *Weighing In: Obesity, Food Justice and the Limits of Capitalism.* Berkeley, CA: University of California Press.

———. 2008. "If They Only Knew": Color Blindness and Universalism in California Alternative Food Institutions." *The Professional Geographer* 60(3): 387–97.

Halse, C. 2009. Bio-Citizenship. Virtue and Discourses and the Birth of the Bio-Citizen. In *Biopolitics and the "Obesity Epidemic:" Governing Bodies*, edited by J. Wright and V. Harwood, 45–59. New York: Routledge.

Haraway, Donna. 1988. Situated Knowledges: The Science Question in Feminism and the Privilege of Partial Perspective. *Feminist Studies* 14(3): 575–99.

Hardin, Garret. 1968. The Tragedy of the Commons. *Journal of Natural Resources Policy Research* 162(13): 3.

Harper, A. Breeze. 2010. *Sistah Vegan: Black Female Vegans Speak on Food, Identity, Health, and Society.* New York: Lantern Books.

Harris, Francesca, Cami Moss, Edward J. M. Joy, Ruth Quinn, Pauline F. D. Scheelbeek, Alan D. Dangour, and Rosemary Green. 2020. The Water Footprint of Diets: A Global Systematic Review and Meta-analysis. *Advances in Nutrition* 11(2): 375–86.

Hartwick, Elaine R. 2000. Towards a Geographical Politics of Consumption. *Environment and Planning A* 32(7): 1177–92.

Harvey, David. 1990. Between Space and Time: Reflections on the Geographical Imagination. *Annals of the Association of American Geographers* 80(3): 418-434.

Hassan, Hawa, and Julia Turshen. 2020. *In Bibi's Kitchen: The Recipes and Stories of Grandmothers from Eight African Countries That Touch the Indian Ocean.* New York: Ten Speed Press.

Details about our refund and Exchange policies are on the bookstore's website. We look forward to serving you again soon.

SHOP ONLINE with your bookstore website. We are headquarters for textbooks, college gear, school supplies and gifts. Be sure to sign up for emails to find out about special offers.

Rental Agreement

-This rental agreement is a contract between you and Barnes and Noble College Booksellers, LLC (BNC) and applies to your rental of textbooks and/or course related materials from us. This agreement sets forth your rights and obligations and should be read carefully. Please also visit our online rental service on www.yourschoolyourbookstore.com (the Website).contacting the bookstore.

-When accessing the Services on the website, by clicking "I Agree" or "I Accept" you agree to the terms and conditions of this Agreement, our privacy policy, and our terms of use, and any other documents incorporated into the website from which you accessed the Service. You agree that this Agreement is legally binding between you and the Company.

-We may modify this agreement from time to time as posted on the Website. It is important that you review the Website regularly to ensure you are aware of any changes.

-For questions regarding this Agreement please contact your campus bookstore.

Terms And Conditions

• You must be 18 years of age or older.

• All information provided by you in connection with this agreement must be accurate and complete.

• You must have a valid personal credit card on file with us at all times.

• Rented materials remain the full property of BNC. Your acceptance of rented materials and paying rental fees entitles you to use the property of

• BNC for a limited amount of time. At the end of the rental period, this agreement will terminate and you will lose all rights to the rented materials.

• Rented materials can be purchased during the first two weeks of classes only.

• Rented materials must be returned to the bookstore from which they were originally rented by the rental return date designated by us at the time of rental in salable condition. Salable condition will be determined by us in our sole discretion, but generally means book spine intact, no excessive damage or water damage to cover or contents, all original pages intact, all original components present, and no excessive highlighting, writing or other markings. Normal use highlighting and writing is permitted.

• You are responsible for loss or theft of all rented materials. Rented materials not returned by the rental return date or returned on or before that date not in salable condition will be subject to non-return fees equal to 75% of the new book price (at the time of rental) plus a 7.5% processing fee. Nonreturn fees will be automatically charged to the credit card on file for this agreement. In the event that the credit card on file for this agreement is no longer valid or if the purchasing limit on such credit card has been exceeded, we will contact you for, and you agree to promptly pay in full, the nonreturn fees.

• If you have not returned the rented materials by the rental return date and we are unable to charge your credit/debit card, it is your responsibility to pay the non-return fees immediately. You will be notified via email if your credit/debit card was declined and have 15 days to pay the fees before your account and any information, including personally identifiable information, you have provided to us is turned over to a third party collection agency ('agency').

• You agree that BNC and any agency it hires to collect non-return fees may contact you via e-mail and you confirm that you are the only person who opens e-mail at the address you have provided or that if anyone else opens e-mail at the address you have provided, you waive any claims of a violation of your privacy or of potential third party disclosure if persons other than you view your e-mail. Additionally, you agree that BNC and any agency it hires may contact you via U.S. Mail, telephone or cellular telephone should such contact information be provided by you or obtained as provided below regarding your failure to return rented materials or pay applicable non-return fees.

• You authorize BNC to share details of your rental transactions with the College, University, or School at which you are enrolled and you acknowledge that your College, University, or School may provide BNC with contact information, including but not limited to, your e-mail address, student address, home address, home telephone number and cellular telephone number, which may be different than the information you provided to us, for the purpose of contacting you regarding failure to return Rented Materials or open non-return fees.

• Returns shipped UPS, USPS or other carriers must be postmarked and shipped on or before the rental return date.

• Standard tax rates apply and vary by state.

CUSTOMER COPY

Shipment #

108722743

Order #

1031939508

Florida International University

Shipment No:	108722743
Box Number:	20020-1-S23-PCK
Order Number:	1031939508
Placed On:	20-Jan-2023
Invoice Date:	03-Feb-2023

BILL TO

Emma MacDonald

1014 Garden Circle

Winter Garden, FL

34787, US

Email:emacd008@fiu.edu

Phone: (407) 867-0793

SHIP TO

Pickup Name: Emma MacDonald

Shipping Method : Store Pickup

Pickup Location: 11000 SW 8th Street

Florida International University (785)

Phone: (305) 348-2691

SHIP FROM

Florida International University

11000 SW 8th Street

B&N BKSTR #785

33174, US, Miami, FL

Email:SM785@bncollege.com

Phone: (305) 348-2691

Website: https://fiu.bncollege.com

ORDER DETAILS

Item	Description	Price	QTY	Discount	Total
785_800007705_new	FOOD GEOGRAPHIES: SOCIAL, POLITICAL, AN - RENTAL_NEW Non-Return Fee: $3.70 , Replacement Fee: $36.75 , Due Date:Apr-29-2023	$43.35	1	$0.00	$43.35

SubTotal	$43.35
Discount	$0.00
Shipping Discount	$0.00
Tax	$3.03
Shipping Tax	$0.00
Shipping	$0.00
Total Amount 1 Item(s)	$46.38

PROCESSED PAYMENTS

Payment Type	Account Number	Date	Amount
DISCOVER	XXXXXXXXXXXX8023	03-Feb-2023	$46.38

Total Payments	$46.38
Order Total	$46.38
Outstanding	$0.00

Terms & Instructions:

Questions about your order? For fastest service, refer to your Box Number (see top of page) when contacting the bookstore.

Hassberg, Analena H. 2020. Nurturing the Revolution: The Black Panther Party and the Early Seeds of the Food Justice Movement. In *Black Food Matters: Racial Justice in the Wake of Food Justice*, edited by Hannah Garth and Ashanté Reese, 82–106. Minneapolis: University of Minnesota Press.

Hayden, Dolores. 1982. *The Grand Domestic Revolution*. Cambridge, MA: MIT Press.

Hayes-Conroy, Allison, and Jessica Hayes-Conroy. 2008. Taking Back Taste: Feminism, Food and Visceral Politics. *Gender, Place and Culture* 15(5): 461–73.

———— (Eds.). 2013. *Doing Nutrition Differently: Critical Approaches to Diet and Dietary Intervention*. Surrey, UK: Ashgate.

————. 2015. Political Ecology of the Body: A Visceral Approach. In *The International Handbook of Political Ecology*, edited by Raymond L. Bryant, 659–72. Cheltenham, UK: Edward Elgar Publishing.

Hayes-Conroy, Jessica, Adele Hite, Kendra Klein, Charlotte Biltekoff, and Aya H. Kimura. 2014. Doing Nutrition Differently. *Gastronomica: The Journal of Food and Culture* 14(3): 56–66.

Held, David, Anthony McGrew, David Goldblatt, and Jonathan Perraton. 1999. *Global Transformations: Politics, Economics and Culture*. Stanford, CA: Stanford University Press.

Henley, David. 2011. Swidden Farming as an Agent of Environmental Change: Ecological Myth and Historical Reality in Indonesia. *Environment and History* 17(4): 525–54.

Herod, Andrew. 2001. *Labor Geographies: Workers and the Landscapes of Capitalism*. New York: Guilford Press.

Heynen, Nik. 2009. Bending the Bars of Empire from Every Ghetto for Survival: The Black Panther Party's Radical Antihunger Politics of Social Reproduction and Scale. *Annals of the Association of American Geographers* 99(2): 406–22.

Hightower, J. 1972. *Hard Tomatoes, Hard Times: Failure of the Land Grant College Complex*. Washington, DC: Agribusiness Accountability Project.

Hitchcock, Robert K. 2020. The Plight of the Kalahari San: Hunter-Gatherers in a Globalized World. *Journal of Anthropological Research* 76(2): 164–84.

Hislop, Rasheed Salaam. 2014. *Reaping Equity: A Survey of Food Justice Organizations in the U.S.A.* MS thesis, University of California, Davis.

Hochschild, Arlie, with Anne Machung. 2003. *The Second Shift: Working Families and the Revolution at Home*. New York: Penguin Books.

Hoekstra, Arjen Y., and Ashok K. Chapagain. 2007. The Water Footprints of Morocco and the Netherlands: Global Water Use as a Result of Domestic Consumption of Agricultural Commodities. *Ecological Economics* 64(1): 143–51.

Hollander, G., 2003. Re-naturalizing Sugar: Narratives of Place, Production and Consumption. *Social and Cultural Geography* 4(1), 60–74.

Holt-Giménez, Eric. 2019. *Can We Feed the World without Destroying It?* Hoboken, NJ: John Wiley & Sons.

Holt Giménez, Eric, and Annie Shattuck. 2011. Food Crises, Food Regimes and Food Movements: Rumblings of Reform or Tides of Transformation? *Journal of Peasant Studies* 38(1): 109–44.

hooks, bell. 2008. *Belonging: A Culture of Place*. New York: Routledge.

————. 2004. *The Will to Change: Men, Masculinity, and Love*. New York: Washington Square Press.

Hormel, L. M., and K. M. Norgaard. 2009. Bring the Salmon Home! Karuk Challenges to Capitalist Incorporation. *Critical Sociology* 35(3): 343–66.

Hollows, Joanne. 2003. Feeling Like a Domestic Goddess: Postfeminism and Cooking. *European Journal of Cultural Studies* 6(2): 179–202.

Human Rights Watch. 2004. *Blood, Sweat, and Fear: Workers' Rights in U.S. Meat and Poultry Plants*. https://www.hrw.org/reports/2005/usa0105/usa0105.pdf.

————. 2010. *Fields of Peril: Child Labor in US Agriculture*. https://www.hrw.org/report/2010/05/05/fields-peril/child-labor-us-agriculture.

IAASTD. 2009. *Agriculture at a Crossroads*. Global Report. International Assessment of Agricultural Knowledge, Science, and Technology for Development. Washington, DC: Island Press.

ILO. 2020. *Child Labor in Agriculture*. Geneva: International Labor Organization. https://www.ilo.org/ipec/areas/Agriculture/lang—en/index.htm.

———. 2017. *Global Estimates of Child Labour: Rules and Trends, 2012–2016*. Geneva: International Labour Office. https://www.ilo.org/wcmsp5/groups/public/@dgreports/@dcomm /documents/publication/wcms_575499.pdf.

———. 2013. *Caught at Sea: Forced Labour and Trafficking in Fisheries*. Geneva: International Labor Organization. https://www.ilo.org/wcmsp5/groups/public/—-ed_norm/—-declara tion/documents/publication/wcms_214472.pdf.

———. 2003. *Decent Work in Agriculture*. Background Paper. Geneva: International Labour Office. https://www.ilo.org/wcmsp5/groups/public/—-ed_dialogue/—-sector/documents /publication/wcms_161567.pdf.

Imhoff, Daniel. 2019. *The Farm Bill: A Citizen's Guide*. Washington, DC: Island Press.

Inness, Sherrie A. (Ed.). 2001. *Kitchen Culture in America: Popular Representations of Food, Gender, and Race*. Philadelphia: University of Pennsylvania Press.

IPCC. 2014: *Climate Change 2014: Synthesis Report. Contribution of Working Groups I, II and III to the Fifth Assessment Report of the Intergovernmental Panel on Climate Change*, edited by R. K. Pachauri and L. A. Meyer. Geneva, Switzerland: IPCC.

———. 2019. *Climate Change and Land: An IPCC Special Report on Climate Change, Desertification, Land Degradation, Sustainable Land Management, Food Security, and Greenhouse Gas Fluxes in Terrestrial Ecosystems*. Edited by P. R. Shukla, J. Skea, E. Calvo Buendia, V. Masson-Delmotte, H.-O. Pörtner, D. C. Roberts, P. Zhai, R. Slade, S. Connors, R. van Diemen, M. Ferrat, E. Haughey, S. Luz, S. Neogi, M. Pathak, J. Petzold, J. Portugal Pereira, P. Vyas, E. Huntley, K. Kissick, M. Belkacemi, and J. Malley. Geneva, Switzerland: IPCC.

IPC Global Partners. 2008. *Integrated Food Security Phase Classification Technical Manual*. Version 1.1. Rome: FAO.

IUCN. 2020. Red Lists of Threatened Species: Summary Tables. International Union for Conservation of Nature. https://www.iucnredlist.org/resources/summary-statistics#Summary%20 Tables.

———. 2018. Palm Oil and Biodiversity. Issues Brief. International Union for Conservation of Nature. https://www.iucn.org/sites/dev/files/iucn_issues_brief_palm_oil_and_biodiversity.pdf.

Jackson, P. 2002. Commercial Cultures: Transcending the Cultural and the Economic. *Progress in Human Geography* 26: 3–18.

James, Ian. 2015. The Cost of Peru's Farming Boom. *Desert Sun*. December 10. https:// www.desertsun.com/story/news/environment/2015/12/10/costs-perus-farming-boom /76605530/.

Jarosz, Lucy. 2011. Defining World Hunger: Scale and Neoliberal Ideology in International Food Security Policy Discourse." *Food, Culture & Society* 14(1): 117–139.

———. 2009. Energy, Climate Change, Meat, and Markets: Mapping the Coordinates of the Current World Food Crisis. *Geography Compass* 3(6): 2065–83.

Jayaraman, Saru. 2013. *Behind the Kitchen Door*. Ithaca, NY: Cornell University Press.

Jayne, Mark, Gill Valentine, and Sarah L. Holloway. 2016. *Alcohol, Drinking, Drunkenness: (Dis) Orderly Spaces*. New York: Routledge.

Jefferson, Thomas. 1787. Letter to George Washington, Paris August 14. National Archives: Founders Online. Accessed at https://founders.archives.gov/documents/Jeffer son/01-12-02-0040.

Jennings, J., S. Aitken, S. L. Estrada, and A. Fernandez. 2006. Learning and Earning: Relational Scales of Children's Work. *Area* 38(3): 231–39.

Jennings, Michael. 2017. Ujamaa. In *Oxford Research Encyclopedia of African History*, edited by Thomas Spear. Oxford University Press, online.

Joassart-Marcelli, Pascale. 2021. *The $16 Taco: Contested Geographies of Food, Ethnicity, and Gentrification*. Seattle: University of Washington Press.

Joassart-Marcelli, P. and F. J. Bosco. 2018. *Food and Place: A Critical Exploration*. Lanham, MD: Rowman & Littlefield.

Johnston, Josée, and Shyon Baumann. 2014. *Foodies*. New York: Routledge.

Joyce, C. 2006. Ancient Figs May Be First Cultivated Crops. *All Things Considered*. National Public Radio. June 2. https://www.npr.org/templates/story/story.php?storyId=5446137.

Jung-a, Song, Christian Oliver, and Tom Burgis. 2008. Daewoo to Cultivate Madagascar Land for Free. *Financial Times*. November 19.

Kaiser, Paul J. 1996. Structural Adjustment and the Fragile Nation: The Demise of Social Unity in Tanzania, *The Journal of Modern African Studies* 34(2): 227–37.

Kaplan, D. M. (Ed.). 2012. *The Philosophy of Food*. Berkeley, CA: University of California Press.

Kaplan, David. 2000. The Darker Side of the "Original Affluent Society." *Journal of Anthropological Research* 56(3): 301–24.

Katz, Cindi. 2004. *Growing up Global: Economic Restructuring and Children's Everyday Lives*. Minneapolis: University of Minnesota Press.

Katz, Claudio J. 1992. Marx on the Peasantry: Class in Itself or Class in Struggle? *The Review of Politics* 54(1): 50–71.

Kimball, Ann Marie. 2016. *Risky Trade: Infectious Disease in the Era of Global Trade*. New York: Routledge.

Kipple, K. F. 2007. *A Movable Feast: Ten Millenia of Food Globalization*. Cambridge: Cambridge University Press.

Koeppel, Dan. 2008. *Banana-the Fate of the Fruit that Changed the World*. New York: Penguin Group.

Kosoy, Nicolás, and Esteve Corbera. 2010. Payments for Ecosystem Services as Commodity Fetishism. *Ecological Economics* 69(6): 1228–36.

Kurlansky, Mark. 1998. *Cod: A Biography of the Fish that Changed the World*. Ontario: Penguin.

Kwate, Naa Oyo A. 2008. Fried Chicken and Fresh Apples: Racial Segregation as a Fundamental Cause of Fast Food Density in Black Neighborhoods. *Health & Place* 14(1): 32–44.

Lanchester, John. 2017. The Case against Civilization: Did Our Ancestors Hunter-Gatherers Have It Better? *New Yorker*. September 18.

Land Matrix. 2020. Global Dataset. https://landmatrix.org/data/.

Lappé, Frances Moore. 2011. The City That Ended Hunger. In *Food and Democracy: Introduction to Food Sovereignty*, edited by Marcin Gerwin. Kraków: Polish Green Network.

Lappé, Frances Moore, and Joseph Collins. 2015. *World Hunger: Ten Myths*. New York: Food First.

Latour, Bruno. 2005. *Reassembling the Social: An Introduction to Actor-Network-Theory*. Oxford: Oxford University Press.

Laudan, Rachel. 2013. *Cuisine and Empire: Cooking in World History*. Berkeley: University of California Press.

Laurence, Felicity. 2010. How Peru's Wells Are Being Sucked Dry by British Love of Asparagus. *The Guardian*. September 14. https://www.theguardian.com/environment/2010/sep/15/peru-asparagus-british-wells.

La Via Campesina. Declaration of Nyéléni. Nyéléni Forum for Food Sovereignty, Mali, 2007. https://nyeleni.org/spip.php?article290.

LeBesco, Kathleen. 2010. Fat Panic and the New Morality. In *Against Health*, edited by Metzl and Kirkland, 72–82. New York: New York University Press.

Le Billon, Karen. 2012. *French Kids Eat Everything*. New York: Harper Collins.

Lee, Helen. 2012. The Role of Local Food Availability in Explaining Obesity Risk among School-Aged Children. *Social Science & Medicine* 74(8): 1193–203.

Le Heron, R., G. Penny, M. Paine, G. Sheath, J. Pedersen, and N. Botha, 2001. Global Supply Chains and Networking: A Critical Perspective on Learning Challenges in the New Zealand Dairy and Sheepmeat Commodity Chains. *Journal of Economic Geography* 1(4): 439–56.

Leslie, D., and S. Reimer. 1999: Spatializing Commodity Chains. *Progress in Human Geography* 23: 401–20.

Lévi-Strauss, C. 1964. *Mythologiques: Le Cru et le Cuit*. Paris: Plon.

Lewis, P., M. A. Monem, and A. Impiglia. 2018. *Impacts of Climate Change on Farming Systems and Livelihoods, with a Special Focus on Small-Scale Family Farming*. Cairo: Food and Agriculture Organization of the United Nations. http://www.fao.org/3/ca1439en/CA1439EN.pdf.

Lipinski, Brian, Craig Hanson, Richard Waite, Tim Searchinger, James Lomax, and Lisa Kitinoja. 2013. *Reducing Food Loss and Waste*. Working paper. World Resource Institute. https://www.wri.org/research/reducing-food-loss-and-waste.

Little, Jo. 2017. *Gender and Rural Geography*. New York: Routledge.

Liu, Yvonne Yen, Patrick Burns, and Daniel Flaming. 2015. Sidewalk Stimulus: Economic and Geographic Impacts of Los Angeles Street Vendors. Los Angeles Economic Roundtable. https://papers.ssrn.com/sol3/papers.cfm?abstract_id=3380029.

Lockie, Stewart, and Simon Kitto. 2000. Beyond the Farm Gate: Production-Consumption Networks and Agri-Food Research. *Sociologia Ruralis* 40(1): 3–19.

Loh, Penn, and Julian Agyeman. Boston's Emerging Food Solidarity Economy. In *The New Food Activisim: Opposition, Cooperation, and Collective Action*, edited by Alison Hope Alkon and Julie Guthman. Berkeley: University of California Press. 257–83.

Londoño, Ernesto, and Letícia Casado. 2020. As Bolsonaro Keeps Amazon Vows, Brazil's Indigenous Fear "Ethnocide." *New York Times*. April 20. https://www.nytimes.com/2020/04/19/world/americas/bolsonaro-brazil-amazon-indigenous.html.

Long, J. W., R. W. Goode, R. J. Gutteriez, J. J. Lackey, and M. K. Anderson. 2017. Managing California Black Oak for Tribal Ecocultural Restoration. *Journal of Forestry* 115(5): 426–34.

Longhurst, R. 2005. Fat Bodies: Developing Geographical Research Agendas. *Progress in Human Geography* 29(3):247–59.

Longhurst, R., and Johnston, L. 2012. Embodied Geographies of Food, Belonging and Hope in Multicultural Hamilton, Aotearoa New Zealand. *Geoforum* 43: 325–31.

Lovell, Edward R., Taratau Kirata, and Tooti Tekinaiti. 2000. *Status Report for Kiribati's Coral Reefs*. Coral Reefs in the Pacific: Status and Monitoring, Resources and Management. Nouméa: International Coral Reef Initiative.

Lustgarten, Abrahm. 2020. Refugees from the Earth. *New York Times Magazine*. The Climate Issue. July 26, 8–23.

Lyman, B. 2012. *A Psychology of Food: More Than a Matter of Taste*. New York: Springer Science & Business Media.

Macarow, Aaron. 2016. What the Average School Lunch Cost in Your State. *Attn:* March 19. https://archive.attn.com/stories/6365/school-lunch-cost-by-state.

Malthus, Thomas. 1798. *An Essay on the Principle of Population, as It Affects the Future Improvement of Society*. Originally published in London by J. Johnson. Available online https://archive.org/details/essayonprincipl00malt/page/44/mode/2up.

Mandelblatt, Bertie. 2012. Geography of Food. In *The Oxford Handbook of Food History*, edited by Jeffrey Pilcher, 154-71. Oxford: Oxford University Press.

Mannur, A. 2007. Culinary Nostalgia: Authenticity, Nationalism, and Diaspora. *Melus* 32(4): 11–31.

Mansfield, B. 2003. "Imitation Crab" and the Material Culture of Commodity Production. *Social and Cultural Geographies* 10: 176–95.

Martin, Philip. 2003. *Promise Unfulfilled: Unions, Immigration, and Farm Workers*. Ithaca, NY: Cornell University Press.

Marx, Karl. 1867. The Fetishism of the Commodity and Its Secret. In *The Consumer Society Reader*, edited by Juliet B. Schor and Douglas B. Holt, 331–42. New York: The New Press.

Mason, Fred. 2002. The Newfoundland Cod Stock Collapse: A Review and Analysis of Social Factors. *Electronic Green Journal* 1(17).

Massey, Doreen. 2005. *For Space*. Thousand Oaks, CA: Sage.

Matta, Raúl. 2016. Food Incursions into Global Heritage: Peruvian Cuisine's Slippery Road to UNESCO. *Social Anthropology* 24(3): 338–52.

McClintock, Nathan. 2011. From Industrial Garden to Food Desert. In *Cultivating Food Justice: Race, Class, and Sustainability*, edited by Alison Alkon and Julian Agyeman, 89–120. Cambridge, MA: MIT Press.

McCullough, Ellen B., Prabhu L. Pingali, and Kostas G. Stamoulis. (Eds.). 2008. *The Transformation of Agri-food Systems: Globalization, Supply Chains and Smallholder Farmers*. London: Earthscan

McKittrick, K. 2013. Plantation Futures. *Small Axe: A Caribbean Journal of Criticism* 17(3): 1–15.

Meijer, Seline S., Gudeta W. Sileshi, Godfrey Kundhlande, Delia Catacutan, and Maarten Nieuwenhuis. 2015. The Role of Gender and Kinship Structure in Household Decision-Making for Agriculture and Tree Planting in Malawi. *Journal of Gender, Agriculture and Food Security* 1(1): 54–76.

Mekonnen, Mesfin M. and Arjen Y. Hoekstra. 2012. A Global Assessment of the Water Footprint of Farm Animal Products. *Ecosystems* 15: 401–15

Millennium Ecosystem Assessment. 2005. *Ecosystems and Human Well-Being: Synthesis*. Washington, DC: Island Press.

Mintz, Sideny Wilfred. 1986. *Sweetness and Power: The Place of Sugar in Modern History*. New York: Penguin.

Mitchell, Don. 2011. Labor's Geography: Capital, Violence, Guest Workers and the Post-World War II Landscape. *Antipode* 43(2): 563–95.

Mkwela, H. S. 2013. Urban Agriculture in Dar es Salaam: A Dream or Reality? *Sustainable Development and Planning* 173: 161–72.

Monterey Bay Aquarium. 2020. Seafood Watch. Consumer Guides. https://www.seafoodwatch.org/seafood-recommendations/consumer-guides.

Mordor Intelligence. 2020. Dietary Supplements Market: Growth, Trends, Covid-19 Impact, and Forecasts (2021–2026). https://www.mordorintelligence.com/industry-reports/dietary-supplement-market.

Morrow, Oona, and Brenda Parker. 2020. Care, Commoning and Collectivity: From Grand Domestic Revolution to Urban Transformation. *Urban Geography* 41(4): 607–24.

Morse, Stephen S., et al. 2012. Prediction and Prevention of the Next Pandemic Zoonosis. *Lancet* 380(9857): 1956–65.

Mougeot, L. 1999. *Urban Agriculture: Definition, Presence, Potential and Risks, Main Policy Challenges*. CFP Report No. 31. International Development Research Center, Canada.

Muñoz, Lorena. 2016. Agency, Choice and Restrictions in Producing Latina/o Street-Vending Landscapes in Los Angeles. *Area* 48(3): 339–45.

Murcott, Anne. 2012. Lamenting the "Decline of the Family Meal" as a Moral Panic? Methodological Reflections. *Recherches Sociologiques et Anthropologiques* 43(1): 97–118.

Murphy, Mollie K., and Tina M. Harris. 2018. White Innocence and Black Subservience: The Rhetoric of White Heroism in *The Help*. *Howard Journal of Communications* 29(1): 49–62.

National Gardening Association. 2014. *Garden to Table: A 5-Year Look at Food Gardening in America*. A special Report. https://garden.org/special/pdf/2014-NGA-Garden-to-Table.pdf.

National Marine Protected Areas Center. 2020. Marine Protected Areas 2020. https://nmsmarineprotectedareas.blob.core.windows.net/marineprotectedareas-prod/media/docs/2020-mpa-building-effective-conservation-networks.pdf.

Nellemann, C., M. MacDevette, T. Manders, B. Eickhout, B. Svihus, A. G. Prins, and B. P. Kaltenborn. (Eds.). 2009. *The Environmental Food Crisis: The Environment's Role in Averting Future Food Crises*. A UNEP rapid response assessment. United Nations Environment Programme.

Nestle, M., 2013. *Food Politics: How the Food Industry Influences Nutrition and Health*. Berkeley, CA: University of California Press.

NOAA, 2020. *FishWatch U.S. Seafood Facts*. https://www.fishwatch.gov/sustainable-seafood/the-global-picture.

Nourishlife. 2020. Food System Map. https://www.nourishlife.org/teach/food-system-tools/.

Norgaard, K. M., R. Reed, and C. Van Horn. 2011. A Continuing Legacy: Institutional Racism, Hunger, and Nutritional Justice on the Klamath. In *Cultivating Food Justice: Race, Class, and Sustainability*, edited by A. H. Alkon, and J. Agyeman, 23–46. Cambridge, MA: MIT Press.

Novo, Mario Gonzalez, and Catherine Murphy. 2000. Urban Agriculture in the City of Havana: A Popular Response to a Crisis. In *Growing Cities, Growing Food: Urban Agriculture on the Policy Agenda*, edited by Bakker N., Dubbeling M., Gündel S., Sabel-Koshella U., de Zeeuw H. Feldafing, 329–46. Germany: Zentralstelle für Ernährung und Landwirtschaft.

Oakland Institute. 2015. Biointensive Agriculture Training Program in Kenya. Agroecology Case Studies. Oakland Institute and Alliance for Food Sovereignty in Africa. https://www.oaklandinstitute.org/sites/oaklandinstitute.org/files/Biointensive_Agriculture_Kenya.pdf.

Olivero, J., J. E. Fa, R. Real, A. L. Márquez, M. A. Farfán, J. M. Vargas, D. Gaveau, M. A. Salim, D. Park, J. Suter, and S. King. 2017. Recent Loss of Closed Forests Is Associated with Ebola Virus Disease Outbreaks. *Scientific Reports* 7(1): 1–9.

Orsini, Francesco, Remi Kahane, Remi Nono-Womdim, and Giorgio Gianquinto. 2013. Urban Agriculture in the Developing World: A Review. *Agronomy for Sustainable Development* 33(4): 695–720.

Oxfam. 2016. *Unearthed: Land, Power and Inequality in Latin America*. Oxfam International. https://www-cdn.oxfam.org/s3fs-public/file_attachments/bp-land-power-inequality-latin -america-301116-en.pdf.

Paarlberg, R., 2013. *Food Politics: What Everyone Needs to Know*. New York: Oxford University Press.

Padoch, C., and T. Sunderland. 2013. Managing Landscapes for Greater Food Security and Improved Livelihoods. *Unasylva* 64(241): 3–13.

Passel, Jeffrey. 2015. Written Testimony Submitted to U.S. Senate Committee on Homeland Security and Governmental Affairs. Hearing on *Securing the Border: Defining the Current Population Living in the Shadows and Addressing Future Flows*. Pew Research Center, March 25. https:// www.pewresearch.org/hispanic/2015/03/26/testimony-of-jeffrey-s-passel-unauthorized-im migrant-population/ph_2015-03-26_unauthorized-immigrants-testimony-23/.

Patel, Raj. 2012. *Stuffed and Starved: The Hidden Battle for the World Food System*. Brooklyn, NY: Melville House.

———. 2009. Food Sovereignty. *The Journal of Peasant Studies* 36(3): 663–706.

Pauly, Daniel, Villy Christensen, Johanne Dalsgaard, Rainer Froese, and Francisco Torres Jr. 1998. Fishing Down Marine Food Webs. *Science* 279(5352): 860–63.

Pazo, P. T. 2016. *Diasporic Tastescapes*. Münster, Germany: LIT Verlag.

Pearce J., and K. Witten (Eds.). 2010. *Geographies of Obesity: Environmental Understandings of the Obesity Epidemic*. Farnham, UK: Ashgate

Peet, R. 1998. *Modern Geographical Thought*. Oxford: Blackwell.

Philo, Chris. 1992. Neglected Rural Geographies: A Review. *Journal of Rural Studies* 8(2): 193–207.

Pietrorazio, Gabriel. 2021. The Seneca Nation is Building Food Sovereignty, One Bison at a Time. Civil Eats. January 14. https://civileats.com/2021/01/14/from-bison-to -syrup-the-seneca-nation-is-making-strides-in-food-sovereignty/.

Pilcher, Jeffrey M. 1998. *Que vivan los tamales! Food and the Making of Mexican Identity*. Albuquerque: University of New Mexico Press.

Pollan, Michael. 2006. *The Omnivore's Dilemma: A Natural History of Four Meals*. New York: Penguin.

Poore, Joseph, and Thomas Nemecek. 2018. Reducing Food's Environmental Impacts through Producers and Consumers. *Science* 360(6392): 987–92.

Poppendieck, Janet. 1998. *Sweet Charity: Emergency Food and the End of Entitlement*. New York: Viking.

Powell, Lisa M., Sandy Slater, Donka Mirtcheva, Yanjun Bao, and Frank J. Chaloupka. 2007. Food Store Availability and Neighborhood Characteristics in the United States. *Preventive Medicine* 44(3): 189–95.

Probyn, Elspeth. 2000. *Carnal Appetites: FoodSexIdentities*. London: Routledge.

Progressio. 2010. *Drop by Drop: Understanding the Impacts of the UK's Water Footprint through a Case Study of Peruvian Asparagus*. London: Progressio, Centro Peruano de Estudios Sociales, and Water Witness International. https://www.progressio.org.uk/sites/default/files/Drop -by-drop_Progressio_Sept-2010.pdf.

Pulido, Laura. 2000. Rethinking Environmental Racism: White Privilege and Urban Development in Southern California. *Annals of the Association of American Geographers* 90(1): 12–40.

Radi, S. M., N. A. El-Sayed, L. M. Nofal, and Z. A. Abdeen. 2013. Ongoing Deterioration of the Nutritional Status of Palestinian Preschool Children in Gaza under the Israeli Siege. *Eastern Mediterranean Health Journal* 19(3): 234–41.

Ramírez, Margaret Marietta. 2015. The Elusive Inclusive: Black Food Geographies and Racialized Food Spaces. *Antipode* 47(3): 748–69.

Ranganathan, Janet, Daniel Vennard, Richard Waite, Partice Dumas, Brian Lipinski, Tim Searchinger, and Globagri-WRR model authors. 2016. *Shifting Diets for a Sustainable Food Future*. Working Paper. World Resources Institute. https://files.wri.org/s3fs-public/Shift ing_Diets_for_a_Sustainable_Food_Future_1.pdf.

Rapley, T. 2011. Some Pragmatics of Data Analysis. *Qualitative Research* 3: 273–90.

Ratta, Annu, and Joe Nasr. 1996. Urban Agriculture and the African Urban Food Supply System. *African Urban Quarterly* 11(2): 154–61.

Rautner, M., M. Leggett, and F. Davis. (2013). *The Little Book of Big Deforestation Drivers*. Oxford: Global Canopy Programme.

Rawson, Katie, and Elliott Shore. 2019. *Dining Out: A Global History of Restaurants*. London: Reaktion Books.

Ray, Deepak K., Navin Ramankutty, Nathaniel D. Mueller, Paul C. West, and Jonathan A. Foley. 2012. Recent Patterns of Crop Yield Growth and Stagnation. *Nature Communications* 3(1): 1–7.

Raynolds, L. T. 2002. Consumer/Producer Links in Fair Trade Coffee Networks. *Sociologia Ruralis* 42(4): 404–24.

Reese, Ashanté M. 2019. *Black Food Geographies: Race, Self-Reliance, and Food Access in Washington, DC*. Chapel Hill: University of North Carolina Press.

Riches, Graham. 2018. *Food Bank Nations: Poverty, Corporate Charity, and the Right to Food*. New York: Routledge.

Rigg, Jonathan, Albert Salamanca, and Eric C. Thompson. 2016. The Puzzle of East and Southeast Asia's Persistent Smallholder. *Journal of Rural Studies* 43: 118–33.

Ritchie, Hannah, and Max Roser. 2020. Meat and Dairy Production. OurWorldInData.org. https://ourworldindata.org/meat-production.

———. 2018. Micronutrient Deficiency. *OurWorldInData.org*. https://ourworldindata.org/micronutrient-deficiency.

Ritchie, Hannah, Max Roser, and Edouard Mathieu. 2020. Data on CO2 and Greenhouse Gas Emissions. *OurWorldInData.org*. Accessed at https://ourworldindata.org/greenhouse-gas-emissions.

Ritzer, George. 2013. *The McDonaldization Thesis: Explorations and Extensions*. London: Sage.

Robinson, Guy. 2014. *Geographies of Agriculture: Globalisation, Restructuring and Sustainability*. New York: Routledge.

Rocha, Cecilia, and Iara Lessa. 2009. Urban Governance for Food Security: The Alternative Food System in Belo Horizonte, Brazil. *International Planning Studies* 14(4): 389–400.

Roggeveen, K. 2014. Tomato Journeys from Farm to Fruit Shop. *Local Environment* 19(1): 77–102.

Roos, Christopher I., María Nieves Zedeño, Kacy L. Hollenback, and Mary M. H. Erlick. 2018. Indigenous Impacts on North American Great Plains Fire Regimes of the Past Millennium. *Proceedings of the National Academy of Sciences* 115(32): 8143–48.

Rosales, Rocio. 2020. *Fruteros: Street Vending, Illegality, and Ethnic Community in Los Angeles*. Oakland: University of California Press.

Rousseau, S. 2012. *Food and Social Media: You Are What You Tweet*. Lanham, MD: Altamira.

Rousseau, Yannick, Reg A. Watson, Julia L. Blanchard, and Elizabeth A. Fulton. 2019. Evolution of Global Marine Fishing Fleets and the Response of Fished Resources. *PNAS* 116(25): 12238–43.

Rudd Center. 2017. Food Industry Self-Regulation after 10 Years: Progress and Opportunities to Improve Food Advertising to Children. University of Connecticut. http://www.uconnruddcenter.org/facts2017.

Ruetschlin, Catherine. 2014. Fast Food Failure: How CEO-to-Worker pay Disparity Undermines the Industry and the Overall Economy. *Demos*. https://www.demos.org/sites/default/files/publications/Demos-FastFoodFailure.pdf

Sachs, Carolyn E. 2018. *Gendered Fields: Rural Women, Agriculture, and Environment*. New York: Routledge.

Sage, Colin. 2011. *Environment and Food*. New York: Routledge.

Sahlins, M. 1968. Notes on the Original Affluent Society. In *Man the Hunter*, edited by R. Lee and I. DeVore, 85. Chicago: Aldine Publishing.

Saksena, Michelle, Abigail M. Okrent, and Karen S. Hamrick. (Eds.). 2018. *America's Eating Habits: Food Away from Home*, Economic Information Bulletin 196, US Department of Agriculture, Economic Research Service. https://www.ers.usda.gov/webdocs/publications/90228/eib-196.pdf?v=7125.2.

San Diego Food System Alliance. 2021. *San Diego County Food Vision 2030*. https://www.sdfsa.org/vision.

Sartori, Martina, George Philippidis, Emanuele Ferrari, Pasquale Borrelli, Emanuele Lugato, Luca Montanarella, and Panos Panagos. 2019. A Linkage between the Biophysical and the Economic: Assessing the Global Market Impacts of Soil Erosion. *Land Use Policy* 86: 299–312.

Schlosser, Eric. 2001. *Fast Food Nation: The Dark Side of the All-American Meal*. Boston: Houghton Mifflin.

Schmidt, Stephan. 2011. Urban Agriculture in Dar es Salaam, Tanzania. Case Study #7-12. Food Policy for Developing Countries: The Role of Government in the Global Food System. Cornell University.

Schultz, T. W., 1964. *Transforming Traditional Agriculture*. New Haven, CT: Yale University Press.

Searchinger, Tim, Richard Waite, Craig Hanson, Janet Ranganathan. 2019. *Creating a Sustainable Food Future: A Menu of Solutions to Feed Nearly 10 Billion People by 2050*. World Resources Report. https://www.wri.org/publication/creating-sustainable-food-future.

Sen, Amartya. 1992. Missing Women. *BMJ: British Medical Journal* 304(6827): 587–88.

———. 1984. *Resources, Values and Development*. Oxford: Basil Blackwell.

———. 1982. *Poverty and Famines: An Essay on Entitlement and Deprivation*. New York: Oxford University Press.

Sherman, Taylor C. 2013. From "Grow More Food" to "Miss a Meal": Hunger, Development and the Limits of Post-Colonial Nationalism in India, 1947–1957. *South Asia: Journal of South Asian Studies* 36(4): 571–88.

Shierholz, Heidi. 2014. Low Wages and Few Benefits Mean Many Restaurant Workers Can't Make Ends Meet. Briefing Paper #383. Economic Policy Institute. https://www.epi.org/publication/restaurant-workers/.

Shiva, Vandana. 1991. *The Violence of the Green Revolution: Third World Agriculture, Ecology and Politics*. New York: Zed Books.

Shortbridge, B. G., and J. R. Shortbridge. 1998. *The Taste of American Place: A Reader on Regional and Ethnic Food*. Lanham, MD: Rowman & Littlefield.

Singh, Nirvikar, and Deepali Singhal Kohli. 1997. The Green Revolution in Punjab, India: The Economics of Technological Change. *Journal of Punjab Studies* 12(2): 285–06.

Slocum, Rachel. 2007. Whiteness, Space and Alternative Food Practice. *Geoforum* 38(3): 520–33.

Slow Food. 2020. Food and Health. https://www.slowfood.com/what-we-do/themes/food-and-health/.

Smaal, Aad, Marnix van Stralen, and Johan Craeymeersch. 2005. Does the Introduction of the Pacific Oyster *Crassostrea Gigas* Lead to Species Shifts in the Wadden Sea? In *The Comparative Roles of Suspension-Feeders in Ecosystems*, edited by R. F. Dame and S. Olenin, 277–89. NATO Science Series IV: Earth and Environmental Sciences. Dordrecht, Netherlands: Springer.

Smil, Vaclav. 2001. *Enriching the Earth: Fritz Haber, Carl Bosch, and the Transformation of World Food Production*. Cambridge, MA: MIT Press.

Smith, Bruce D. 2001. Low-Level Food Production. *Journal of Archaeological Research* 9(1): 1–43.

Smith, D. M., and S. Cummins. 2008. Obese Cities: How Our Environment Shapes Overweight. *Geography Compass* 3: 518–35.

Snowdon, Wendy, and Anne Marie Thow. 2013. Trade Policy and Obesity Prevention: Challenges and Innovation in the Pacific Islands. *Obesity Reviews* 14 (S2): 150–58.

Solway, Jacqueline. 2009. Human Rights and NGO "Wrongs": Conflict Diamonds, Culture Wars and the "Bushman Question." *Africa* 79(3): 321–46.

Sova, Chase, Kimberly Flowers, and Christian Man. 2019. Climate Change and Food Security: A Test of US Leadership in a Fragile World. CSIS Brief. Center for Strategic and International Studies. http://csis-website-prod.s3.amazonaws.com/s3fs-public/publication/191015_Flowers_ClimateChangeFood_WEB.pdf.

Spang, Rebecca L., and Adam Gopnik. 2019. *The Invention of the Restaurant: Paris and Modern Gastronomic Culture, with a New Preface*. Cambridge, MA: Harvard University Press.

Stassart, P., and S. Whatmore. 2003. Metabolizing Risk: Food Scares and the Un/Remaking of Belgian Beef. *Environment and Planning A* 35: 449–62.

Statista. 2020a. U.S. Diets and Weight Loss—Statistics and Facts. Statista Research Department. https://www.statista.com/topics/4392/diets-and-weight-loss-in-the-us/#dossier Summary__chapter2.

———. 2020b. Distribution of Global Corn Production in 2019/2020, by Country. https://www.statista.com/statistics/254294/distribution-of-global-corn-production-by-country-2012/.

Stief, Colin. 2019. Slash and Burn Agriculture Explained. Geography. ThoughtCo. https://www.thoughtco.com/slash-and-burn-agriculture-p2-1435798.

Stocked, Kathryn. 2009. *The Help*. New York: G. P. Putnam.

Stokstad, Eric. 2019. Fishing Fleets Have Doubled Since 1950—But They're Having a Harder Time Catching Fish. *Science*. May 27. https://www.sciencemag.org/news/2019/05/fishing-fleets-have-doubled-1950-theyre-having-harder-time-catching-fish.

Stone, Daniel. 2016. Why Are Bananas So Cheap? *National Geographic*, June 8. https://www.nationalgeographic.com/people-and-culture/food/the-plate/2016/08/bananas-are-so-cool/.

Strings, Sabrina. 2019. *Fearing the Black Body: The Racial Origins of Fat Phobia*. New York: New York University Press.

Sugawara, Kazuyoshi. 2002. Optimistic Realism or Opportunistic Subordination? The Interaction of the G/wi and G//ana with Outsiders. In *Ethnicity, Hunter-Gatherers, and the "Other,"* edited by Susan Kent, 93–126. Washington, DC: Smithsonian Institution Press.

———. 2002. Voices of the Dispossessed. *Cultural Survival Quarterly* 26(1): 28–29.

Survival International. 2019. *The Bushmen*. Accessed on 5/3/2019 at https://www.survivalinternational.org/tribes/bushmen.

Suzman, James. 2017. *Affluence without Abundance: The Disappearing World of the Bushmen*. New York: Bloomsbury Publishing.

Telfer, E. 2012. *Food for Thought: Philosophy and Food*. New York: Routledge.

Third World Network and SOCLA. 2015. *Agroecology: Key Concepts, Principles and Practices*. Penang, Malaysia: Third World Network and Berkeley, CA: Sociedad Científica Latinoamericana de Agroecología (SOCLA).

Tickler, David, Jessica J. Meeuwig, Katharine Bryant, Fiona David, John A. H. Forrest, Elise Gordon, Jacqueline Joudo Larsen, et al. 2018. Modern Slavery and the Race to Fish. *Nature Communications* 9(1): 1-9.

Tornaghi, Chiara. 2014. Critical Geography of Urban Agriculture. *Progress in Human Geography* 38(4): 551–67.

Treuhaft, Sarah, Michael J. Hamm, and Litjens, Charlotte Hamm. 2009. Healthy Food for All: Building Equitable and Sustainable Food Systems in Detroit and Oakland. Policy Link and Michigan State University. https://www.policylink.org/resources-toolshealthy-food-for-all-building-equitable-and-sustainable-food-systems-in-detroit-and-oakland.

Trinh, T. M. 1999. Write Your Body and the Body in Theory. In *Feminist Theory and The Body. A Reader*, edited by J. Price and M. Shildrick, 258–66. New York: Routledge.

Tuomainen, Helena Margaret. 2009. Ethnic Identity, (Post)Colonialism and Foodways: Ghanaians in London. *Food, Culture & Society* 12(4): 525–54.

UNCTAD. 2018. Merchandise Trade Matrix–Product Groups, Exports in Thousands of United States Dollars, Annual. United Nations Conference on Trade and Development. https://unctadstat.unctad.org.

UNEP. 2014. *The Importance of Mangroves to People: A Call to Action*. Cambridge: World Conservation Monitoring Centre, United Nations Environment Programme.

UNICEF. 2020. Malnutrition Data. https://data.unicef.org/topic/nutrition/malnutrition/.

United Nations. 2020. Water Action Decade. https://wateractiondecade.org.

Urbina, Ian. 2015. "Sea Slaves:" The Human Misery That Feeds Pets and Livestock. *New York Times*. July 27. https://www.nytimes.com/2015/07/27/world/outlaw-ocean-thailand-fishing-sea-slaves-pets.html.

USDA (US Department of Agriculture). 2020. Farm Labor. Economic Research Service. https://www.ers.usda.gov/topics/farm-economy/farm-labor/.

———. 2019a. *2017 Census of Agriculture. United States Summary and State Data*. United States Department of Agriculture. National Agriculture Statistics Service. https://www.nass.usda.gov/Publications/AgCensus/2017/Full_Report/Volume_1,_Chapter_1_US/usv1.pdf.

———. 2019b. Farm Structure. Economic Research Service. United States Department of Agriculture. https://www.ers.usda.gov/topics/farm-economy/farm-structure-and-organization/farm-structure/

———. 2018. America's Diverse Family Farms. Economic Research Service. United States Department of Agriculture. Economic Information Bulletin Number 203. https://www.ers.usda.gov/webdocs/publications/90985/eib-203.pdf?v=9520.4.

———. 2017. *Food Access Research Atlas.* Washington, DC: United States Department of Agriculture, Economic Research Service. https://www.ers.usda.gov/data-products/food-access-research-atlas/documentation/.

———. 2005. *Structure and Finances of U.S. Farms: 2005 Family Farm Report.* Washington, DC: United States Department of Agriculture, Economic Research Service. at https://www.ers.usda.gov/publications/pub-details/?pubid=43824

US Census. 2020. *American Community Survey Five-Year Estimates, 2014–2018.* Summary File Data. Washington, DC: US Census Bureau.

US Department of the Interior. 2012. *Klamath Dam Removal Overview Report for the Secretary of the Interior.* An Assessment of Science and Technical Information. https://klamathrestoration.gov/sites/klamathrestoration.gov/files/2013%20Updates/Final%20SDOR%20/0.Final%20Accessible%20SDOR%2011.8.2012.pdf.

US Department of State. 2013. Trafficking in Persons Report. https://2009-2017.state.gov/j/tip/rls/tiprpt/2013/index.htm.

US Food and Drug Administration. 2018. Environmental Assessment of Factors Potentially Contributing to the Contamination of Romaine Lettuce Implicated in a Multi-State Outbreak of E. coli O157:H7. https://www.fda.gov/food/outbreaks-foodborne-illness/environmental-assessment-factors-potentially-contributing-contamination-romaine-lettuce-implicated.

Valentine, G., 1999. Eating In: Home, Consumption and Identity. *Sociological Review* 47: 491–524.

Vild Mad. 2019. Web application. Accessed on 4/2/2019 at https://vildmad.dk/en/om-vild-mad.

Viviano, Frank. 2017. This Tiny Country Feeds the World. *National Geographic*, September. https://www.nationalgeographic.com/magazine/2017/09/holland-agriculture-sustainable-farming/.

Voigt, Maria, Serge A. Wich, Marc Ancrenaz . . . , Jessie Wells, Kerrie A. Wilson, Hjalmar S. Kühl. 2018. Global Demand for Natural Resources Eliminated More Than 100,000 Bornean Orangutans. *Current Biology* 28: 761–69.

Wahlbeck, Östen. 2012. Entrepreneurship as Social Status: Turkish Immigrants' Experiences of Self-Employment in Finland. *Migration Letters* 5(1): 53–62.

Walk Free Foundation. 2019. *The Global Slavery Index 2018.* https://www.globalslaveryindex.org/resources/downloads/.

Walker, R. E., C. R. Keane, and J. G. Burke. 2010. Disparities and Access to Healthy Food in the United States: A Review of Food Deserts Literature. *Health & Place* 16(5): 876–84.

Wallace, Rob. 2016. *Big Farms Make Big Flu: Dispatches on Influenza, Agribusiness, and the Nature of Science.* New York: New York University Press.

Warshansky, Daniel. 2018. Food Banks and the Devolution of Anti-Hunger Policy. In *Food & Place: A Critical Exploration*, edited by Pascale Joassart-Marcelli and Fernando J. Bosco, 166–84. Lantham, MD: Rowman and Littlefield.

Watson, James L., ed. 1997. *Golden Arches East: McDonald's in East Asia.* Stanford, CA: Stanford University Press.

Watts, M. 1983. *Silent Violence: Food, Famine and Peasantry in Northern Nigeria.* Berkeley: University of California Press.

———. 2004. Are Hogs like Chickens? Enclosure and Mechanization in Two "White Meat" Filieres. In *Geographies of Commodity Chains*, edited by A. Hughes and S. Reimer, 39–62. London: Routledge.

Western Klamath Restoration Partnership. 2018. Website homepage. https://www.wkrp.network.

WFP. 2020. Palestine. World Food Programme. https://www.wfp.org/countries/state-palestine.

————. 2017. Food Security and Emigration: Why People Flee and the Impact on Family Members Left Behind in El Salvador, Guatemala and Honduras. United Nations' World Food Programme. https://docs.wfp.org/api/documents/WFP-0000022124 /download/?_ga=2.137485751.62933729.1599142250-1604877483.1599142250

Whatmore, Sarah. 1991. *Farming Women: Gender, Work, and Family Enterprise.* London: Macmillan Academic and Professional Ltd.

Whatmore, Sarah, and Lorraine Thorne. 1997. Nourishing Networks: Alternative Geographies of Food. In *Globalising Food: Agrarian Questions and Global Restructuring,* edited by D. Goodman and M. Watts, 222–35. London: Routledge.

White, Monica M. 2018. *Freedom Farmers: Agricultural Resistance and the Black Freedom Movement.* Chapel Hill: NC: University of North Carolina Press, 2018.

Whittlesey, D. (1936). Major Agricultural Regions of the Earth. *Annals of the Association of American Geographers* 26(4): 199–240.

WHO. 2020a. Global Health Observatory Data Repository. World Health Organization. https://apps.who.int/gho/data/view.main.CTRY2430A?lang=en.

————. 2020b. Food Safety. Fact Sheets. World Health Organization. https://www.who.int /news-room/fact-sheets/detail/food-safety

Williams-Forson, Psyche A. 2006. *Building Houses out of Chicken Legs: Black Women, Food, and Power.* Chapel Hill: University of North Carolina Press.

Wills, A. B. 2003. Pilgrims and Progress: How Magazines Made Thanksgiving. *Church History* 72(1): 138–58.

Wilson, Rachel. 2013. Cocina Peruana Para El Mundo: Gastrodiplomacy, the Culinary Nation Brand, and the Context of National Cuisine in Peru. *Exchange: The Journal of Public Diplomacy* 2(1): 13–20.

WinklerPrins, Antoinette. 2017. *Global Urban Agriculture.* Boston: CABI.

Winter, M. 2003. Geographies of Food: Agro-Food Geographies—Making New Connections. *Progress in Human Geography* 27: 505–13.

Wise, Timothy. 2020. Replacing Hunger with Malnutrition: Former UN Official calls out failing African Green Revolution. Institute for Agriculture and Trade Policy. August 19. https://www.iatp.org/blog/202008/replacing-hunger-malnutrition-former-un-official-calls -out-failing-african-green.

————. 2019. *Eating Tomorrow: Agribusinesses, Family Farmers, and the Battle for the Future of Food.* New York: The New Press.

————. 2009. *Agricultural Dumping under NAFTA: Estimating the Costs of US Agricultural Policies to Mexican Producers.* Working Paper No. 09-08. Global Development and Environment Institute. Somerville, MA: Tufts University.

Wolfe, Julia, Jori Kandra, Lora Engdahl, and Heidi Shierholz. 2020. *Domestic Workers Chartbook.* Washington, DC: Economic Policy Institute. http://www.epi.org/194214.

World Bank. 2017. *Future of Food: Shaping the Food System to Deliver Jobs.* Washington, DC: International Bank for Reconstruction and Development/The World Bank.

————. 2020a. World Development Indicators. Washington, DC: The World Bank. http://data .worldbank.org/data-catalog/world-development-indicators.

————. 2020b. Arable Land Hectares. https://data.worldbank.org/indicator/AG.LND.ARBL .HA.

WOR. 2013. *World Ocean Review 2: The Future of Fish: The Fisheries of the Future.* Hamburg: Maribus. https://worldoceanreview.com/wp-content/downloads/wor2/WOR2_en.pdf.

WWF. 2015. *Living Blue Planet Report: Species, Habitat, and Human Well-Being.* Gland: WWF International and London: Zoological Society of London. https://c402277.ssl.cf1. rackcdn.com/publications/817/files/original/Living_Blue_Planet_Report_2015_Final _LR.pdf?1442242821.

————. 2016. *Fishing for Proteins: How Marine Fisheries Impact on Global Food Security up to 2050.* Hamburg: International WWF Centre for Marine Conservation. https://c402277.ssl.cf1.rack cdn.com/publications/982/files/original/Report_food_and_fish_Final.pdf?1484256747.

Young, E. M. 2012. *Food and Development.* London: Routledge.

————. 1997. *World Hunger.* London: Routledge.

Index

About the Author

Pascale Joassart-Marcelli is professor of geography and director of the Interdisciplinary Food Studies Program at San Diego State University, where she teaches courses such as *Geography of Food*; *Food, Place, and Culture*; *Food Justice*; and *Feeding the World*. Her research focuses on the relationships between food, place, and ethnicity, including the role of food in creating just and sustainable cities. She has published over fifty peer-reviewed articles and book chapters and is the author of *The $16 Taco: Contested Geographies of Food, Ethnicity, and Gentrification* (2021) and the coeditor of *Food and Place: A Critical Exploration* (2018). Her research has been funded by the National Science Foundation and other private and public funding agencies.

Exploring Geography

Series Editor: David H. Kaplan

CPSIA information can be obtained
at www.ICGtesting.com
Printed in the USA
BVHW052022140222
628718BV00004B/13